AF361845

Hydrology in Water Resources Management

Hydrology in Water Resources Management

Hydrology in Water Resources Management

Editors

Andrzej Wałęga
Tamara Tokarczyk

MDPI • Basel • Beijing • Wuhan • Barcelona • Belgrade • Manchester • Tokyo • Cluj • Tianjin

Editors
Andrzej Wałęga
University of Agriculture in Krakow
Poland

Tamara Tokarczyk
Institute of Meteorology and Water
Management National Research Institute
(IMGW-PIB)
Poland

Editorial Office
MDPI
St. Alban-Anlage 66
4052 Basel, Switzerland

This is a reprint of articles from the Special Issue published online in the open access journal *Water* (ISSN 2073-4441) (available at: https://www.mdpi.com/journal/water/special_issues/hydrol_water_resour_manage).

For citation purposes, cite each article independently as indicated on the article page online and as indicated below:

LastName, A.A.; LastName, B.B.; LastName, C.C. Article Title. *Journal Name* **Year**, *Volume Number*, Page Range.

ISBN 978-3-0365-3266-0 (Hbk)
ISBN 978-3-0365-3267-7 (PDF)

Contents

Article

Performance Evaluation of a Two-Parameters Monthly Rainfall-Runoff Model in the Southern Basin of Thailand

Pakorn Ditthakit [1,2,*], Sirimon Pinthong [1,2], Nureehan Salaeh [1,2], Fadilah Binnui [1,2], Laksanara Khwanchum [2,3], Alban Kuriqi [4], Khaled Mohamed Khedher [5,6] and Quoc Bao Pham [7,*]

[1] School of Engineering and Technology, Walailak University, Nakhon Si Thammarat 80161, Thailand; Sirimon.pi@mail.wu.ac.th (S.P.); Nureehan.sa@mail.wu.ac.th (N.S.); Fadilah.bi@mail.wu.ac.th (F.B.)
[2] Center of Excellence in Sustainable Disaster Management, Walailak University, Nakhon Si Thammarat 80161, Thailand; laksanara.kh@wu.ac.th
[3] School of Languages and General Education, Walailak University, Nakhon Si Thammarat 80161, Thailand
[4] CERIS, Instituto Superior Técnico, Universidade de Lisboa, 1049-001 Lisbon, Portugal; alban.kuriqi@tecnico.ulisboa.pt
[5] Department of Civil Engineering, College of Engineering, King Khalid University, Abha 61421, Saudi Arabia; kkhedher@kku.edu.sa
[6] Department of Civil Engineering, High Institute of Technological Studies, Mrezgua University Campus, Nabeul 8000, Tunisia
[7] Institute of Applied Technology, Thu Dau Mot University, Binh Duong 75000, Vietnam
* Correspondence: dpakorn@mail.wu.ac.th (P.D.); phambaoquoc@tdmu.edu.vn (Q.B.P.)

Abstract: Accurate monthly runoff estimation is crucial in water resources management, planning, and development, preventing and reducing water-related problems, such as flooding and droughts. This article evaluates the monthly hydrological rainfall-runoff model's performance, the GR2M model, in Thailand's southern basins. The GR2M model requires only two parameters: production store (X_1) and groundwater exchange rate (X_2). Moreover, no prior research has been reported on its application in this region. The 37 runoff stations, which are located in three sub-watersheds of Thailand's southern region, namely; Thale Sap Songkhla, Peninsular-East Coast, and Peninsular-West Coast, were selected as study cases. The available monthly hydrological data of runoff, rainfall, air temperature from the Royal Irrigation Department (RID) and the Thai Meteorological Department (TMD) were collected and analyzed. The Thornthwaite method was utilized for the determination of evapotranspiration. The model's performance was conducted using three statistical indices: Nash–Sutcliffe Efficiency (NSE), Correlation Coefficient (r), and Overall Index (OI). The model's calibration results for 37 runoff stations gave the average NSE, r, and OI of 0.657, 0.825, and 0.757, respectively. Moreover, the NSE, r, and OI values for the model's verification were 0.472, 0.750, and 0.639, respectively. Hence, the GR2M model was qualified and reliable to apply for determining monthly runoff variation in this region. The spatial distribution of production store (X_1) and groundwater exchange rate (X_2) values was conducted using the IDW method. It was susceptible to the X_1, and X_2 values of approximately more than 0.90, gave the higher model's performance.

Keywords: GR2M; inverse distance weighting; rainfall-runoff model; sensitivity analysis

Citation: Ditthakit, P.; Pinthong, S.; Salaeh, N.; Binnui, F.; Khwanchum, L.; Kuriqi, A.; Khedher, K.M.; Pham, Q.B. Performance Evaluation of a Two-Parameters Monthly Rainfall-Runoff Model in the Southern Basin of Thailand. *Water* **2021**, *13*, 1226. https://doi.org/10.3390/w13091226

Academic Editor: Andrzej Walega

Received: 28 March 2021
Accepted: 25 April 2021
Published: 28 April 2021

Publisher's Note: MDPI stays neutral with regard to jurisdictional claims in published maps and institutional affiliations.

1. Introduction

A tropical climate characterizes Thailand's southern region since it is close to the equator. Consequently, many areas have been experiencing flooding problems leading to a vast majority of devastation to human beings' lives and properties that hindered economic growth and development. Each year, during a dry spell of approximately two months, this region usually faces a drought situation due to increasing water demand from all activities and insufficient water supply and storage. Accurate estimation of runoff quantity and time variation benefits urban water management, e.g., planning for urban water supply and distribution infrastructure. Besides, it helps water resources management-related issues personnel for effective disaster response planning, preventing and reducing the adverse impact [1,2]. Hence, it is fundamentally imperative to obtain hydrological information since the water supply is in demand from all activities, including domestic consumption, agriculture, and various industries [3,4].

Although runoff is essential, most hydrologists cannot access it due to insufficient runoff measuring stations than rainfall measuring stations equipped throughout the country's regions [5]. Many research topics regarding the rainfall-runoff model have been studied, developed, and applied by hydrologists and irrigation engineers to investigate different water management and planning issues. For example, Chen et al. [6], Kabiri et al. [7], and Lin et al. [8] applied the rainfall-runoff model to assess runoff impacts due to climate and land-use change. Kwak et al. [9] also used the rainfall-runoff model to reconstruct the missing runoff time-series information. Similarly, Ballinas-González et al. [10] studied the sensitivity analysis of the rainfall-runoff modeling parameters in the data-scarce urban catchment. Lerat et al. [11] proposed the alternative method for calibrating daily rainfall-runoff models to monthly streamflow data when no daily streamflow data recorded. Likewise, Abdessamed and Abderrazak [12] utilized a coupling HEC-RAS and HEC-HMS modeling for evaluating floodplain inundation maps in arid environments. Zhang et al. [13] tested the performance of the shuffled complex evolution (SCE-UA) as a global optimization method to calibrate the Xinanjiang (XAJ) model. Lastly, Khazaei et al. [14] applied a simple genetic algorithm to automatically calibrate the ARNO conceptual rainfall-runoff model.

The Rural Genius model (GR2M) model has recently been successfully applied as a rainfall-runoff relationship model to comprehend the variation of watershed's hydrological characteristics and determine alleviation measures of unexpected hydrological situations in many regions throughout the world. Dezetter et al. [15] applied the GR2M model for study runoff in West Africa due to climate variability on hydrologic regimes for large-scale water resources management and planning. Okkan and Fistikoglu [16] evaluated the effects of climate change on runoff in the Izmir-Tahtali watershed, Turkey, using statistical downscaling under the AR5 scheme GR2M model. They recommended that it immediately took on the drought alleviating water supply and agriculture measures on a national scale. Lyon et al. [17] utilized the GR2M model as the first step for screening hydrologic data for evaluating the changes of hydrological response across the Lower Mekong Basin. Zamoum and Souag-Gamane [18] developed regionalized parameters of the GR2M model for predicting monthly runoff in the ungauged catchment of northern Algeria. Boulariah, et al. [19] conducted a comparative study between two conceptual non-linear models, i.e., the GR2M and the ABCD. The results showed that the GR2M model outperformed the ABCD in the validation phase. Topalović et al. [20] compared four monthly rainfall-runoff models based on the water balance concept, i.e., abcd, Budyko, GR2M, and the Water and Snow Balance Modelling System (WASMOD), to simulate runoff in the Wimmera catchment under changing climate conditions. Hadour et al. [21] applied the GR2M model to study the effects of climate scenario on monthly river runoff in the Cheliff, Tafna, and Macta in North-West Algeria. Rintis and Setyoasri [22] compared the GR2M model's performance to two well-known rainfall-runoff models in Indonesia: Mock and NRECA. Using the Bah Bolon Basin in Indonesia as a studied area, they found the GR2M model's performance was comparable to Mock and NRECA methods requiring fewer parameters. O'Connor, et al. [23] applied the GR2M hydrological model and an Artificial Neural Network for reconstructing monthly river flow for Irish catchments.

The spatiotemporal characteristic with a hydrological analysis of Southern Basins of Thailand constitutes a vital platform for understanding the hydrological behavior. Furthermore, it gives particular interest to the valorization of the hydraulic potential of the region. Hydrological modeling is essential for studying the development and management of water resources in the watershed. The main reason for choosing GR2M in this study is that it requires little hydrological information (i.e., rainfall data, potential evapotranspiration, and flow rates). Only two model parameters need to be calibrated.

This article mainly focused on investigating the monthly hydrological rainfall-runoff variation using the GR2M model in Thailand's southern basins, namely, Songkhla Lake Basin, West basin, and the Eastern Basin. The study's novelty is that it is the first attempt to apply a two-parameters monthly rainfall-runoff model, namely the GR2M model, in Thailand's southern basins. It is also drastically useful for water resources planning and management in this region. This article is organized as follows: Section 1 reviews the study area's dominant characteristic and data analysis for model input. In Section 2, the GR2M theory is briefly explained. The model's calibration and verification are delineated in Section 3. The performance criteria for evaluating the applicability of the GR2M Model is depicted in Section 4. Our result findings and discussion are portrayed in Section 5. Finally, in Section 6, we concluded significant contributions from our research work.

2. Study Area and Data Analysis

This research was conducted in Thailand's southern basin. It encompasses five major river basins, including the Peninsula-East Coast. Peninsula-West Coast, Mae Nam Tapi, Thale Sap Songkhla, and Mae Nam Pattani, as shown in Figure 1. When investigating monthly rainfall, evapotranspiration, and runoff data, we found only three river basins, i.e., the Peninsula-East Coast, Peninsula-West Coast, Thale Sap Songkhla. Thus, we focused our analysis on these three basins. These river basins have an area of approximately in the range of 13 to 6713 km^2. Geographically, this portion is the peninsula between the Andaman Sea, which is on the western side, and the South China Sea, which is on the eastern side. The long western mountain range in the northern and central regions also extends to this portion. The Phuket ridge along the west coast and the Nakhon Si Thammarat ridge at the center of the lower portion of the ridge's southern part is divided into two regions: the east and the west coasts. Climate variability on both sides of the river basins is mainly dominated by the north-eastern monsoon and the south-western monsoon winds. The southwest monsoon wind typically starts in mid-May and ends in mid-October. In contrast, the northeast monsoon typically begins in mid-October and ends in mid-February.

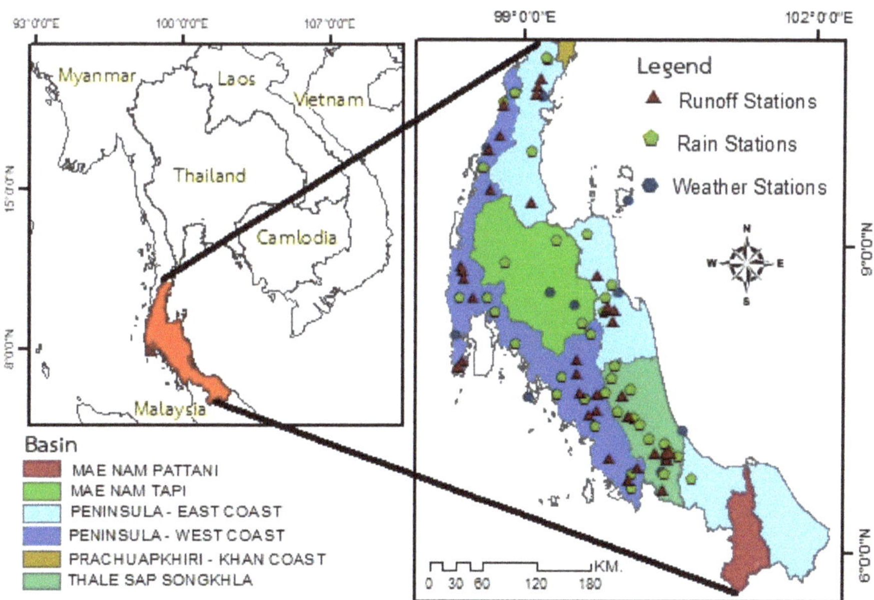

Figure 1. Location of rainfall, runoff, and weather stations selected in the southern basin of Thailand.

The Peninsula-East Coast watershed covers an area of 26,023.91 km^2 and encompasses 11 provinces. It also consists of areas covering all parts of Chumphon, Trang, Nakhon Si Thammarat, Narathiwat, Prachuap Khiri Khan, Pattani, Phatthalung, Yala, Ranong, Songkhla, and Surat Thani. The flat coast has a small plain from Chumphon to Narathiwat. Additionally, most rivers are short rivers with approximately 150 km flowing into the Gulf of Thailand. There are nine runoff stations in the Peninsula-East Coast watershed. The Peninsula-West Coast Watershed, 18,841.20 km^2, consists of seven provinces: Ranong, Phang Nga, Phuket, Krabi, Nakhon Si Thammarat, Trang, and Satun. It also includes Chumphon, Surat Thani, Phatthalung, and Songkhla, with similar topography to the Peninsula-East Coast Watershed. It is a coastal area next to the Andaman Sea. The Phuket Mountains go from Ranong Province to Phang Nga Province, the origin of various rivers and streams. They are generally not long, and they flow mainly to the Andaman Sea in the west and southwest directions. The nineteen runoff stations were used for our analysis. Thale Sap Songkhla watershed, an area of 8484.35 km^2, primarily covers three provinces, the province of Nakhon Si Thammarat (Some portions of the district of Cha-Uat and the district of Hua Sai), the province of Phatthalung, both provinces, and the province of Songkhla, except for the district of Nathawi, the district of Chana, the district of Thepha and the district of Saba Yoi). Thus, 147 sub-districts and 26 districts, with nine runoff stations, were our study setting. Figure 1 shows the rainfall location, runoff, and weather stations selected in Thailand's southern basin.

We collected the monthly meteorological and hydrological data from the Royal Irrigation Department (RID) and the Thai Meteorological Department (TMD), including runoff (37 stations), rain (38 stations), and air temperature (13 stations) as shown in the statistical values in Figure 2. We also investigated and analyzed the time corresponding among those three meteorological and hydrological data to select the suitable periods of model's calibration and verification, as shown in Table 1. The Thiessen polygon was used to determine the mean areal precipitation in the considered basin from rain gauge observations. The monthly evapotranspiration, which is one of the input data for the GR2M model, was calculated from the average monthly air temperature (Ti) data by Thornthwaite [24], as shown below:

- Monthly values of the heat index

$$I_i = \left(\frac{T_i}{5}\right)^{1.514} \tag{1}$$

- Annual temperature efficiency index

$$J = \sum_{i=1}^{12}(I_i) \tag{2}$$

- Evapotranspiration

$$PET_i(0) = 1.6\left(\frac{10T_i}{J}\right)^C \tag{3}$$

- The C value can be obtained from:

$$C = 0.000000675J^3 - 0.0000771J^2 + 0.01792J + 0.49239 \tag{4}$$

- Potential Evapotranspiration

$$PET_i(L) = K \times PET_i(0) \tag{5}$$

where T_i = Monthly average temperature ($^\circ$C), K = PET constants at different latitudes.

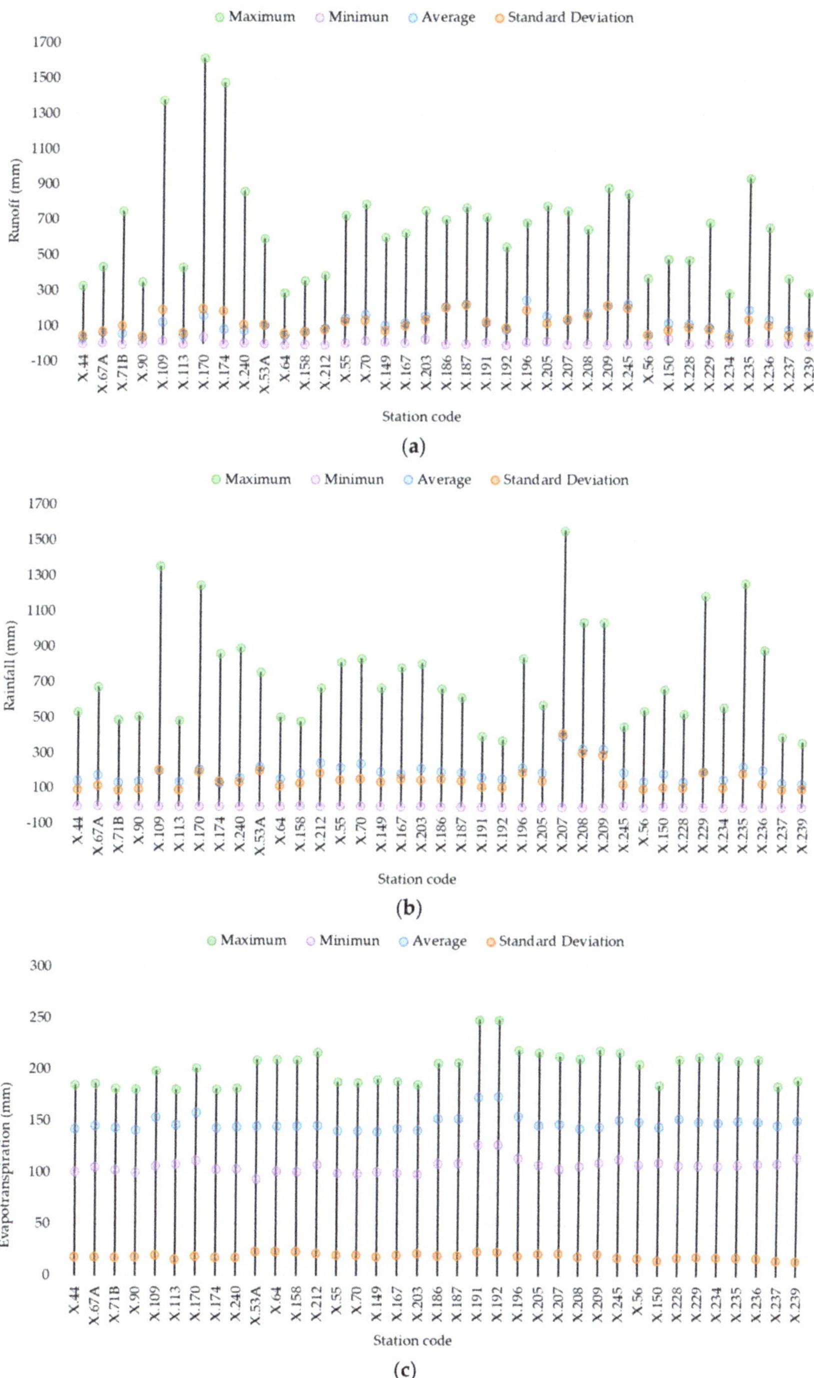

Figure 2. Statistical values of monthly (**a**) runoff, (**b**) rainfall, and (**c**) evapotranspiration data used in this analysis.

Table 1. The periods of data used for the GR2M model's calibration and verification.

No.	Code	Basin Name	Period			
			All	Warm-Up	Calibration	Validation
1	X.44	TSS	April 2004–March 2009	April 2004–September 2004	October 2004–February 2007	March 2007–March 2009
2	X.67A	TSS	April 2005–March 2009	April 2005–October 2005	November 2005–September 2007	October 2007–March 2009
3	X.71B	TSS	April 2004–March 2009	April 2004–October 2004	November 2004–April 2007	May 2007–March 2009
4	X.90	TSS	April 2003–March 2009	April 2003–October 2003	November 2003–July 2007	August 2007–March 2009
5	X.109	TSS	April 2003–March 2008	April 2003–October 2003	November 2003–December 2006	January 2007–March 2008
6	X.113	TSS	April 2003–March 2009	April 2003–October 2003	November 2003–November 2006	December 2006–March 2009
7	X.170	TSS	April 2003–March 2009	April 2003–October 2003	November 2003–February 2007	March 2007–March 2009
8	X.174	TSS	April 2003–March 2009	April 2003–October 2003	November 2003–January 2007	February 2007–March 2009
9	X.240	TSS	April 2004–March 2009	April 2004–September 2004	October 2004–February 2007	March 2007–March 2009
10	X.53A	PEC	April 2003–March 2010	April 2003–July 2003	August 2003–December 2006	January 2007–March 2010
11	X.64	PEC	April 2004–March 2009	April 2004–September 2004	October 2004–September 2007	October 2007–March 2009
12	X.158	PEC	April 2004–March 2009	April 2004–August 2004	September 2004–September 2007	October 2007–March 2009
13	X.212	PEC	April 2005–March 2012	April 2005–July 2005	August 2005–June 2009	July 2009–March 2012
14	X.55	PEC	April 2005–March 2009	April 2005–October 2005	November 2005–November 2007	December 2007–March 2009
15	X.70	PEC	April 2005–March 2009	April 2005–September 2005	October 2005–May 2008	June 2008–March 2009
16	X.149	PEC	April 2005–March 2009	April 2005–October 2005	November 2005–April 2008	May 2008–March 2009
17	X.167	PEC	April 2003–March 2009	April 2003–October 2003	November 2003–October 2006	November 2006–March 2009
18	X.203	PEC	April 2005–March 2009	April 2005–October 2005	November 2005–December 2007	January 2008–March 2009
19	X.186	PWC	April 2003–March 2009	April 2003–September 2003	October 2003–December 2006	January 2007–March 2009
20	X.187	PWC	April 2003–March 2009	April 2003–August 2003	September 2003–August 2006	September 2006–March 2009
21	X.191	PWC	April 2003–March 2009	April 2003–August 2003	September 2003–May 2007	June 2007–March 2009
22	X.192	PWC	April 2003–March 2009	April 2003–September 2003	October 2003–August 2006	September 2006–March 2009
23	X.196	PWC	April 2003–March 2009	April 2003–August 2003	September 2003–August 2006	September 2006–March 2009
24	X.205	PWC	April 2005–March 2012	April 2005–August 2005	September 2005–May 2009	June 2009–March 2012
25	X.207	PWC	April 2003–March 2010	April 2003–August 2003	September 2003–September 2007	October 2007–March 2010
26	X.208	PWC	April 2005–March 2009	April 2005–July 2005	August 2005–September 2007	October 2007–March 2009
27	X.209	PWC	April 2005–March 2011	April 2005–July 2005	August 2005–May 2009	June 2009–March 2011
28	X.245	PWC	April 2005–March 2009	April 2005–July 2005	August 2005–June 2007	July 2007–March 2009
29	X.56	PWC	April 2004–March 2009	April 2004–August 2004	September 2004–August 2007	September 2007–March 2009
30	X.150	PWC	April 2005–March 2009	April 2005–September 2005	October 2005–July 2007	August 2007–March 2009
31	X.228	PWC	April 2003–March 2009	April 2003–October 2003	November 2003–September 2006	October 2006–March 2009
32	X.229	PWC	April 2003–March 2009	April 2003–October 2003	November 2003–March 2007	April 2007–March 2009
33	X.234	PWC	April 2004–March 2009	April 2004–September 2004	October 2004–August 2007	September 2007–March 2009
34	X.235	PWC	April 2004–March 2009	April 2004–September 2004	October 2004–September 2007	October 2007–March 2009
35	X.236	PWC	April 2004–March 2009	April 2004–July 2004	August 2004–December 2006	January 2007–March 2009
36	X.237	PWC	April 2004–March 2009	April 2004–August 2004	September 2004–January 2007	February 2007–March 2009
37	X.239	PWC	April 2004–March 2009	April 2004–July 2004	August 2004–December 2007	January 2007–March 2009

Remark: TSS = Thale Sap Songkhla; PWC = Peninsular-West Coast; PEC = Peninsular-East Coast.

3. GR2M Model

The GR2M, a conceptual model, was first introduced by Demagref in the late 1980s and it has been widely applied for water resources management [25]. The model aims to simulate the relationship between monthly rainfall and runoff and reproduce the hydrological system's response. It has been continuously being developed to improve its efficiency by Kabouya [26], Makhlouf and Michel [27], Mouelhi [28] until Mouelhi et al. [29]. The model selected in this study was the latest version, GR2M 2006. It is the most popular and efficient compared to other models [13]. The GR2M model's advantage requires only two parameters: production store: X_1 (mm) and groundwater exchange rate (X_2). Additionally, it needs only three monthly meteorological and hydrological data, i.e., rainfall, runoff, and evapotranspiration [30,31]. The GR2M model results give runoff hydrograph and other elements such as soil moisture content, surface runoff, the groundwater flow.

The structure of the GR2M model consisted of two reservoirs, as presented in Figure 3. The first reservoir represents soil moisture (S) of the basin-controlled production store: X_1 (mm). Furthermore, the second reservoir is water flow through the river (R). Its capacity is up to 60 mm and is regulated by the groundwater exchange rate (X_2). This model starts with the precipitation infiltrated into the soil, causes soil moisture at the level: S_1 (mm). When the soil reaches a saturation point, the remnants of infiltration rain become rainfall excess: P_1 (mm). The soil moisture loss from evapotranspiration: E until the remaining moisture level: S_2 (mm).

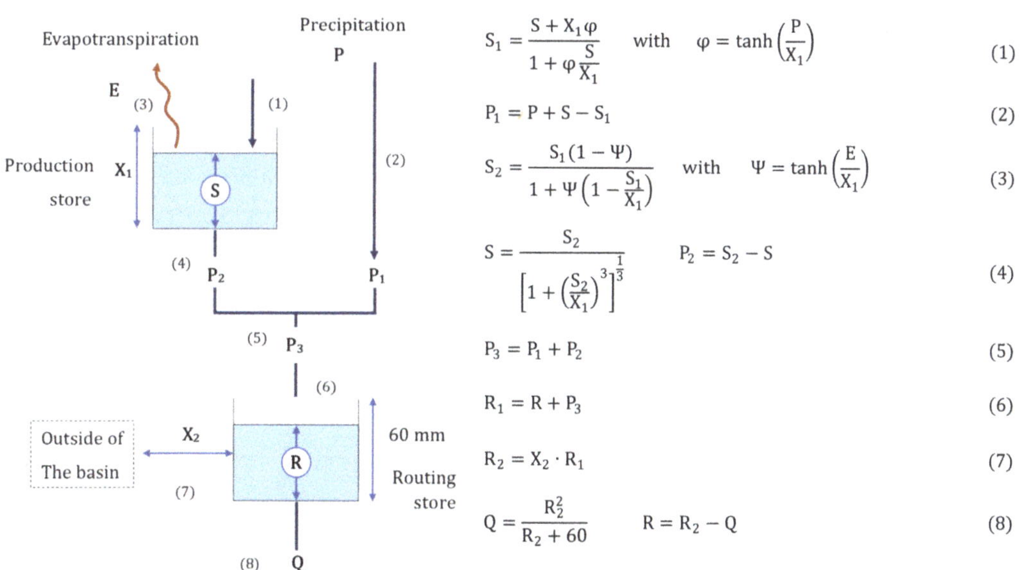

$$S_1 = \frac{S + X_1\varphi}{1 + \varphi\frac{S}{X_1}} \quad \text{with} \quad \varphi = \tanh\left(\frac{P}{X_1}\right) \tag{1}$$

$$P_1 = P + S - S_1 \tag{2}$$

$$S_2 = \frac{S_1(1 - \Psi)}{1 + \Psi\left(1 - \frac{S_1}{X_1}\right)} \quad \text{with} \quad \Psi = \tanh\left(\frac{E}{X_1}\right) \tag{3}$$

$$S = \frac{S_2}{\left[1 + \left(\frac{S_2}{X_1}\right)^3\right]^{\frac{1}{3}}} \qquad P_2 = S_2 - S \tag{4}$$

$$P_3 = P_1 + P_2 \tag{5}$$

$$R_1 = R + P_3 \tag{6}$$

$$R_2 = X_2 \cdot R_1 \tag{7}$$

$$Q = \frac{R_2^2}{R_2 + 60} \qquad R = R_2 - Q \tag{8}$$

Figure 3. Structure of the GR2M model. (Source: Adapted from Bachir et al. [31]; Rwasoka et al. [32]).

Additionally, some moisture content is released as surface water: P_2 (mm) and gradually released with rainfall excess. This water section is called surface runoff or net rainfall: P3 (mm), which moved into the flow path combined with the remaining water from the initial or existing water in the river: R (mm). It causes the water content at level R1 (mm). The water volume movement may change because some water may be lost, causing the residual water volume at the level: R_2 (mm). Ultimately, the total amount of water discharge into the runoff streamflow gauging station conducted the assessment.

4. Model's Calibration and Verification

In achieving our aims in evaluating a Two-Parameters Monthly Rainfall-Runoff Model's performance, the GR2M model applied in Thailand's southern basin was calibrated and verified. It included two steps, i.e., the warm-up period and calibrating and verifying the GR2M Model.

4.1. Warm-Up Period

In this process, the appropriate initial parameters of X_1 and X_2 are determined. It enables the model to mimic the basin's existing hydrological behavior at the considered runoff stations before conducting the model's calibration and verification. The R-value, the initial or existing water capacity in the river, is varied between 10 mm and 60 mm to determine the suitable warm-up period. In our study, we found the warm-up periods of approximately 4 to 7 months.

4.2. Calibrating and Verifying the GR2M Model

As widely known, the calibration and verification processes are imperative for applying the mathematical model to find the most suitable model's parameters. The model can simulate the behavior of our concerning water system. For the GR2M model, only two parameters: the production store (X_1) and the groundwater exchange rate (X_2), must be calibrated and validated. Microsoft Excel solvers help by giving an objective function and practical constraints, which can automatically solve the fair values of X_1 and X_2 parameters for each runoff station. The GR2M model was calibrated and verified for 37 different runoff stations in the Southern Basins in this study. The details of the intervals for the calibration and verification of the model are presented in Table 1. The lowest and the highest periods used for running the GR2M model are 41 and 80 months. The used range of the calibration and verification periods consists of 22 and 48, and 10 and 39 months, respectively.

5. Performance Criteria for Evaluating the Applicability of the GR2M Model

In this study, three performance criteria were used for evaluating the performance and applicability of the GR2M Model. They included Nash–Sutcliffe Efficiency (NSE), Correlation Coefficient (r), and Overall Index (OI). The details for each performance criteria can be delineated as shown the following:

Nash-Sutcliffe Efficiency (NSE) [32] is a popular index used to tell model accuracy or efficiency-effectiveness of the model (Model Performance) in estimating the desired value. As the equation below:

$$\text{NSE} = 1 - \frac{\sum_{i=1}^{n}\left(Q_{cal} - Q_{obs}\right)^2}{\sum_{i=1}^{n}\left(Q_{obs} - \acute{Q}_{obs}\right)^2} \tag{6}$$

NSE is between $-\infty$ to 1. Suppose the Nash values are close to 1. In that case, the model results and the measurement results are similar. They are considered the model of efficiency or accuracy in forecasting [33].

Correlation Coefficient (r) is a simple linear regression equation. It is a simple linear regression equation that can be used to estimate the Y as well. If X and Y are correlated well. The correlation coefficient between X and Y can be calculated from the following equation.

$$r = \frac{\sum_{i=1}^{n}\left(Q_{obs} - \overline{Q}_{obs}\right)\left(Q_{cal} - \overline{Q}_{cal}\right)}{\sqrt{\sum_{i=1}^{n}\left(Q_{obs} - \overline{Q}_{obs}\right)^2} \cdot \sqrt{\sum_{i=1}^{n}\left(Q_{cal} - \overline{Q}_{cal}\right)^2}} \tag{7}$$

The r-value is between -1 and 1. The squares of r or R2 will always be between 0–1, and in this sense, if R2 is 0, then the two variables have no linear correlation. If R2 is equal to 1, then there is an entirely linear correlation. If the r-value approaches 1, the model results and the measurement results are related. The plus sign (+) or minus sign can also tell the direction of the data set's relationship. The plus sign (+) means the dataset is related. Suppose the data obtained from the model is precious. The data obtained from the measurement is also precious. The minus sign ($-$) means the dataset is in the opposite relationship. If the information is valuable More information will be less [34–36].

Overall Index:

$$OI = \frac{1}{2}\left[2 - \frac{RMSE}{Q_{obs,max} - Q_{obs,min}} - \frac{\sum_{i=1}^{n}(Q_{obs} - Q_{cal})^2}{\sum_{i=1}^{n}(Q_{obs} - \overline{Q}_{obs})^2}\right] \qquad (8)$$

The OI value is a criterion that indicates model performance. It is between $-\infty$ to 1. If the higher OI is closer to 1, the model's performance is favorable [37,38]. where; Q_{obs} is the amount of runoff obtained from the measurement, Q_{cal} is the amount of runoff obtained from the calculation, $\overline{Q}_{obs}$ is the average runoff from the measure, $\overline{Q}_{cal}$ is the average runoff from the calculation, $Q_{obs,max}$ is the runoff from the highest measurement $Q_{obs,min}$ is the runoff from the lowest measurements, and n is the amount of information.

6. Results and Discussion

6.1. The Results of Calibrating and Verifying the GR2M Model

Table 2 shows the results of the model's calibration and verification. It explicitly indicated that the GR2M model could be applied for modeling monthly rainfall-runoff in the southern region of Thailand.

Table 2. Results of calibrating and verifying the GR2M model.

No.	Code	Performance Criteria						No.	Code	Performance Criteria					
		Calibration			Validation					Calibration			Validation		
		NSE	r	OI	NSE	r	OI			NSE	r	OI	NSE	r	OI
1	X.44	0.942	0.973	0.949	0.465	0.705	0.657	20	X.187	0.563	0.756	0.668	0.349	0.654	0.548
2	X.67A	0.978	0.99	0.974	0.719	0.852	0.795	21	X.191	0.177	0.492	0.505	0.664	0.831	0.749
3	X.71B	0.688	0.954	0.793	0.605	0.797	0.733	22	X.192 [b]	0.165	0.462	0.493	0.167	0.670	0.451
4	X.90	0.772	0.887	0.85	0.468	0.502	0.478	23	X.196	0.333	0.691	0.544	0.283	0.691	0.505
5	X.109	0.925	0.987	0.94	0.577	0.849	0.696	24	X.205 [c]	0.518	0.755	0.693	−0.119	0.663	0.289
6	X.113	0.736	0.91	0.821	0.479	0.796	0.648	25	X.207	0.758	0.878	0.798	0.836	0.920	0.856
7	X.170	0.805	0.903	0.867	0.038	0.451	0.392	26	X.208	0.796	0.906	0.838	0.751	0.894	0.808
8	X.174	0.725	0.975	0.821	0.385	0.731	0.61	27	X.209 [a]	0.880	0.943	0.896	0.870	0.935	0.883
9	X.240	0.975	0.993	0.973	0.511	0.735	0.687	28	X.245	0.476	0.715	0.636	0.199	0.503	0.457
10	X.53A	0.822	0.908	0.868	0.714	0.847	0.793	29	X.56	0.813	0.911	0.868	0.676	0.866	0.746
11	X.64 [a]	0.787	0.888	0.838	0.941	0.970	0.942	30	X.150	0.833	0.915	0.871	0.226	0.623	0.461
12	X.158	0.573	0.759	0.714	0.752	0.869	0.818	31	X.228	0.111	0.527	0.450	0.346	0.702	0.554
13	X.212 [b]	0.383	0.668	0.594	0.173	0.431	0.467	32	X.229 [c]	0.564	0.794	0.730	−0.437	0.407	0.120
14	X.55	0.654	0.903	0.761	0.987	0.996	0.980	33	X.234	0.854	0.934	0.890	0.713	0.882	0.773
15	X.70 [a]	0.780	0.943	0.845	0.923	0.976	0.916	34	X.235	0.430	0.758	0.648	0.404	0.679	0.596
16	X.149	0.557	0.912	0.702	0.957	0.986	0.950	35	X.236	0.801	0.896	0.857	0.513	0.719	0.647
17	X.167	0.892	0.973	0.912	0.278	0.803	0.534	36	X.237 [c]	0.734	0.277	0.411	−0.305	0.497	0.202
18	X.203	0.732	0.970	0.809	0.897	0.973	0.904	37	X.239	0.376	0.754	0.602	0.055	0.673	0.388
19	X.186 [b]	0.400	0.660	0.580	0.405	0.680	0.620								
	Maximum									0.978	0.993	0.974	0.987	0.996	0.980
	Minimum									0.111	0.277	0.411	−0.437	0.407	0.120
	Average									0.657	0.825	0.757	0.472	0.750	0.639
	Standard Deviation									0.233	0.170	0.153	0.350	0.166	0.213

Remark: TSS = Thale Sap Songkhla; PWC = Peninsular-West Coast; PEC = Peninsular-East Coast. [a] the green text shows the best top-three model performance stations, [b] the red text shows the worst top-three model performance stations, [c] the blue text shows stations having the overfitting models.

The average performance criteria gave NSE, r, and OI values for the calibration stage of 0.657, 0.825, and 0.757. Those values for the verification stage of 0.472, 0.750, and 0.639, respectively. Lian, et al. [39] suggested that the model had a good prediction since NSE was in the range of 0.36 to 0.75. By obtaining an r-value of more than 0.70, it indicated a strong positive linear relationship between simulated and observed runoff [36]. Moreover, the OI value of more than 0.60 showed the model had relatively high forecasting accuracy. The three performance criteria previously mentioned emphasized a strong consistency between the runoff data obtained from the measurements and model-simulated for our study.

Considering the best top-three model performance stations obtaining from X.64, X.70, and X.209, NSE, r, and OI values for both calibration and verification processes gave more than 0.76, it showed the GR2M model performed quite satisfactorily for simulating monthly runoff. Conversely, the worst top-three model performance stations were X.212, X.186, and X.192. They gave NSE, r, and OI values for both calibration and verification processes less than 0.690. However, some runoff stations, i.e., X.205, X.229, and X.237, had a negative NSE value. It represented overfitting models for those three runoff stations and could not be generally applied. Although many attempts were being made for the model's calibration and verification processes, the quality and accuracy of measured hydrological and meteorological data are the most important things to concern and check the consistency. Figure 4 illustrates the relationship between rainfall and runoff obtained from running the GR2M model. Herein present six examples of runoff stations, i.e., X.64, X.70, X.209, X.212, X186, and X.192. The best top-three and the worst top-three model performance stations are presented.

Likewise, the bar chart in blue represents rainfall time-series variation. The line graphs in orange and green also show the observed and simulated runoff time-series variation, respectively. For both runoff time-series variations, the solid and dot lines mean calibration and validation periods, respectively. A slight difference runoff time-series value was observed for the best top-three model performance stations. A significant difference was observed among runoff time-series values for the worst top-three model performance stations. However, both cases underestimated runoff value; that is, the simulated runoff was lower than the observed runoff. It could realize when using the calibrated and verified GR2M model, especially for water resources management and planning for rainy and dry seasons.

6.2. The Optimal Values of Production Store Capacity (X_1) and Groundwater Exchange Rate (X_2)

Figure 5 shows suitable X_1 and X_2 parameters of the GR2M model for each runoff station obtained from the model's calibration and verification.

The production store (X_1) value results ranged from 2.00 mm to 10.00 mm. It showed a spatial variation of X_1 value, and its values ranged from the minimum (2.00 mm) and maximum (10.00 mm) values. The average and standard deviation values of X_1 were 5.71 mm, and 2.49 mm, respectively. Furthermore, the skewness and kurtosis values of X_1 were -0.52 and -1.03, respectively. It could physically explain river basin characteristics in terms of production store (X_1). It had left skew, platykurtic, and non-symmetric distributions. The groundwater exchange rate (X_2) value results ranged from 0.54 to 1.00. Those X_2 values mostly reached the maximum value (1.00). The average and standard deviation values of X_2 were 0.93 and 0.12, respectively. Moreover, the skewness and kurtosis values of X_2 were -2.01 and 3.69, respectively. It could physically explain river basin characteristics in terms of the groundwater exchange rate (X_2). It had left skew, leptokurtic, and non-symmetric distributions. The positive value of groundwater exchange rate (X_2) displayed no groundwater flows outside the basin.

6.3. The Spatial Distribution of X_1 and X_2 Values Using the Inverse Distance Weighting (IDW) Method

Figure 6 shows the spatial distribution of X_1 and X_2 values using the IDW method. As seen from Figure 5a, the low production store (X_1) value (yellow and green color) was generally located on the Peninsular-West Coast. The significant area roughly was covered by the average production store (X_1) value (5.71 mm). Most areas were a light blue color. Only the northern part of Surat Thani province shows the high production store (X_1) value, which shows the dark blue zone. For the groundwater exchange rate (X_2), as depicted in Figure 5b, most areas were governed by the dark blue zone. It indicated that most areas in the southern basin, Thailand, had a high groundwater exchange rate (X_2).

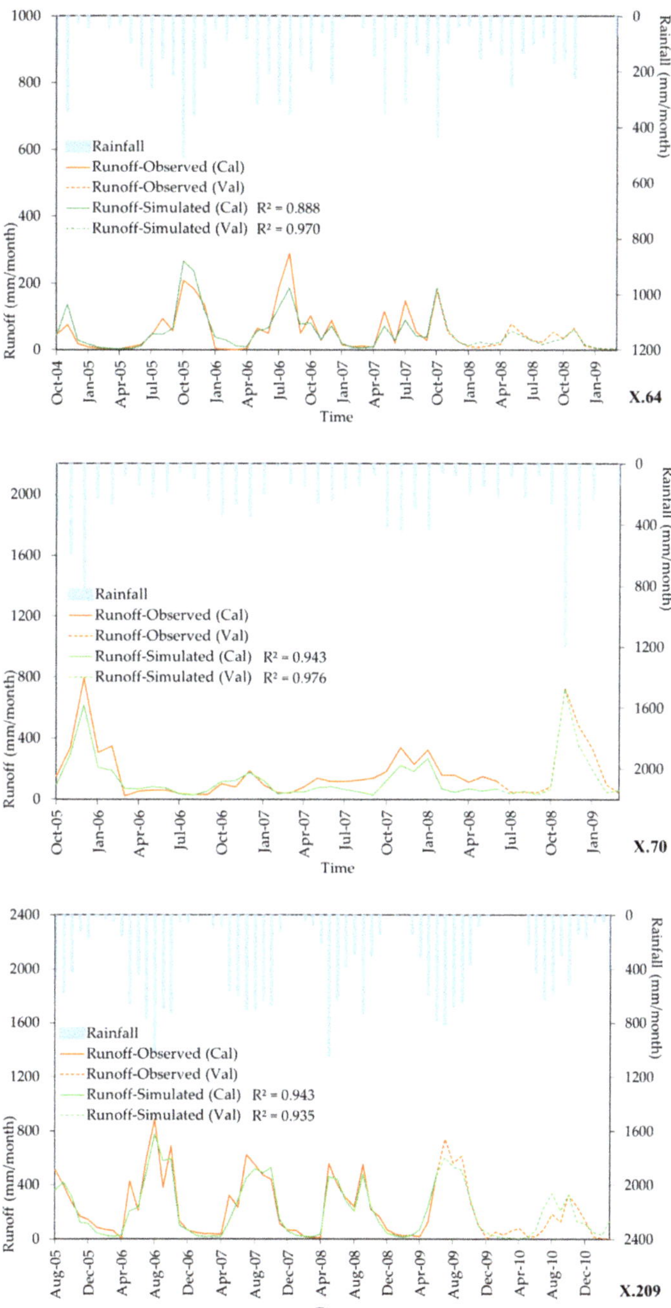

Figure 4. *Cont.*

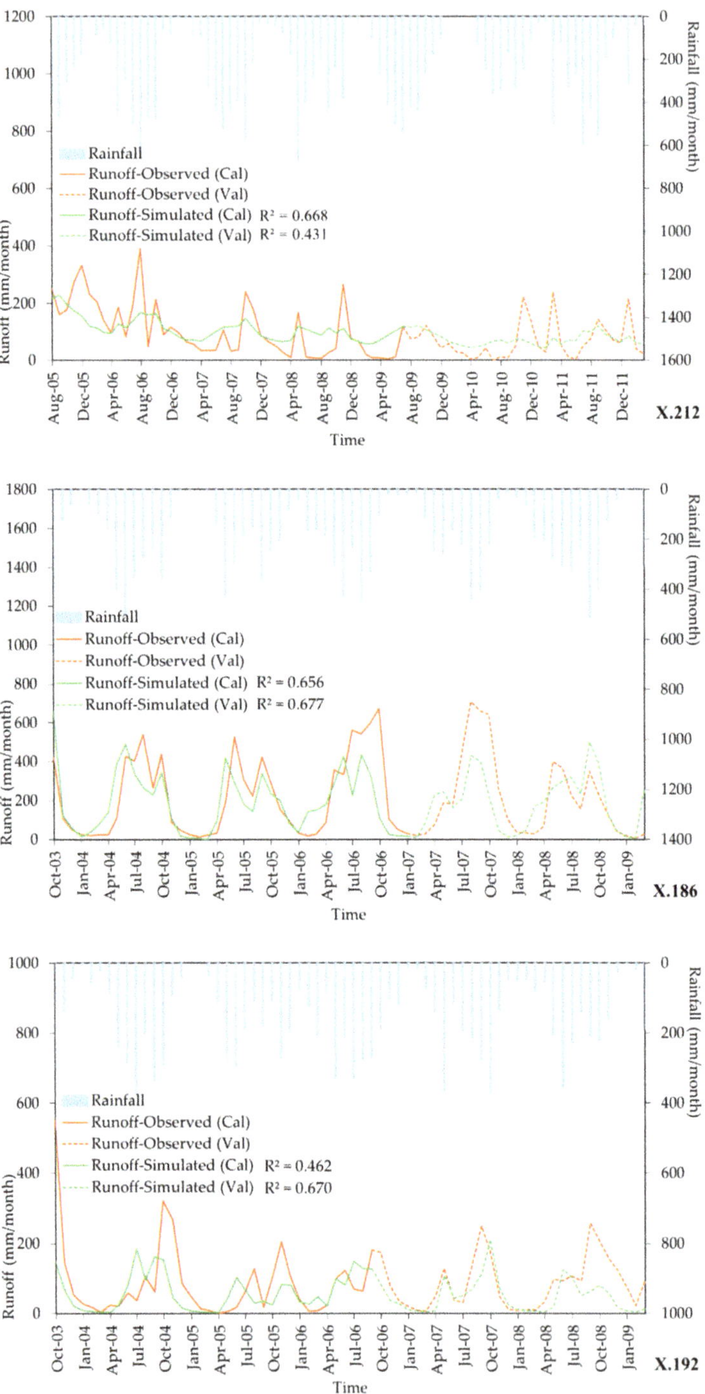

Figure 4. The relationship between rainfall and runoff of models (GR2M) stations.

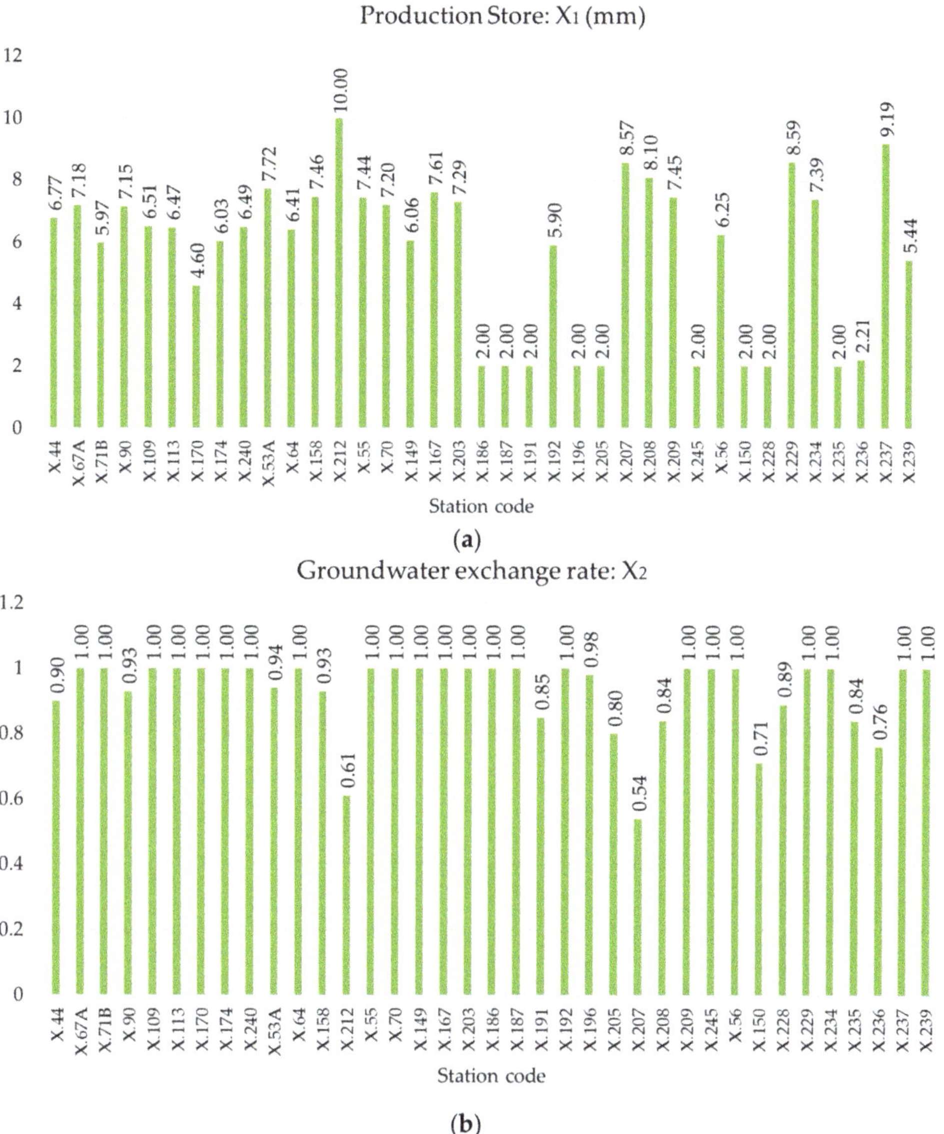

Figure 5. The suitable X_1 and X_2 parameters of the GR2M model: (**a**) Production Store: X_1, and (**b**) Groundwater exchange rate: X_2.

Furthermore, it agreed to the average X_2 value of 0.93. The northern part of Surat Thani province and some Chumporn, Trang, and Satun provinces show the low groundwater exchange rate (X_2) value, as portrayed in the yellow and green zone. Suppose we do not have a measured gauged or ungauged. In that case, we can use these figures to determine the values of X_1 and X_2 roughly. If we know areal rainfall and evaporation, we can also estimate the runoff via the GR2M model.

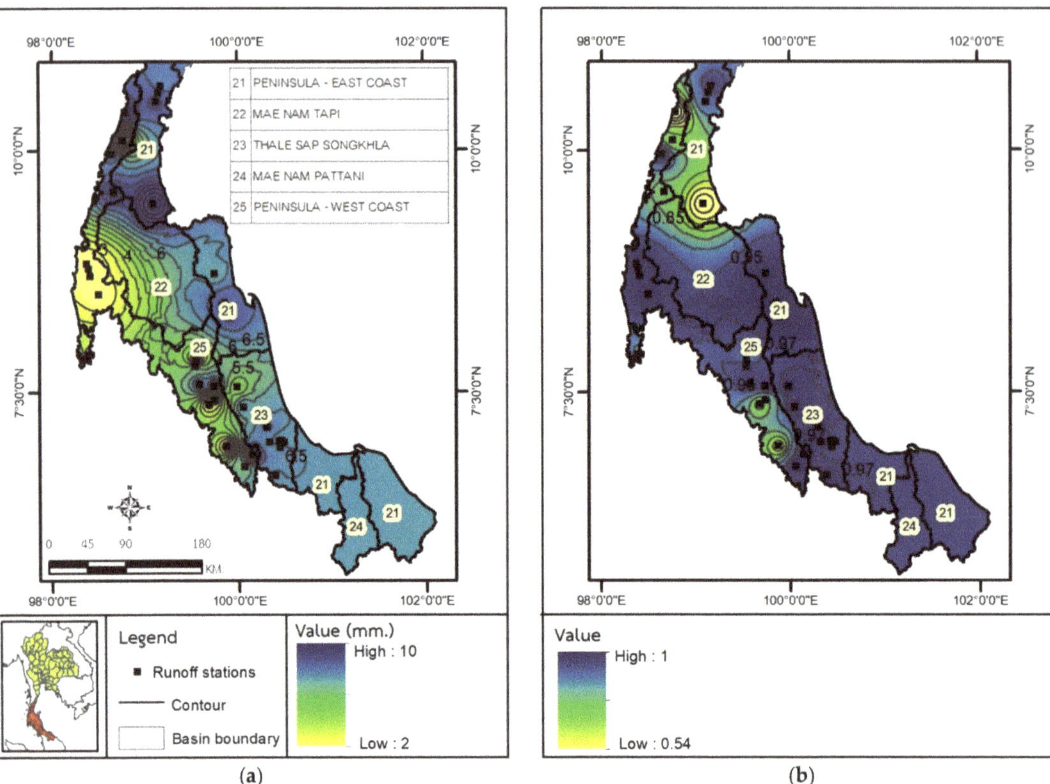

Figure 6. The spatial distribution of X_1 and X_2 values using IDW method: (**a**) Production Store: X_1, and (**b**) Groundwater exchange rate: X_2.

7. Sensitivity Analysis

The sensitivity analysis [10] was conducted in this study to understand the effects of the two model parameters (i.e., X_1 and X_2). We randomly selected three runoff stations (X.44, X.64, and X.240) as the representative for all 37 runoff stations due to the analysis sensitivity. By fixing the optimal X_2 value obtained from calibration and verification stages and then varying the X_1 value in it ranges from the minimum to maximum (2 mm to 10 mm) [31,36], we received the results of X_1's sensitivity analysis. Similarly, by fixing the optimal X_1 value obtained from calibration and verification stages and then varying the X_1 value in it ranges from the minimum to maximum (-1 to 1) [31,34], we got the results of X_2's sensitivity analysis. It was rarely reported about the sensitivity analysis for the GR2M model's two parameters to our best knowledge. Thus, it was the early attempt to conduct their sensitivity analysis. As evidentially presented in Figure 7, the X_1 value was sensitive. Apart from the optimal value obtained from the calibration and verification stages, the other value gave a lower model's performance. Considering the X_2 value, we found that the higher value (approximately more than 0.90) was trial, it gave the higher model's performance. It also confirmed and corresponded with the results, as found in Figures 5 and 6.

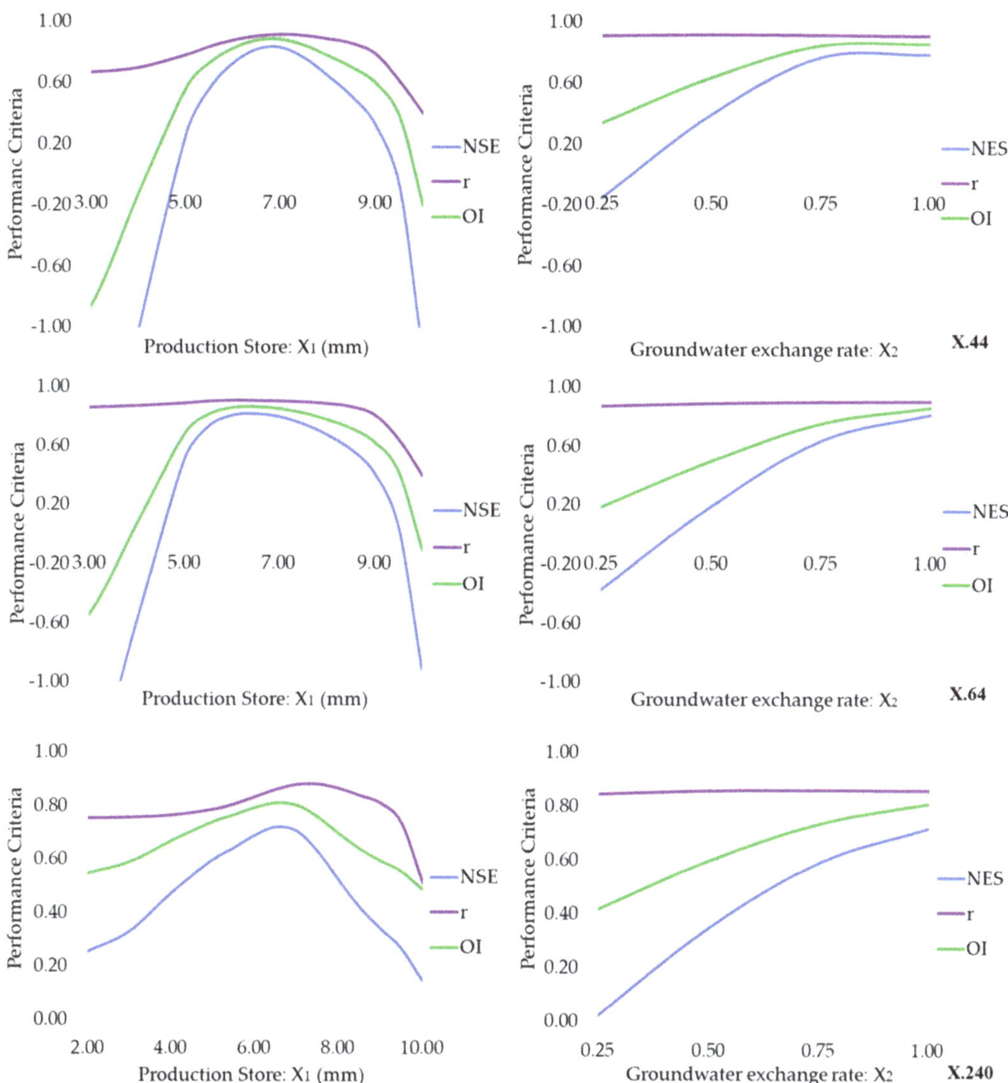

Figure 7. The sensitivity analysis of the GR2M model's two parameters: X_1 and X_2.

Rainfall-runoff modeling is among the most challenging task for hydrologists, particularly in regions with scarce rainfall and runoff data records. The complexity of the rainfall-runoff modeling also comes from the non-stationary features of its components, such as seasonality, potential trend, and the non-linear behavior of the variables involved in the modeling process [11,40]. Geomorphological features characterizing the watershed influence significantly the runoff regime; namely, in urban areas, high imperviousness areas cause increased runoff by originating floods while the same behavior is not observed in fewer imperviousness areas [1,3]. Thus, it is crucial to know the sensitivity of parameters in the rainfall-runoff modeling, especially in the urban areas, making the calibration process more efficient by focusing only on the parameters for which the modeling results are more sensitive [10]. The findings resulted from this study contribute to enhance the understanding of the hydrological parameters and processes that govern a watershed system. Also, it offers new insights on the application of the GR2M model in regions characterized by a similar climate and geomorphological conditions to support decision-makers and optimize the planning and operation rules of water resources systems [21,40]. Last, for areas, especially large basins suffering from a lack of hydrometeorological data records it is important to assess the areal inhomogeneity of the investigated gauging station network [41,42]. In that regard, knowing the fractal dimension of the hydrometeorological network and its limits of validity is the key to understanding the limits of reliability of an inhomogeneous distribution of gauging stations [42].

8. Conclusions

With only two parameters, namely, the production store (X_1) and the groundwater exchange rate (X_2), our research work explicitly indicated GR2M model could be applied for modeling monthly rainfall-runoff in the southern region of Thailand. The model's calibration results for 37 runoff stations gave the average NSE, r, and OI of 0.657, 0.825, and 0.757. Those values for verification of 0.472, 0.750, and 0.639, respectively. The range of X_1 was between 2.00 and 10.00, and the range of X_2 was between 0.54 and 1.00. It was sensitive to the X_1 value. The other value indicates lower model efficiency, apart from the optimum value obtained from the calibration and verification phases. We also found that the higher value of X_2 (approximately more than 0.90) gave the higher model's performance. Personnel concerning water resources planning and management can apply our work for a guideline for utilizing the GR2M model to determine monthly runoff in other runoff stations located in the southern region, Thailand. It is because there are similar hydrological, geological, and topological basin characteristics. However, to further enhance the GR2M model's reliability, a more extended period of recorded hydrological data is required. Also, more runoff gauging station installation will cover the variety of existing watershed characteristics.

Author Contributions: Conceptualization, methodology, supervision: P.D. and Q.B.P.; data curation, formal analysis, investigation: S.P., N.S., and F.B.; writing-original draft preparation: S.P., N.S., F.B., and L.K.; writing-review and editing: L.K., P.D., Q.B.P., K.M.K., and A.K.; proofread the text and helped in structuring the publication: P.D., Q.B.P., K.M.K., and A.K. All authors have read and agreed to the published version of the manuscript.

Funding: This research work was supported by the Deanship of Scientific Research at King Khalid University under Grant number RGP. 2/173/42.

Acknowledgments: This research was partially supported by the new strategic research (P2P) project, Walailak University, Thailand, under contract number CGS-P2P-2564-038. The authors also extend their thanks to the Deanship of Scientific Research at King Khalid University for funding this work through the large research groups under grant number RGP. 2/173/42.

Conflicts of Interest: The authors declare no conflict of interest.

References

1. Vaze, J.; Chiew, F.; Perraud, J.; Viney, N.; Post, D.; Teng, J.; Wang, B.; Lerat, J.; Goswami, M. Rainfall-runoff Modelling across Southeast Australia: Datasets, Models and Results. *Australas. J. Water Resour.* **2010**, *14*, 101–116. [CrossRef]
2. Suwal, N.; Kuriqi, A.; Huang, X.; Delgado, J.; Młyński, D.; Walega, A. Environmental Flows Assessment in Nepal: The Case of Kaligandaki River. *Sustainability* **2020**, *12*, 8766. [CrossRef]
3. Ahmad, M.-U.-D.; Peña-Arancibia, J.L.; Stewart, J.P.; Kirby, J.M. Water balance trends in irrigated canal commands and its implications for sustainable water management in Pakistan: Evidence from 1981 to 2012. *Agric. Water Manag.* **2021**, *245*, 106648. [CrossRef]
4. López-Lambraño, A.; Martínez-Acosta, L.; Gámez-Balmaceda, E.; Medrano-Barboza, J.; López, J.R.; López-Ramos, A. Supply and Demand Analysis of Water Resources. Case Study: Irrigation Water Demand in a Semi-Arid Zone in Mexico. *Agriculture* **2020**, *10*, 333. [CrossRef]
5. Zhang, X.; Guo, P.; Zhang, F.; Liu, X.; Yue, Q.; Wang, Y. Optimal irrigation water allocation in Hetao Irrigation District con-sidering decision makers' preference under uncertainties. *Agric. Water Manag.* **2021**, *246*, 106670. [CrossRef]
6. Chen, Q.; Chen, H.; Wang, J.; Zhao, Y.; Chen, J.; Xu, C. Impacts of Climate Change and Land-Use Change on Hydrological Extremes in the Jinsha River Basin. *Water* **2019**, *11*, 1398. [CrossRef]

7. Kabiri, R.; Bai, V.R.; Chan, A. Assessment of hydrologic impacts of climate change on the runoff trend in Klang Watershed, Malaysia. *Environ. Earth Sci.* **2015**, *73*, 27–37. [CrossRef]

8. Lin, B.; Chen, X.; Yao, H.; Chen, Y.; Liu, M.; Gao, L.; James, A. Analyses of landuse change impacts on catchment runoff using different time indicators based on SWAT model. *Ecol. Indic.* **2015**, *58*, 55–63. [CrossRef]

9. Kwak, J.; Lee, J.; Jung, J.; Kim, H.S. Case Study: Reconstruction of Runoff Series of Hydrological Stations in the Nakdong River, Korea. *Water* **2020**, *12*, 3461. [CrossRef]

10. Ballinas-González, H.A.; Alcocer-Yamanaka, V.H.; Canto-Rios, J.J.; Simuta-Champo, R. Sensitivity Analysis of the Rainfall–Runoff Modeling Parameters in Data-Scarce Urban Catchment. *Hydrology* **2020**, *7*, 73. [CrossRef]

11. Lerat, J.; Thyer, M.; McInerney, D.; Kavetski, D.; Woldemeskel, F.; Pickett-Heaps, C.; Shin, D.; Feikema, P. A robust approach for calibrating a daily rainfall-runoff model to monthly streamflow data. *J. Hydrol.* **2020**, *591*, 125129. [CrossRef]

12. Abdessamed, D.; Abderrazak, B. Coupling HEC-RAS and HEC-HMS in rainfall–runoff modeling and evaluating floodplain inundation maps in arid environments: Case study of Ain Sefra city, Ksour Mountain. SW of Algeria. *Environ. Earth Sci.* **2019**, *78*, 586. [CrossRef]

13. Zhang, C.; Wang, R.-B.; Meng, Q.-X. Calibration of Conceptual Rainfall-Runoff Models Using Global Optimization. *Adv. Meteorol.* **2015**, *2015*, 545376. [CrossRef]

14. Khazaei, M.R.; Zahabiyoun, B.; Saghafian, B.; Ahmadi, S. Development of an Automatic Calibration Tool Using Genetic Algorithm for the ARNO Conceptual Rainfall-Runoff Model. *Arab. J. Sci. Eng.* **2013**, *39*, 2535–2549. [CrossRef]

15. Dezetter, A.; Girard, S.; Paturel, J.; Mahé, G.; Ardoin-Bardin, S.; Servat, E. Simulation of runoff in West Africa: Is there a single data-model combination that produces the best simulation results? *J. Hydrol.* **2008**, *354*, 203–212. [CrossRef]

16. Okkan, U.; Fistikoglu, O. Evaluating climate change effects on runoff by statistical downscaling and hydrological model GR2M. *Theor. Appl. Clim.* **2014**, *117*, 343–361. [CrossRef]

17. Lyon, S.W.; King, K.; Polpanich, O.-U.; Lacombe, G. Assessing hydrologic changes across the Lower Mekong Basin. *J. Hydrol. Reg. Stud.* **2017**, *12*, 303–314. [CrossRef]

18. Zamoum, S.; Souag-Gamane, D. Monthly streamflow estimation in ungauged catchments of northern Algeria using regionalization of conceptual model parameters. *Arab. J. Geosci.* **2019**, *12*, 342. [CrossRef]

19. Boulariah, O.; Longobardi, A.; Meddi, M. Statistical comparison of nonlinear rainfall-runoff models for simulation in Africa North-West semi-arid areas. In Proceedings of the 15th International Conference on Environment Science and Technology, Rhodes, Greece, 31 August–2 September 2017.

20. Topalović, Ž.; Todorović, A.; Plavšić, J. Evaluating the transferability of monthly water balance models under changing climate conditions. *Hydrol. Sci. J.* **2020**, *65*, 928–950. [CrossRef]

21. Hadour, A.; Mahé, G.; Meddi, M. Watershed based hydrological evolution under climate change effect: An example from North Western Algeria. *J. Hydrol. Reg. Stud.* **2020**, *28*, 100671. [CrossRef]

22. Rintis, H.; Suyanto; Setyoasri, Y.P. Rainfall-Discharge Simulation in Bah Bolon Catchment Area by Mock Method, NRECA Method, and GR2M Method. *Appl. Mech. Mater.* **2016**, *845*, 24–29. [CrossRef]

23. O'Connor, P.; Murphy, C.; Matthews, T.; Wilby, R.L. Reconstructed monthly river flows for Irish catchments 1766–2016. *Geosci. Data J.* **2020**, 1766–2016. [CrossRef]

24. Thornthwaite, C.W. An approach toward a rational classification of climate. *Geogr. Rev.* **1948**, *38*, 55–94. [CrossRef]

25. Paturel, J.E.; Servat, E.; Vassiliadis, A. Sensitivity of conceptual rainfall-runoff algorithms to errors in input data—Case of the GR2M model. *J. Hydrol.* **1995**, *168*, 111–125. [CrossRef]

26. Kabouya, M. Modélisation Pluie-Débit aux Pas de Temps Mensuel et Annuel en Algérie Septentrionale. Ph.D. Thesis, Université Paris Sud Orsay, Orsay, France, 1990.

27. Makhlouf, Z.; Michel, C. A two-parameter monthly water balance model for French watersheds. *J. Hydrol.* **1994**, *162*, 299–318. [CrossRef]

28. Mouelhi, S. Vers Une Chaîne Cohérente de Modèles Pluie-Débit Conceptuels Globaux aux Pas de Temps Pluriannuel, Annuel, Mensuel et Journalier. Ph.D. Thesis, ENGREF Paris, Paris, France, 2003.

29. Mouelhi, S.; Michel, C.; Perrin, C.; Andréassian, V. Stepwise development of a two-parameter monthly water balance model. *J. Hydrol.* **2006**, *318*, 200–214. [CrossRef]

30. Fathi, M.M.; Awadallah, A.G.; Abdelbaki, A.M.; Haggag, M. A new Budyko framework extension using time series SARIMAX model. *J. Hydrol.* **2019**, *570*, 827–838. [CrossRef]

31. Bachir, S.; Nouar, B.; Hicham, C.; Azzedine, H.; Larbi, D. Application of GR2M for rainfall-runoff modeling in Kébir Rhumel Watershed, north east of Algeria. *World Appl. Sci. J.* **2015**, *33*, 1623–1630.

32. Nash, J.E.; Sutcliffe, J.V. River flow forecasting through conceptual models part I—A discussion of principles. *J. Hydrol.* **1970**, *10*, 282–290. [CrossRef]

33. Zolfaghari, M.; Mahdavi, M.; Rezaei, A.; Salajegheh, A. Evaluating GR2M model in some small watersheds of Iran (Case study Gilan and Mazandaran Provinces). *J. Basic Appl. Sci. Res.* **2013**, *3*, 463–472.

34. Kunnath-Poovakka, A.; Eldho, T.I. A comparative study of conceptual rainfall-runoff models GR4J, AWBM and Sacramento at catchments in the upper Godavari river basin, India. *J. Earth Syst. Sci.* **2019**, *128*, 33. [CrossRef]

35. Moriasi, D.N.; Arnold, J.G.; Van Liew, M.W.; Bingner, R.L.; Harmel, R.D.; Veith, T.L. Model Evaluation Guidelines for Systematic Quantification of Accuracy in Watershed Simulations. *Trans. ASABE* **2007**, *50*, 885–900. [CrossRef]

36. Ratner, B. The correlation coefficient: Its values range between +1/−1, or do they? *J. Target. Meas. Anal. Mark.* **2009**, *17*, 139–142. [CrossRef]
37. Sarzaeim, P.; Bozorg-Haddad, O.; Bozorgi, A.; Loáiciga, H.A. Runoff projection under climate change conditions with data-mining methods. *J. Irrig. Drain. Eng.* **2017**, *143*, 04017026. [CrossRef]
38. Alazba, A.A.; Mattar, M.A.; El-Nesr, M.N.; Amin, M.T. Field assessment of friction head loss and friction correction factor equations. *J. Irrig. Drain. Eng. ASCE* **2012**, *138*, 166–176. [CrossRef]
39. Lian, Y.; Chan, I.-C.; Singh, J.; Demissie, M.; Knapp, V.; Xie, H. Coupling of hydrologic and hydraulic models for the Illinois River Basin. *J. Hydrol.* **2007**, *344*, 210–222. [CrossRef]
40. Safari, M.J.S.; Arashloo, S.R.; Danandeh Mehr, A. Rainfall-runoff modeling through regression in the reproducing kernel Hilbert space algorithm. *J. Hydrol.* **2020**, *587*, 125014. [CrossRef]
41. Lovejoy, S.; Schertzer, D.; Ladoy, P. Fractal characterization of inhomogeneous geophysical measuring networks. *Nature* **1986**, *319*, 43–44. [CrossRef]
42. Mazzarella, A.; Tranfaglia, G. Fractal Characterisation of Geophysical Measuring Networks and its Implication for an Optimal Location of Additional Stations: An Application to a Rain-Gauge Network. *Theor. Appl. Climatol.* **2000**, *65*, 157–163. [CrossRef]

Article

Integrating a GIS-Based Multi-Influence Factors Model with Hydro-Geophysical Exploration for Groundwater Potential and Hydrogeological Assessment: A Case Study in the Karak Watershed, Northern Pakistan

Umair Khan [1,2], Haris Faheem [1], Zhengwen Jiang [1], Muhammad Wajid [1], Muhammad Younas [2] and Baoyi Zhang [1,*]

[1] MoE Key Laboratory of Metallogenic Prediction of Nonferrous Metals & Geological Environment Monitoring/School of Geosciences & Info-Physics, Central South University, Changsha 410083, China; umairkhanbku@gmail.com (U.K.); harisgeo1997@gmail.com (H.F.); jiangzhengwen@csu.edu.cn (Z.J.); mw260532@gmail.com (M.W.)

[2] Department of Geology, Bacha Khan University, Charsadda 24600, Pakistan; unas.geo@gmail.com

* Correspondence: zhangbaoyi@csu.edu.cn; Tel.: +86-731-88877676

Citation: Khan, U.; Faheem, H.; Jiang, Z.; Wajid, M.; Younas, M.; Zhang, B. Integrating a GIS-Based Multi-Influence Factors Model with Hydro-Geophysical Exploration for Groundwater Potential and Hydrogeological Assessment: A Case Study in the Karak Watershed, Northern Pakistan. *Water* 2021, 13, 1255. https://doi.org/10.3390/w13091255

Academic Editors: Andrzej Wałęga and Tamara Tokarczyk

Received: 5 April 2021
Accepted: 29 April 2021
Published: 30 April 2021

Publisher's Note: MDPI stays neutral with regard to jurisdictional claims in published maps and institutional affiliations.

Abstract: The optimization of groundwater conditioning factors (GCFs), the evaluation of groundwater potential (GW_{pot}), the hydrogeological characterization of aquifer geoelectrical properties and borehole lithological information are of great significance in the complex decision-making processes of groundwater resource management (GRM). In this study, the regional GW_{pot} of the Karak watershed in Northern Pakistan was first evaluated by means of the multi-influence factors (MIFs) model of optimized GCFs through geoprocessing tools in geographical information system (GIS). The distribution of petrophysical properties indicated by the measured resistivity fluctuations was then generated to locally verify the GW_{pot}, and to analyze the hydrogeological and geoelectrical characteristics of aquifers. According to the weighted overlay analysis of MIFs, GW_{pot} map was zoned into low, medium, high and very high areas, covering 9.7% (72.3 km^2), 52.4% (1307.7 km^2), 31.3% (913.4 km^2), and 6.6% (44.8 km^2) of the study area. The GW_{pot} accuracy sequentially depends on the classification criteria, the mean rating score, and the weights assigned to GCFs. The most influential factors are geology, lineament density, and land use/land cover followed by drainage density, slope, soil type, rainfall, elevation, and groundwater level fluctuations. The receiver operating characteristic (ROC) curve, the confusion matrix, and Kappa (K) analysis show satisfactory and consistent results and expected performances (the area under the curve value 68%, confusion matrix 68%, Kappa (K) analysis 65%). The electrical resistivity tomography (ERT) and vertical electrical sounding (VES) data interpretations reveals five regional hydrological layers (i.e., coarse gravel and sand, silty sand mixed lithology, clayey sand/fine sand, fine sand/gravel, and clayey basement). The preliminary interpretation of ERT results highlights the complexity of the hydrogeological strata and reveals that GW_{pot} is structurally and proximately constrained in the clayey sand and silicate aquifers (sandstone), which is of significance for the determination of drilling sites, expansion of drinking water supply and irrigation in the future. Moreover, quantifying the spatial distribution of aquifer hydrogeological characteristics (such as reflection coefficient, isopach, and resistivity mapping) based on Olayinka's basic standards, indirectly and locally verify the performance of the MIF model and ultimately determine new locations for groundwater exploitation. The combined methods of regional GW_{pot} mapping and hydrogeological characterization, through the geospatial MIFs model and aquifer geoelectrical interpretation, respectively, facilitate decision-makers for sustainable GRM not only in the Karak watershed but also in other similar areas worldwide.

Keywords: multi-influencing factors (MIF); vertical electrical sounding (VES); electrical resistivity tomography (ERT); groundwater resource management (GRM); hydro-stratigraphy; well logs

1. Introduction

Increasing anthropogenic repression, climate change, and environmental problems are affecting the supply and demand of domestic and irrigation water. The efficient and innovative use of geospatial and geophysical datasets for understanding groundwater management and hydrological processes in various climatic and vegetation regimes under topographical, geological, hydrological, and land-covered influence has become an important challenge, which offers a wide range of research opportunities [1–5]. There are several conventional geological, geophysical, and hydrogeological methods, and the most commonly used methods are geophysical, but they are time-consuming and mainly applicable on a small scale [6,7]. However, remote sensing (RS) and geographical information system (GIS) provide spatial, temporal, and spectral data availability that can cover large and inaccessible areas within a short period and serve as a useful tool for assessing and managing groundwater resources [8–12].

The groundwater potential (GW_{pot}) is influenced by multiple geological, hydrological, and land-covered processes [10,12,13]. Usually, the occurrence and movement of surface water and groundwater could be assessed by optimized groundwater conditioning factors (GCFs), i.e., rainfall, lineament density, slope, soil types, drainage density, land use/land cover, lithology, elevation, and groundwater level fluctuation. [14,15]. GIS and RS analysis are useful for large-scale estimates of surface water and groundwater. Several methods have been employed to monitor GW_{pot}, such as cumulative rainfall departure (CRD), Monte Carlo (MC) simulation, frequency ratio (FR), certainty factor (CF), weights-of-evidence (WoE), fuzzy logic index models, logistic regression (LR) model, analytical hierarchy process (AHP), and multi-influence factors (MIFs) [8,16–23]. The CRD is a water balance method which defines groundwater level fluctuations in shallow aquifers as a function of rainfall. The statistical methods (e.g., FR, LR, WoE) estimates the coefficient for each GCF by defining the relationship between the dependent variable and independent variables, while the AHP assigns a score to each conditioning factor based on expert's opinion. The MC simulation is considered to be the main tool to quantify the uncertainty in groundwater predictions. To reduce the mathematical complexity by incorporating a decision-making reasoning process based on expert system judgment, the MIF technique has become a useful GW_{pot} modeling approach, that can quickly, accurately, cost-effectively, and consequently monitor GW_{pot} [23–25]. MIFs constitute a GIS-based multi-criteria decision-making (MCDM) technique that enumerates the spatial relationships between dependent and independent variables according to scores assigned based on major and minor GCFs influencing GW_{pot} [24,26]. This method is economical as it relatively simple and useful for practical applications before starting an expensive field survey [3,9,20]. It helps in narrowing down the potential areas for conducting detailed hydrogeological and geophysical surveys and ultimately locating the drilling sites [7,27].

Hydro-stratigraphy and hydrogeology are essential for characterizing aquifer potentiality and developing hydrological models to predict groundwater resources for future availability [28,29]. For geoscientists, finding and locating the source and availability of the groundwater in a complex area with multiple hydrogeological features is a vital task. Although surface geophysical measurements can provide effective spatial coverage services [30,31], these measurements depend on the area extent to be investigated, cost, geological condition, and the acquired data readability. They contribute information on groundwater levels, hydrogeological behaviors, and corresponding lithology, ensuring a higher positioning accuracy for groundwater resources [32–34]. With the proper GW_{pot} and hydrogeological evaluation, geophysical techniques can be combined to improve efficiency. Specifically, the electrical resistivity techniques are well established and commonly used to solve numerous geological and environmental problems [35,36], which are considered as the most effective geophysical methods for the characterization of GW_{pot} and hydrologic stratigraphy. These methods are widely used to scrutinize high-resistance and low-resistance layers, and are, therefore, valuable tools for studying aquifer vulnerability [32,37]. The quantification of the aquifer potential analyzed by VES-based reflection coefficient, isopach, and resistivity mapping can directly verify the predicted result of the MIF model and its performance. The combination of vertical electrical sounding (VES) and electrical resistivity tomography (ERT) methods produces a high ratio of 90% compared to 82% for the VES method and 85% for the ERT method [33,38]. The spatial distribution of aquifer hydrologic characteristics, such as resistivity, reflection coefficient, overburden thickness, hydraulic conductivity, and specific productivity, plays an essential role in assessing and managing GW_{pot} [39,40]. Apparent resistivity and reflection coefficients are the most critical hydrogeological data needed to manage groundwater resources [41]. These parameters also outline variances in the hydrological strata that help to explain aquifer models for GW_{pot} modeling. In addition, geophysical well logging also generates useful information about the geological structure and the formations' lithology [22]. The feasibility study of resistivity surveys through boreholes has been used worldwide and is supported by general hydrogeological studies. Drilling (machinery type deployed subsurface soil/rock conditions) and electrical logs record the true location of the aquifer and corresponding lithology.

The phenomenon of surface water resource depletion and irregular spatial-temporal distribution of precipitation have made groundwater a vital natural resource for the reliable and economic provision of potable water supply in low- and mid-income regions of the Karak watershed, Northern Pakistan. In this context, this study addresses the applicability of comprehensive MCDM-MIFs model with optimized GCFs for GW_{pot} assessment and hydro-geophysical investigation for hydrogeological characterization. As the GW_{pot} mapping depends on the suitable GCFs and the weights assigned to them, various GCFs, such as geology, lineament density, land use/land cover, drainage density, slope, soil type, rainfall, elevation, and groundwater level fluctuations, were processed and optimized through geospatial analysis in GIS environment. The predicted GW_{pot} results using the MIF model were then analyzed by the receiver operating characteristic (ROC) curve and the confusion matrix, and Kappa (K) analysis. However, groundwater is an invisible resource that is difficult to measure or quantify directly. Therefore, the interpretation of VES and ERT data was employed to predict hydrogeological properties, aquifer hydraulic characteristics and GW_{pot} zones for future exploitation and installing tube wells for its utilization. Moreover, our methodology not only improves the reliability of the integrated geospatial and geoelectrical modeling and bridges the gap of GW_{pot} evaluation and hydrogeological characterization in the Karak watershed, but also provides an optional solution of groundwater assessment in other similar areas worldwide.

2. Study Area

2.1. Physical Geographical Background

The study area is located at geodetic coordinates between the latitudes of 32°46′ and 33°22′ N and between the longitudes of 70°43′ and 71°33′ E, covering an area of approximately 2372 km^2 (Figure 1b). A 123 km road from Peshawar on the Indus Highway leads to Karachi and is easily accessible from various parts of the country via mettled roads (Figure 1a). Geographically, the Karak watershed is located in the southern part of the Kohat Plateau of the upper Indus basin Pakistan. The Kohat Plateau itself lies between 70°–74° E and 32°–34° N, covering an area of approximately 10,000 km^2. Most of the region's climate is semi-arid, with two major seasons, i.e., the rainy season and the dry season. Precipitation is the primary source of groundwater replenishment where the average precipitation is 450 mm/year, and the minimum and maximum average temperatures in the Karak (at an altitude of 706 m) are 10.3 °C and 43.5 °C, respectively, varying by altitude. The harvest depends on the amount of precipitation or pipeline well supply. Annual precipitation in the northeast ranges from 500 to 750 mm. In the study area, rainfall from June to November is 68% and is 32% from December to May. During the short rainy season, rainfall is scarce, unstable, and concentrated, and it is relatively or absolutely dry for the rest of the time. High temperature and rainfall intensity cause large amounts of precipitation loss due to evaporation and runoff, respectively [42]. The highest elevation area of the Karak watershed is in the eastern Surghar Shinghar ranges (Figure 1b), where elevations typically exceed 1415 m above sea level. The lowest elevation area is associated with the Bannu boundary, where the river level is below 305 m.

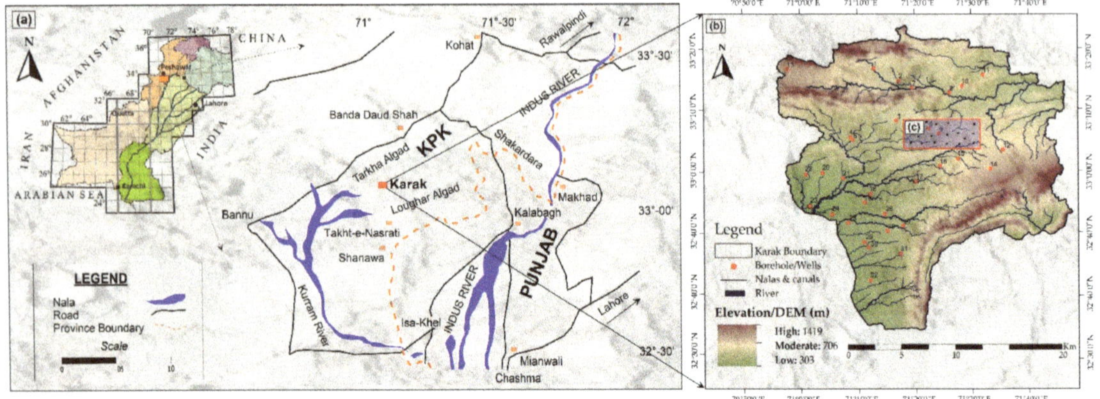

Figure 1. (**a**) Generalized physical geographical features of the Potwar region; and (**b**) Location map of the Karak watershed with the surveyed boreholes and vertical electrical sounding (VES) and electrical resistivity tomography (ERT) measurements.

2.2. Geological Background

A regional geological map of the study area was prepared to plot major geological structures and lithological units (Figure 2). The Karak watershed is part of a large intermontane basin where sedimentation has taken place from weathering and erosion of the surrounding Bannu mountain belts [43,44]. The Bannu basin is located in a depression behind the Trans-Indus current uplift boundary, which leads to the formation of the Bhittani, Khisor/Marwat, and Shinghar mountains. The basin is formed by the uplift boundary from the Kohat mountain range to the Bhittanni and Marwat/Khisor mountain ranges [43], as shown in Figure 2. In the Potwar Plateau and the adjacent Kohat Plateau, the exposed sedimentary formations are Eocene limestone, evaporite, and red beds [45]. Subsurface deposits of the area widely vary from very coarse sediments (such as gravel and boulders) to very fine sediments (such as silt and clay). There are three types of sediments in this region, including alluvial fans, floodplains, and basin-filled sediments [46,47]. An alluvial fan is composed of various proportions of boulders, gravel, sand, silt, and clay. The sediments in the floodplain are mainly clay and silt, with minor amount of sand. Sandy sediments were primarily formed in the Marwat range, mainly due to erosion [48]. The ages of the exposed strata in the study area range from the Precambrian to the Quaternary. The lithological distributions of the Karak watershed are illustrated in Table 1.

Table 1. Lithological characteristics of the Karak watershed.

Product	Formation Names	Lithology Characteristics
(C Fm)	Chinji formation	Sandstones and shales (abundant quartz with subordinate feldspars)
(DS)	Darzinda shale	Dark-brown to gray claystone and subordinate fossiliferous marl beds
(K Fm)	Kamlial formation	Mainly composed of sandstone (subordinate feldspars, lithic grains, micas)
(J/T)	Jurassic or Triassic rocks	Sandstone, siltstone, shale and dolomite
DP Fm	Dhok Pathan formation	Equal amount of sandstone and clay
(K Fm)	Kohat formation	Mainly composed of limestone and divided into three members
(N Fm)	Nagri formation	Primary sandstone and minor number of clays
(Q)	Quaternary alluvium	Mainly composed of sand, gravel, silt and clay

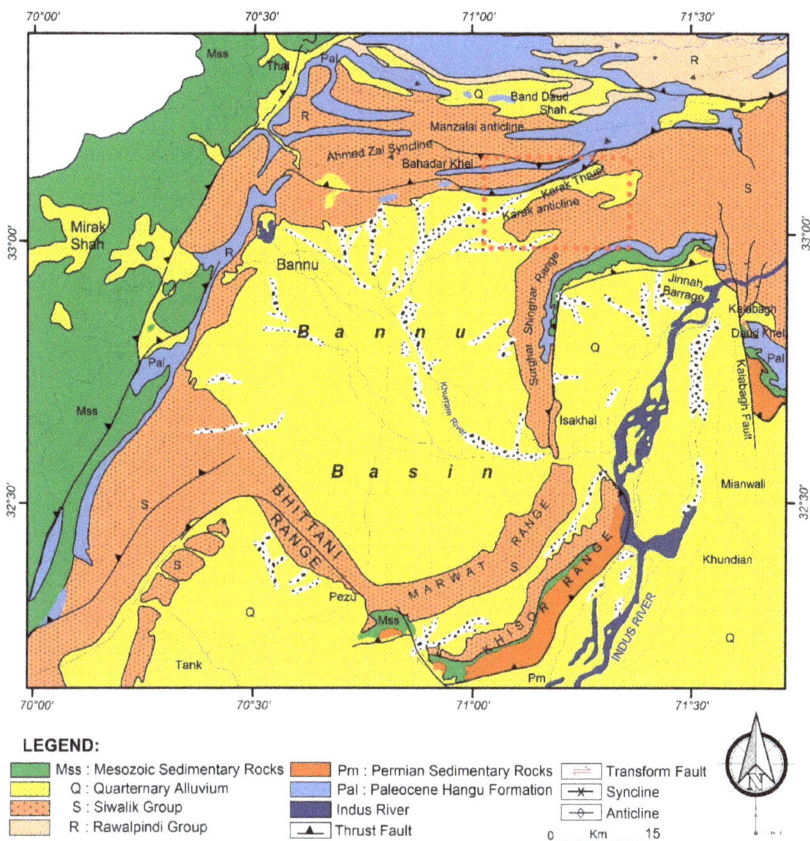

Figure 2. Regional geology map of the study area illustrates main structural and lithological units.

2.3. Hydrological Background

In the study area, the estimated thickness of semi-confined aquifers ranges from 10 to 30 m. The groundwater quality in the northeastern part of the northwest catchment is inferior [42]. This situation occurs due to the salt rock in the northern mountainous region, which is dissolved by runoff water and polluted groundwater due to deep infiltration. Under diving conditions, groundwater flows through weathered layers and fault zones. The alluvial filling is very uneven and contains high level of silt and clay. Locally, sand and gravel beds were encountered in boreholes. The flow rate through the open wells is calculated to be 0.035 mm^3/year. The alluvial aquifer's average annual recharge is approximately equal to the average annual discharge, which is 2.7 mm^3/year. The groundwater level is between 29.03 and 238.66 m. This indicates that a fuzzy groundwater boundary exists corresponding to a surface water boundary [49]. A small dam (Chambia dam) was constructed to maintain the groundwater level in the Karak watershed. The soil texture of the study area is predominantly medium clay, pure sand, cultivable soil and crops.

3. GCF Analysis and Optimization

The evaluation of groundwater condition factors (GCFs) is essential to effectively determine an accurate groundwater potential (GW$_{pot}$) index [50]. GCFs should be considered in terms of regional topographical, geological, hydrological, and land use/land cover characteristics influencing the GW$_{pot}$ [15]. Therefore, the identification of the GW$_{pot}$ spatial distribution was performed by multi-criteria decision-making (MCDM) analysis of nine factors, i.e., drainage density, geology, lineament density, slope, soil type, rainfall, elevation, land use/land cover, and groundwater level fluctuations. These GCFs were extracted independently from appropriate remote sensing, geological, and conventional map datasets (Table 2).

Table 2. GCFs used for mapping groundwater potential of the Karak watershed.

GCFs	Data sources	Format	Product
Drainage density	Digital elevation model (DEM) (ASTER 30 m)	Digital	(D_d)
Slope	Digital elevation model (DEM) (30 m spatial resolution)	Digital	(S_L)
Elevation	Shuttle Radar Topography Mission (SRTM) data from United States Geological Survey (USGS), resolution: 30 m	Digital	(E_L)
Rainfall	Annual rainfall data from Pakistan Meteorological Department (PMD)	numbers	(R_F)
Land use/cover	Forest Management Center Peshawar (FCMP), KPK, Pakistan	Digital	(LULC)
Geology	Geological map from National Centre of Excellence in Geology (NCEG), University of Peshawar	Digital	(G_{EO})
Lineament density	Landsat 8 OLI imagery and Shuttle Radar Topography Mission (SRTM)	Digital	(L_D)
Soil type	Directorate General Soil and Water Conservation (DGSC), Khyber Pakhtunkhwa (KPK), Pakistan	Digital	(S_T)
GW fluctuation	Pre-monsoon and post-monsoon groundwater table data (onsite survey)	Points	(GLF)

The drainage density (D_d) is a measure of the total length of all streams per unit area, regardless of the stream networks [51]. The hydrology toolkit in ArcGIS 10.4 was used to extract stream networks from a digital elevation model (DEM). Accordingly, D_d was calculated as the stream's total length divided by the total drainage using Equation (1) [14]. Subsequently, the drainage frequency was classified into five categories using a natural break classification scheme [16]. High drainage frequency is associated with high permeable lithology and accordingly high GW_{pot}. The groundwater favorability is indirectly related to D_d, which is related to surface runoff and permeability [52].

$$DD = \sum_{l=0}^{n} \frac{D_l(km)}{A(km)^2}\left(km^{-1}\right) \tag{1}$$

where DD represents drainage density, D_l is the stream's length, and A is the watershed area (km^2).

Lineaments are surface manifestations of linear or curvilinear features, such as joints, straight streams, and regional vegetation placement, reflecting potential topographical or geological structure [15]. The seven bands of the Landsat 8 image were stacked using ENVI 4.8 (Harris Geospatial, Broomfield, CO, USA), and principal component analysis (PCA) was performed on the stacked image in QGIS (Open Source Geospatial Foundation, Bern and Chur, Switzerland). The thematic layer for L_d can be defined as the total length of all recorded lineaments divided by the catchment area under consideration, as shown in Equation (2) [53]. The higher the L_d, the higher the favorability of GW_{pot}.

$$LD = \sum_{i=0}^{n} \frac{L_i(km)}{A(sq.\,km)}\left(km^{-1}\right) \tag{2}$$

where LD represent the lineament density, L_i is the lineament's length in km, and A is the grid area in square kilometer.

Data from 17 metrological stations were processed using simple arithmetic mean, isometric, and Thiessen polygon interpolation methods to obtain sufficient uniform precipitation in the catchment area. After these three interpolation methods were used for comparison, isometric interpolation (Equation (3)) was considered the best technical interpolation method. The flat and gentle areas, with less runoff, are more favorable for GW_{pot} than steep slopes [54]. In addition to rainfall quantity, other precipitation characteristics (such as duration and intensity) are also important. For example, a 20 mm rainfall in a long period may have a more significant impact on groundwater recharge than a 50 mm rainfall in a short period.

$$P = \frac{\sum_{l=1}^{n} p_l}{N} \tag{3}$$

where P is the average precipitation depth, with $p1$, $p2$, $p3$ up to p_n being the rainfall records of measurement stations 1, 2, 3, up to n, respectively.

The slope is an important factor that directly controls the infiltration of surface water. A 30-m resolution DEM was processed to generate a slope map in the ArcGIS 10.4 spatial analyst toolkit. The slope gradient was reclassified into five classes using the quantile classification scheme presented by [18]. A higher slope is more conducive to runoff but has a smaller impact on groundwater recharge. Elevation or altitude can have an indirectly inverse effect on GW_{pot}, which relates primarily to the occurrence of rainfall and the resulting recharging. However, high altitudes favor more recharge and ensure groundwater availability in low land areas in a watershed. Mountainous regions are often favorable for the recharge of deep-seated confined aquifers situated at low land areas [55].

Stratum lithology influences the porosity and permeability of aquifers and directly affects the GW_{pot}. The porosity of rocks, alluvial/sedimentary layers, sand, silt, and clay beds determine water infiltration and percolation [56]. Therefore, the lithology factor was also considered concerning groundwater characteristics. The lithology map was extracted, digitized, and reclassified from the geological map of Northern Pakistan. Accordingly, different weights were assigned to rock units depending on the infiltration capacity and GW_{pot} as per multiple influence factor criterion.

Vegetation cover areas, such as forests and agriculture traps, retain water by the roots of plants. In contrast, the built-up and rocky land cover decreases groundwater recharge by increasing the runoff during rainfall [24]. Therefore, to conduct GW_{pot} studies, it is necessary to investigate the land use land cover (LU/LC) characteristics of the study area. Therefore, the LU/LC map from the Forest Management Center Peshawar (FCMP) was reclassified with different score values assigned to several subclasses.

The water retention capacity of an area depends on the type of soil and its permeability. Permeability is directly related to the soil effective porosity which is greatly influenced by the particle shape, size, adsorbed water, porosity, saturation, and the presence of impurities in the soil [57]. The soil type map was primarily derived from the Directorate General Soil and Water Conservation (DGSC), KPK, and updated through onsite inspections. Soil mainly influences infiltration and percolation processes that eventually affect the groundwater recharge and then the GW_{pot} of a given area.

4. Methods

In this study, the application of remote sensing (RS) and geographic information system (GIS)-based spatial data and geoelectric data assisted hydrogeological assessment to distinguish the sediments and rock units of groundwater significance. The flowchart developed in this study is shown in Figure 3, which contains four steps:

1. Using RS and GIS toolkits, the database is ready to be input data for the MIF model, after the GCF' analysis and optimization described in Section 3.
2. Once the GCFs are to be optimized, the weights and ranks of each GCFs are assigned for the multi-criteria decision-making (MCDM) MIF model, and the weighted/ranked GCFs are integrated through the weighted overlay analysis (WOA), based on the principle of superposition in a GIS environment to identify regional GW_{pot} zones of the Karak watershed.
3. The hydrogeological characteristics of the aquifers are evaluated by the interpretation of electrical resistivity tomography (ERT) and vertical electrical sounding (VES) data. Furthermore, the aquifer potential is further quantified through quantitative analysis

of the resistivity mapping, overburden thickness mapping, and reflection coefficient mapping.

4. To evaluate the accuracy of GW$_{pot}$ mapping, the performance of the MIF model is assessed based on groundwater level (GWL) data through a confusion matrix, Kappa (K) analysis, and a Receiver Operating Characteristics (ROC) curve. In addition, the quantitative aquifer potential interpreted by VES data indirectly verifies the MIF model's predictive performance. Meanwhile, the hydrologic stratigraphic prediction derived from ERT and VES numerical models is correlated with known boreholes lithological information.

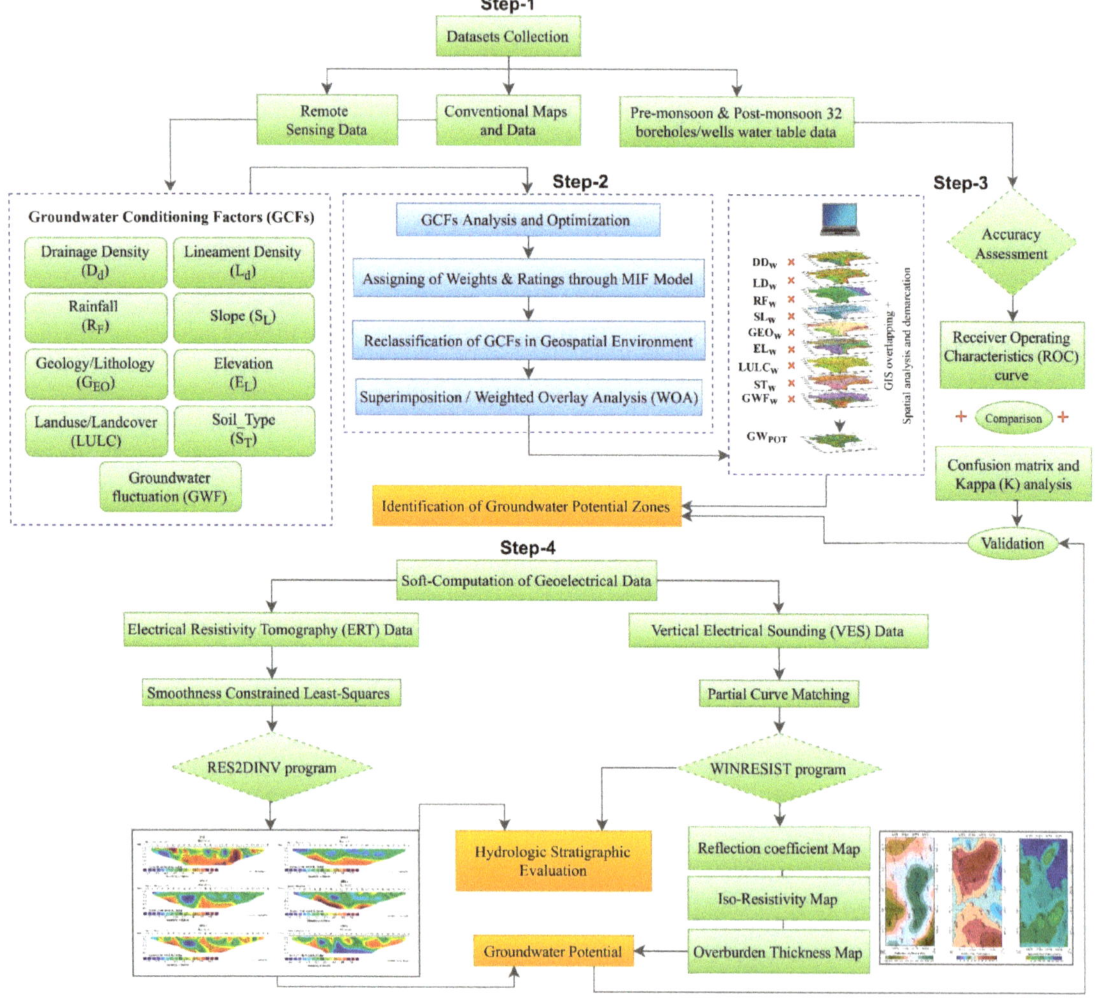

Figure 3. Framework to delineate groundwater potential and to identify hydrogeological characteristics.

4.1. Multi-Influence Factors (MIF) Model

4.1.1. Assigning of Weights and Ranks

The GW$_{pot}$ index is influenced by several hydrological, geological, topographical, environmental, and climatic variables [2]. By means of GCFs analysis and optimization, geology (G_{EO}), lineament density (L_D), drainage density (D_D), slope (S_L), soil type (S_T), rainfall (R_F), elevation (E_L), land use/land cover (LULC), and groundwater level fluctuations (GLF) were identified as the input data of the MIF model. The MIF model involves drawing a graph with correlations between conditioning factors and assigning weights based on the strength of the interrelationships (Figure 4) [2]. In Figure 4, a continuous arrow shows a major influence, and a dashed arrow indicates a minor influence on the other GCFs. The weights and ranks were assigned to each GCFs and different classes based on their relative contribution to GW_{pot} using the heuristic approaches/knowledge-driven method [11,58,59]. Weights of 1.0 and 0.5 were allocated to each major and minor effective variable, respectively. The combined weights of both major (CF_h) and minor (CF_l) were considered for calculating the comparative ranks (Table 3). Since the estimated weight of each GCF is equally distributed and applied to each GCF' category, the final GW_{pot} map is a weighted average. The estimated weight for each conditioning factor was obtained as a percentage using Equation (4).

$$\text{Score} = \left[\frac{(CF_h + CF_l)}{\sum (CF_h + CF_l)} \times 100 \right] \tag{4}$$

where, CF_h is the major weight of the condition factor, and CF_l is the minor weight.

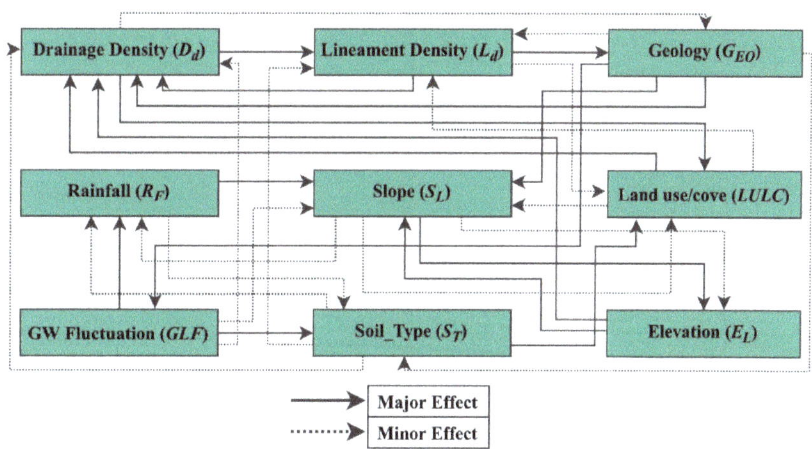

Figure 4. Interrelationship between the GCFs concerning the GW_{pot} index.

Table 3. Effect of GCFs, relative weight and score for each GCFs.

Groundwater Conditioning Factors (GCFs)	Major Effect (GCF$_h$)	Minor Effect (GCF$_l$)	Relative Weights (GCF$_e$ + GCF$_m$)	Proposed Score of GCFs
Rainfall	1	0.5	1.5	06
LU&LC	1	0.5 + 0.5	2	08
Geology	1 + 1 + 1	0.5 + 0.5	5	24
Lineament density	1 + 1	0.5	2.5	10
Drainage density	1 + 1	0.5	2.5	10
Slope	1	0.5 + 0.5 + 0.5	2.5	10
Soil type	1	0.5 + 0.5	2	08
Elevation	1 + 1	0	2	08
GWL fluctuation	1 + 1	0.5 + 0.5	3	16
Total			Σ20.5	100

4.1.2. Weighted Overlay Analysis (WOA)

The GW_{pot} index quality is influenced by the quality and quantity of the input data and the predictive models used [2]. Weighted overlay analysis [60,61] in ArcGIS 10.4 (Environmental System Research Institute, Redlands, California, United States) was used to outline the spatial distribution of the groundwater potential index based on nine GCFs' superimposition and their corresponding percentage effects on the groundwater potential. This work was done by multiplying each factor's category cell value by the factor's weight and summing the resulting cell values to generate a GW_{pot} map, as summarized in Equation (5). A GW_{pot} index is a calculated dimensionless number considering the weight assigned for each GCF and its categories [3]. After the WOA analysis had been completed, the natural break method was used to categorized GW_{pot} into four levels of potentiality (i.e., low, medium, high, and very high).

$$\begin{aligned} GW_{potz} &= \sum\nolimits_{l=0}^{n} W_i \times R_i \\ &= DD_cDD_w + LD_cLD_w + RF_cRF_w + SL_cSL_w + EL_cEL_w + GEO_cGEO_w + \\ &\quad LULC_cLULC_w + ST_cST_w + GLF_cGLF_w \end{aligned} \tag{5}$$

where GW_{potz} is the groundwater potential index, W_i is the weight of each condition factor, R_i is the rank of each GCF's category, DD is the drainage density, LD is the lineament density, RF is the rainfall, SL is the slope variation, EL is the elevation, GEO is the lithology, LULC is land-use/land-cover, ST is the slope type, and GLF is the groundwater level fluctuation. The subscripts c and w indicate a category of a GCF's thematic layer and its corresponding percent influence on GW_{pot}, respectively. This overlay analysis was done by multiplying the rank of each GCF's category (each individual category has a rank) with the weight of each condition factor (each GCF has a unique weight) to obtain the GW_{pot} index at the corresponding position of GCFs.

4.2. Accuracy Assessment of the MIF Model

The pre-monsoon and post-monsoon groundwater table (GWT) data from 32 observed boreholes with global positioning system (GPS) positions were collected for validation purposes. The area under the curve (AUC) based receiver operating characteristic (ROC) curve, the confusion matrix, and Kappa (K) analysis were used to test the performance of the MIF model. The ROC is a mathematical technique developed to explain the efficiency of probabilistic deterministic detection and prediction systems [62,63]. In this study, ROC was used to assess the spatial consistency between real events and to predict the model probability. In the validation phase, pre-monsoon and post-monsoon GWT data of 32 observed boreholes/tube wells were compared with the GW_{pot} result obtained by the MIF model. The ROC curve provides a quantitative evaluation that can determine the uncertainty of modeling and evaluate the spatial model effectiveness. The confusion matrix and Kappa (K) analysis [26] were also used for accuracy evaluation by correlating the GW_{pot} map with the observed GWT data. The overall accuracy was calculated using the following formula [64].

$$OA = \frac{\sum C_{OWL}}{\sum OWL} \tag{6}$$

where, OA is the overall accuracy, C_{OWL} represent the number of correct observation boreholes/well's locations and OWL is the number of observation boreholes/well's locations.

The Kappa (K) analysis is a multivariate approach for MIF accuracy evaluation. It was calculated by the following formula [65].

$$K = \left[\frac{\sum CV\% - CAOV\%}{\sum TC - CAOV\%} \right] \tag{7}$$

where, CV % is the percentage of the correct values, CAOV% is the percentage of the correct agreement to observed values, TC is the total number of class.

4.3. Interpretation of Geoelectrical Data

The geophysical techniques have typically been used to assess hydrogeological structures, hydro-stratigraphic characteristics and the spatial distribution of aquifers [34]. Fundamentally, an electrical current is injected into the ground by two current electrodes and measures the potential difference between the other two pairs of electrodes. In this study, two-dimensional electrical resistivity tomography (ERT) based on dipole–dipole configuration and vertical electrical sounding (VES) based on Schlumberger configuration measurements were performed using essential field equipment (Terameter SAS 100 and SAS 1000 Lund imaging systems and their accessories, ABEM, Sundbyberg, Sweden) (Figure 5).

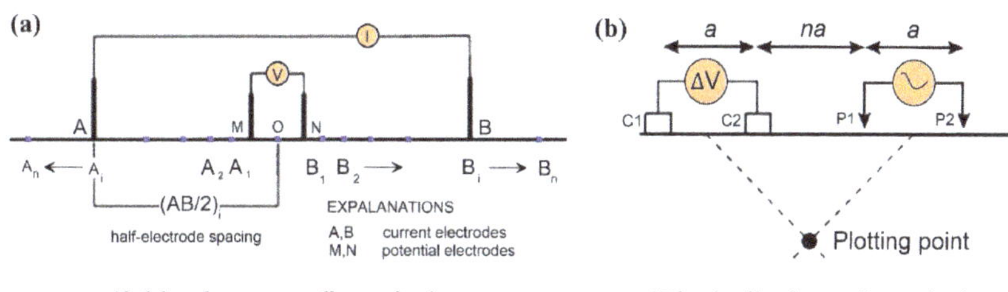

Figure 5. Schematic diagram of (**a**) the Schlumberger array configuration for vertical electrical sounding (VES), and (**b**) dipole–dipole array configuration for electrical resistivity tomography (ERT) techniques.

4.3.1. Electrical Resistivity Tomography (ERT)

The ERT technique was effectively applied in the surveyed area to provide information about subsurface hydrogeological characteristics to fully understand the GW_{pot} and hydro-stratigraphy through vertical and horizontal two-dimensional sections capable of reaching lengths and depths up to 176 m and 30.2 m, respectively. A multi-electrode 2D device (Terameter SAS 100) along a dipole–dipole configuration including electrodes connected to a transmitter/receiver system via a multi-core cable was used to acquire data (Figure 5b). The dipole–dipole configuration exhibits an excellent vertical and horizontal resolution of subsurface geological features, which has great horizontal coverage and penetration depth [66]. The apparent resistivity was calculated for every electrode quadrupole by Equation (8) [34].

$$\rho = K \frac{V}{I} \tag{8}$$

where V is the voltage, I is the current, and K is a geometric factor.

The dipole–dipole configuration data were concatenated to obtain combined apparent resistivity pseudo-sections. The degree of consistency between resistivity and actual subsurface resistivity distribution depends on the combination of acquisition parameters and inversion strategy. The smoothness constrained least-squares technique in the RES2DINV (Landviser, League, Texas, United States) program was used to process the apparent resistivity data [67,68]. This process automatically creates 2D models in a rectangular block by selecting the optimal data inversion parameters (e.g., the damping coefficient, and the vertical and horizontal flatness filter ratio, convergence limit, number of iterations). We used the finite difference method to calculate the module's apparent resistivity and compared it to the measured data. Iteratively, we adjusted the resistivity of the model block until the calculated apparent resistivity value of the model matched the actual measurement. Finally, the program produces a pseudo section (a qualitative method for measuring or calculating resistivity changes) and an inverse model section (slice depth and resistivity tomography image) [68]. As a follow-up to the observation results of ERT lines L1, L2, L3, L4, L5, and L6 were acquired in the E-W, S-W, N-E, E-W, E-W, and E-W directions, respectively. In this study, the ERT technique estimated the spatial subsurface resistivity caused by the lateral and longitudinal inhomogeneities of petrophysical properties. The distribution of petrophysical properties indicated by the measured resistivity fluctuations were generated to guide GW_{pot} and hydro-stratigraphy in the study area.

4.3.2. Vertical Electrical Sounding (VES)

The VES method was used in the surveyed area to evaluate the hydro-stratigraphic structure of the sedimental layer (i.e., the structure of the subsurface sediments), aquifer characteristics (e.g., thickness, resistivity (ρ), overburden thickness, and reflection coefficient), and GW_{pot}. The VES technique is one of the most commonly used conventional resistivity methods to determine the vertical variation of subsurface resistivity parameters [34]. In the surveyed area, 26 VES measurement stations were operated at different positions using the Schlumberger electrode configuration with half-current electrode spacing (AB / 2) ranging from 1.5 to 1000 m in each successive electrode probe to determine the depth to the sediments and apparent resistivity (ρ_a). Meanwhile, using the Schlumberger array (Figure 5a), the adequate penetration depth is typically 20–40% of the external electrode spacing (AB), depending on the subsurface resistivity structure [69]. In this study, we first plotted all resistivity data collected to confirm qualitive and qualitative characteristics. The statistical apparent resistivity (ρ_a) values of the Schlumberger array for each sounding were calculated using Equation (9).

$$\rho_a = \pi \left\{ \frac{\left(\frac{AB}{2}\right)^2 - \left(\frac{MN}{2}\right)^2}{MN} \right\} Ra \tag{9}$$

where, AB represent the distance between two current electrodes, MN is the distance between the potential electrode, and Ra is the apparent electrical resistance.

The preliminary interpretation was performed using Partial Curve Matching (PCM) and auxiliary tools to summarize VES values, i.e., the relationship between the apparent resistivity and corresponding half current electrode spacing (AB/2) on the double logarithmic graph. The results obtained from the exercises were used as an input model for computer-assisted iterations using the WinResist™ (Geotomo Software, Gelugor, Penang, Malaysia) program. The preliminary interpretation of VES data was quantitative, determining the thickness (h) and resistivity (ρ) of different layers, and qualitative inferring lithology was based on the resistivity and reflection coefficient (RC) values of each sounding station. For better depiction, six VES measurements were performed in the two boreholes' immediate vicinity (BH06/BH09) and correlated with known lithological information. The Schlumberger configuration was characterized by tracking and tracing each VES subsurface layer, the vertical changes, and the geoelectric profile with a known borehole/well lithology to horizontally correlate the measured VES to perceive a unified layer model applicable to all field curves. Moreover, geological information of known borehole/wells can improve interpretations that lead to lithological results from VES data, while software analysis can only provide resistivity distinction by depth. The statistical apparent resistivity values of each VES measurements were outlined to create an iso-resistivity map. The RC values for the surveyed area were calculated using the following expression [70].

$$RC = \left\{ \frac{(\rho n - \rho(n-1))}{(\rho n + \rho(n-1))} \right\} \tag{10}$$

where RC represents the reflection coefficient, ρn is the resistivity of the n-th layer, and $\rho(n-1)$ is the resistivity overlying the n-th layer.

4.4. Geophysical Well Logging

Hydrogeological characterization of aquifers using geophysical well/borehole logs has been emphasized in many studies [5,71]. Effective groundwater exploration and well/borehole lithology evaluation require a complete understanding of aquifer hydrogeological characteristics and well/borehole design. In the study area, the drilling sites were selected based on the experience of the MIF model to determine prerequisites for the successful construction of the tube well and evaluate the availability of groundwater supply that can meet the demand for domestic and irrigation water. The GeoLog International (GLI) groundwater and engineering services with reference to Ms. Manahil Engineering & Cons conducted St. Rotary (SR) drilling and geophysical logging in Marwatan Banda, Karak. The borehole's logging survey was conducted using multi-parameter methods, i.e., normal resistivity logs (NRLs) (short and long configuration) and spontaneous potential logs (SPLs). The Geo logger 3030/Mark-2 3433 (GLI, Peshawar, Pakistan) was used for petrophysical property measurements. Through these significant hydrogeological properties, e.g., the formations' lithology, depth, thickness, groundwater water table level, and groundwater quality in total dissolved solids (TDS) were evaluated.

5. Results

5.1. Evaluation of GCFs

The MIF model is an MCDM technique widely used for environmental management and has proven to effectively explain the GW_{pot} influential factors. It can effectively determine GCF weights. Table 4 illustrates the weights and qualitative ranks assigned to each influencing factor described below.

Drainage density (D_d) is a measure of the total length of all streams per unit area, regardless of the stream networks [51]. Subsequently, the drainage frequency was classified into five categories, i.e., very low (1.08–1.61 km/km²), low (1.61–1.86 km/km²), moderate (1.86–2.11 km/km²), high (2.11–2.38 km/km²), and very high (2.38–3.08 km/km²) (Figure 6a), according to a natural break classification scheme. The groundwater favorability is indirectly related to drainage density, as are surface runoff and permeability. Therefore, the highest score was assigned to the 1.08–1.61 km/km² category, indicating high infiltration and low runoff, and the lowest score was assigned to the 2.38–3.08 km/km² category (Table 4).

Lineament density (L_d) of the Karak watershed indirectly indicates the GW_{pot}, as the presence of lineaments usually means a porous zone. The lineaments are spatial distributed in the study area aligned in the directions of E-SW, NNE-SSW, NW-SE, and E-W, and their density was classified into five frequency categories (Figure 6b). The higher the L_d, the higher the probability of GW_{pot}. Therefore, the highest rank was assigned to the 1.46–1.78 km/km² category and the lowest was assigned to the 0.17–0.45 km/km² category.

Rainfall (R_F) interpolated data were reclassified into five categories, i.e., very low (13–281 mm), low (282–577 mm), moderate (578–604 mm), high (605–629 mm), and very high (630–663 mm) (Figure 6c). In addition to the quantity of R_F, other precipitation characteristics, such as duration and intensity, are also important. For example, a long period of 20 mm R_F has a more significant impact on groundwater recharge than a short period of 50 mm R_F.

The slope (S_L) map was reclassified into five categories, i.e., flat (0–5.78°), gentle (5.78°–13.5°), moderate (13.5°–23.1°), steep (23.1°–35.0°), and very steep (35.0°–81.9°) using the quantile classification scheme presented in [18]. The flat and gentle areas are more suitable for GW_{pot} than steep slopes, as a gentle and flat slope allows for less runoff, and a steep slope is more conducive to runoff [54]. The highest rank was assigned to flat area (0–5°.780°), and the lowest was assigned to a very steep area (35.0°–81.9°), which has a smaller impact on recharge in the study area (Figure 6d).

Table 4. Classification of weight and ranks of GCFs.

Groundwater Conditioning Factor (GCF)	Categories within the GCF	Qualitative Rank	Ranks	Weight of GCF
Rainfall	629–664	Very high	06	06
	604–628	High	04	
	577–603	Moderate	03	
	281–576	Low	02	
	13–280	Very low	01	
Land use/ land cover	Agriculture, Rivers/stream	Very high	08	08
	Barren land	High	06	
	Dam/pond	Moderate	05	
	Shrub land	Low	04	
	Built-up	Low–Very low	03	
	Forest	Very low	02	
	Range land	Very low	01	
Geology	QA	Very high	24	24
	K Fm, N Fm	High	18	
	DP Fm, K Fm	Moderate	12	
	J/T rocks	Low	08	
	C Fm, DS	Very low	04	
Lineament density (km/km^2)	1.47–1.78	Very high	10	10
	1.15–1.46	High	08	
	0.82–1.14	Moderate	06	
	0.50–0.81	Low	04	
	0.07–0.49	Very low	02	
Drainage density (km/km^2)	1.08–1.61	Very high	10	10
	1.62–1.86	High	08	
	1.87–2.11	Moderate	06	
	2.12–2.38	Low	04	
	2.39–3.08	Very high	02	
Slope (degree)	0.0–5.78	Flat	10	10
	5.79–13.5	Gentle	08	
	13.6–23.4	Moderate	06	
	23.5–35.3	Steep	04	
	35.4–81.7	Very steep	02	
Soil type	Loamy	High	06	08
	Loamy clay	Moderate	04	
	Mainly loamy	Low	02	
Elevation (meter)	1419	High	06	08
	706	Moderate	04	
	303	Low	02	
Groundwater level fluctuation (meter)	1.57–5.29	High	14	16
	5.3–1.08	Moderate	10	
	1.09–14.6	Low	06	
	14.7–19.3	Very low	02	

The elevation (E_L) map in Figure 6e shows three elevation categories, i.e., high (707–1419 m), moderate (304–706 m), and low (0–303 m).

Geology (G_{EO}) characteristics govern the porosity and permeability of the hydrogeological layer, which in turn influences the formation and distribution of GW_{pot} through physio-mechanical properties that control the water transmitting ability of the hydrogeological layer materials and the rate of groundwater flows. Therefore, the G_{EO} factor was also considered concerning groundwater characteristics. The study area consisted of eight lithological units of formation types and geological ages. The confirmed lithology outcrops are the Quaternary alluvium (Q), Dhok Pathan formation (DP Fm), Chinji formation (C Fm), Jurassic-Triassic rocks (J/T), Kohat formation (K Fm), Nagri formation (N Fm), Kamlial formation (K Fm), and Drazinda shale (DS) (Figure 6f).

Land use/land cover (LULC) greatly influences groundwater occurrence and exploitation. The major portion of the study area is agriculture (62%; 1345 km^2), followed by forest area (15%; 576 km^2), barren land (12%; 292km^2), rangeland (4%; 58.3 km^2), shrubland (3%;

40.7 km^2), built-up (2%; 30 km^2), river/stream (1%; 22 km^2), and dam/pond (1%; 8 km^2) (Figure 6g).

Soil type (S_T) and its permeability decides the water retention capacity of an area. The soil types of the study area include loamy soil, loamy clay, and mainly loamy (Figure 6h). The dominant soil type in the study area is loamy soil. The coverage of the other two soil types (i.e., loam and mainly loam) are relatively low. According to composition and soil water holding capacity, the loam is regarded as the highest grade, and mainly loam is regarded as the lowest grade.

Groundwater level fluctuation (GLF) is of significance in the successful management of GW_{pot}. Pre-monsoon and post-monsoon groundwater levels (GWLs) indicate the degree of saturation and the extent of recharge aquifers. In this study, hydrogeological data of 32 boreholes/wells over 10 years 2009–2019 (from Pakistan Water and Power Development Authority (WAPDA)) was collected through onsite investigation. During the period 2009–2019, the pre-monsoon and post-monsoon water level varies from 5.9 to 15.4 mbgl and from 7.3 to 32.6 mbgl, respectively (Figure 6i). The groundwater fluctuation levels were calculated for the period of 2009–2019, with a minimum of 1.57 m and a maximum of 19.3 m. In the study area, the aquifer is partially saturated due to the inadequate precipitation and other influencing factors. In the northern region, slight fluctuations of groundwater level (about 6 m) were observed, which may have been due to groundwater recharge by surface irrigation. However, groundwater levels fluctuated significantly in the southern and central regions, which may have been caused by topographical influence and the excessive exploitation of groundwater.

5.2. Assessment of GW_{pot}

Using the weighted overlay analysis in the GIS environment, the GW_{pot} zones were evaluated by integrating several conditioning factors (i.e., rainfall, slope, geology, soil type, drainage density, lineament density, land use/cover, elevation and groundwater fluctuation). Based on natural breaks in the histogram of the GW_{pot} index, the GW_{pot} map was categorized into four levels of potentiality, i.e., low, medium, high, and very high (Figure 7a), with the distribution ranges of 9.7% (72.3 km^2), 52.4% (1307.7 km^2), 31.3% (913.4 km^2), and 6.6% (44.8 km^2) of the total area, respectively (Figure 7b,c). The spatial distribution of the various GW_{pot} zones typically shows a mirror reflection of key factors. High and very high GW_{pot} zones confirm their excellent capacities as sedimentary groundwater aquifers. The GW_{pot} map demonstrates that the excellent groundwater is concentrated due to the distribution of Quaternary alluvial and agricultural land with high infiltration ability. Moreover, high drainage densities and low slope gradients can increase groundwater infiltration capacity, which may be related to the evaluated high GW_{pot}. The northwestern, southeastern, and the central part limited regions typically have a low to medium GW_{pot}, accounting for approximately 12.7% of the study area.

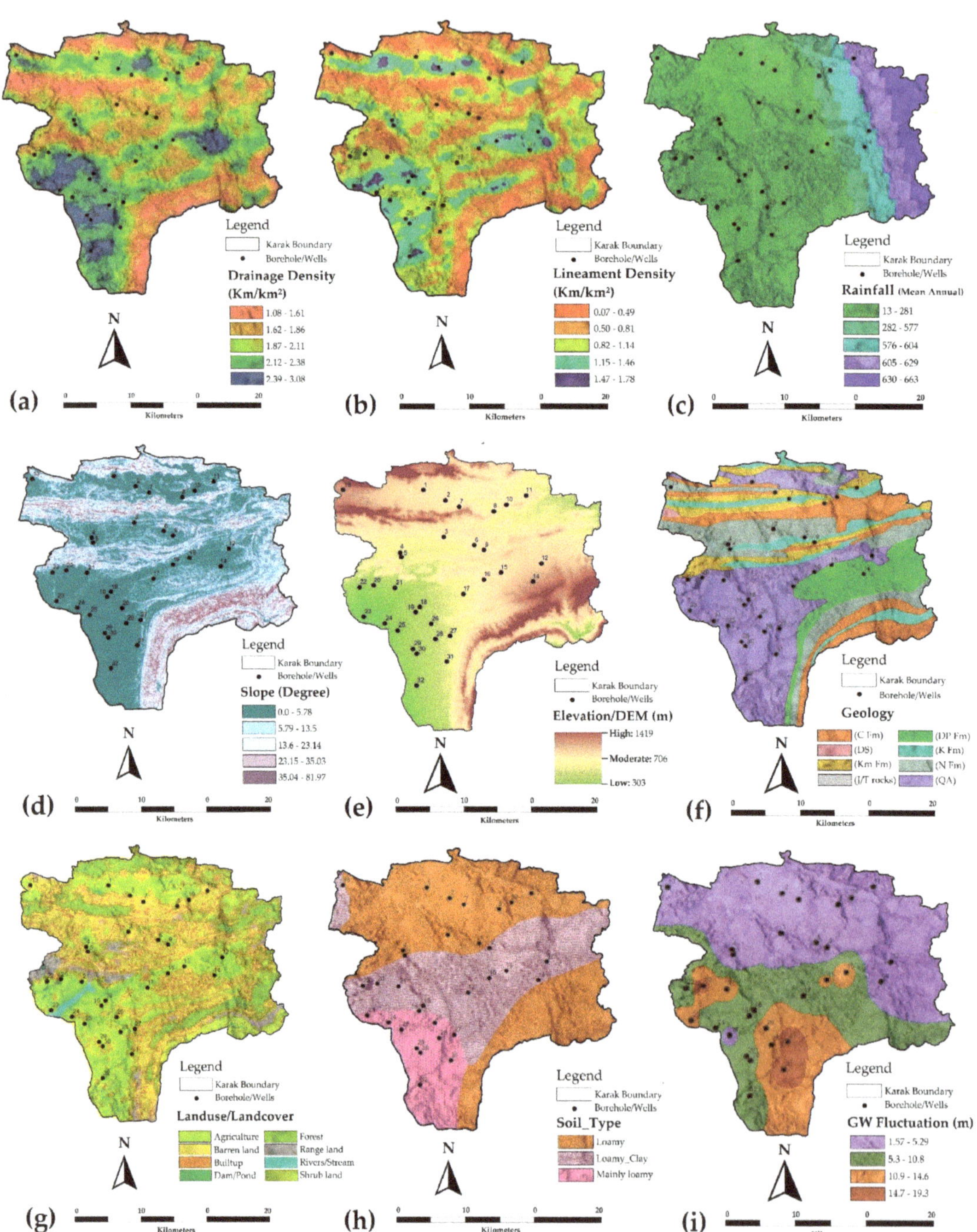

Figure 6. GCFs considered in this study: (**a**) drainage density; (**b**) lineament density; (**c**) rainfall; (**d**) slope; (**e**) soil type; (**f**) land use land cover; (**g**) geology/lithology; (**h**) elevation; (**i**) groundwater level fluctuations.

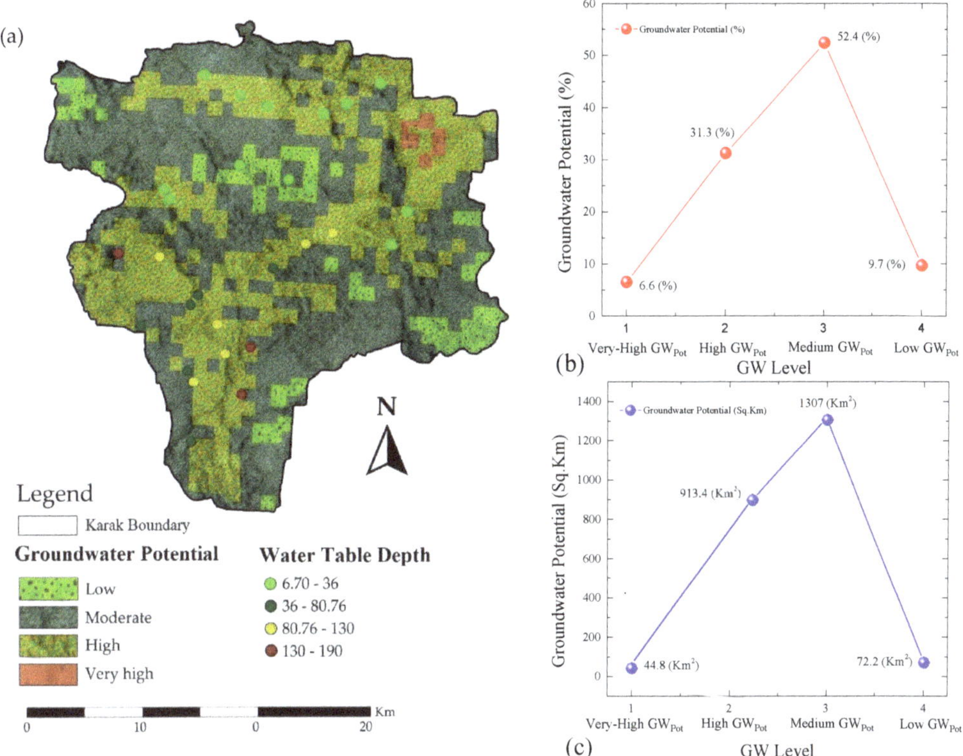

Figure 7. (**a**) GW_{pot} zones and groundwater level depths of boreholes/wells; (**b**) groundwater potentiality in square kilometers and (**c**) in percentage of the Karak watershed.

5.3. MIF Model's Performance

The ROC curve, the confusion matrix, and Kappa (K) analysis were used to evaluate the accuracy of the assessment result and the performance of MIF the model.

ROC graphs are useful tools for visualizing a classifier's performance and for determining the area under the curve (AUC) value to evaluate an algorithm [62]. The ROC curves were implemented in the present study as a goodness of fit, and the success rate can be distinctly visualized. In this study, the predicted GW_{pot} map was examined and compared with 32 pre-monsoon and post-monsoon groundwater level (GWL) fluctuations to evaluate the spatial coincidence between the favorability values (from GW_{pot}) and the actual GWL fluctuation events (Figure 8a). The GWL fluctuations range from 1.57 to 19.3 m (Figure 8b). Since a larger area under the ROC curve indicates that the spatial GW_{pot} mapping is more effective, an AUC value of 1 shows a perfect prediction of the model and indicates that the highest ranked probabilities coincide with the groundwater fluctuation [63]. The result of the ROC chart analysis shows that the AUC value of the presented MIF performance is 68% (Figure 8c) which is consistent with GWL fluctuation.

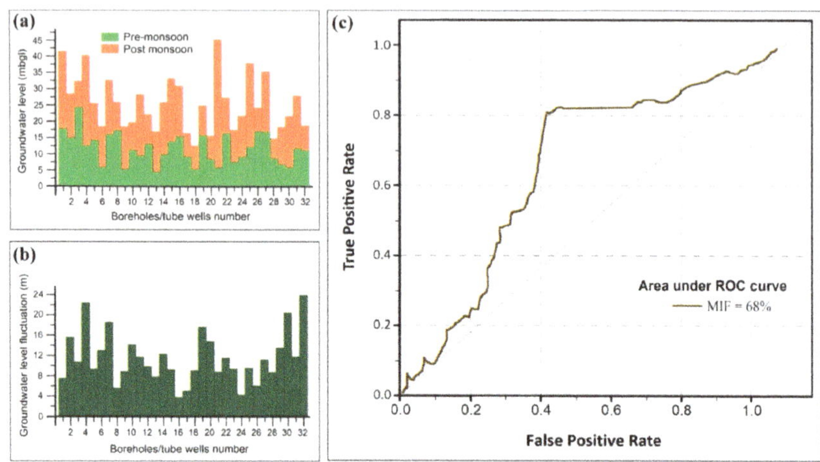

Figure 8. (**a**) The pre-monsoon and post-monsoon groundwater level (mbgl) fluctuations; (**b**) average groundwater level fluctuation (m) of the Karak watershed; (**c**) receiver operating characteristics (ROC) curve of the MIF model.

The confusion matrix and Kappa (K) analysis were performed using the 32 actual groundwater depths from boreholes/wells. The groundwater depth in the study area is between 6.7 and 190 m. These 32 depths were divided into four categories, i.e., 6.7–36 m, 36–80.76 m, 80.76–130 m, and 130–190 m. The groundwater depth data were used to calculate classification accuracy by the confusion matrix and Kappa (K) analysis. Overlay analysis shows that most of the boreholes/wells with higher groundwater levels are located in areas with demarcated higher groundwater potential. The performance evaluation of the MIF model shows that the overall accuracy is 68%, and the Kappa coefficient is 0.65 or 65% (Table 5), which indicates that the estimated potential of groundwater is consistent with the investigated groundwater depths in the study area.

Table 5. Error matrix of the GW_{pot} zone-based confusion matrix and Kappa (K) analysis.

S. No	GWpot zones	Very High	High	Moderate	Low	Total	CS [1]
1	very high	0	0	0	01	1	1
2	high	12	04	06	02	24	16
3	moderate	04	0	01	01	6	4
4	low	1	0	0	0	1	1
	Total	17	04	07	04	32	22

[1] CS refer to the correct sample.

5.4. ERT Interpretation

In this study, the ERT approach with an optimal compromise between the electrode distance and profile length produced a deep characterization of the hydro-stratigraphical layers and groundwater potentiality. The smoothness constrained least-squares outputs by the RES2DINV software show an apparent lateral homogeneity with a gradual increase in resistivity, with depth caused by lateral and longitudinal inhomogeneities of rock physical properties (Figure 9a). Each inversed resistivity section obtained a distribution of petrophysical properties of resistivity variability and possible resistivity anomalies (which may be water-bearing zones). The final depth of the inversed sections ranges from 5 to 30.2 m.

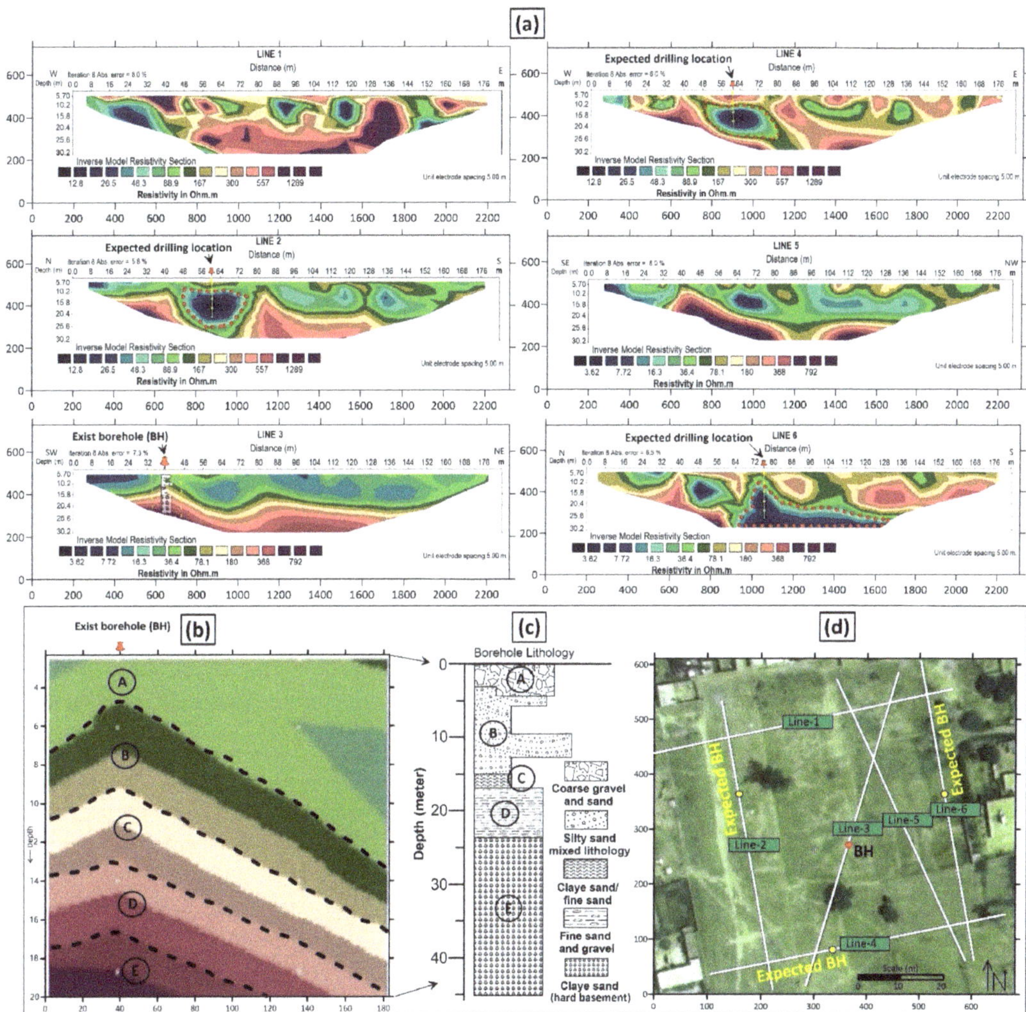

Figure 9. (**a**) Inverse model resistivity section of ERT survey lines (i.e., L1, L2, L3, L4, L5, and L6) containing existing borehole lithological information on L3; (**b**) correlation of acquired ERT hydrologic stratigraphy with (**c**) The existing borehole lithological information; and (**d**) ERT measurements line alignment in the study area, in which the red dot shows the location of the existing borehole, and the yellow dots indicate the proposed locations of wells for groundwater exploitation.

Generally, the root means square (RMS) error at the end of eight iterations of almost every ERT section is less than 8%. The interpretation of ERT sections is based on a standard resistivity range of values. The recommended GW_{pot} zones were based on an understanding of the subsurface sediment/ rock lithology of the study area. Meanwhile, the subsurface lithology related to the resistivity range was derived from the existing standard resistivity chart, which considers other local factors that may cause the resistivity deviation.

In the study area, the L1, L2, L3, L4, L5, and L6 ERT inversed resistivity values area 12.8–189 Ωm, 12.8–189 Ωm, 3.62–792 Ωm, 12.8–189 Ωm, 3.62–792 Ωm, and 3.62–792 Ωm, respectively (Figure 9a). The inverse resistivity models using dipole–dipole configurations on L2, L4, and L6 ultimately revealed the vertical and lateral distribution of subsurface resistivity. According to the predicated GW_{pot} on diffusion and array configuration, the groundwater prospect resistivity values are 12.8–48.3 Ωm, 14–76.4 Ωm, and 3.62–16.3 Ωm on L2, L4, and L6, respectively. Variation of resistivity characteristics within the primary lithological unit ultimately indicates the GW_{pot} prospect adjacent to clayey sand and silicate aquifers (sandstone) (Figure 9a). This result is consistent with the Karak watershed regional geology, which is mainly composed of interlayers of fine sand, sandstone, clay, and gravel. Since the GW_{pot} is structurally controlled, it also needs to locate potential fracture zones, e.g., fractured sandstone, which are considered good aquifer sources. The ERT techniques should be applied with a proper understanding of the hydrogeological background. Therefore, five lithological sequences (i.e., topsoil with coarse gravel and sand, silty sand mixed lithology, clayey sand/fine sand, fine sand/gravel, and clayey basement) of the drilled borehole on L3 at final depths of 45 m were normalized with the ERT model by mean of quantitative quota (Figure 9b). The ERT-predicted hydro-stratigraphy and borehole lithological log signature (Figure 9c) performance analysis shows suitable matches. The marked yellow points on the L3, L4, and L6 sections are considered future prospects for groundwater exploitation (Figure 9d). These high groundwater potential zones will play a vital role in the future expansion of drinking water and irrigation development in the surveyed area.

5.5. VES Interpretation

5.5.1. Hydrogeological Characteristics

The VES technique has been proven efficient in evaluating hydrogeology, aquifer properties, and aquifer potential. In this study, aquifer characteristics (such as thickness, lithology, and resistivity, reflection coefficient, and isopach) were determined, which is an essential factor in hydro-stratigraphic inheritance and GW_{pot} assessment. The apparent resistivity data obtained from the VES positions were plotted against half of the current electrode spacing (AB/2), and a curve matching technique was used to interpret resistivity sounding curves (Figure 10). This technique involves matching small segments of the field curve against the trendline curve to determine the thickness of a particular layer in half-space and the apparent resistivity. As far as the evaluation of the statistical apparent resistivity is concerned, the qualitative interpretation results indicating that the curves, stratification properties, and RMS errors are in complete agreement (Figure 11) (Appendix A). Depending on the shape of the VES curve, the resistivity distributions of various hydro-stratigraphy can be classified into H, K, A, and Q types, which can be mutually combined to generate HA, HK, KH, and QH types [72]. In this study, the type of curves observed include 3-layer H-type (26%), 4-layer HA-type (9%) and KH (52%), and 5-layer HKH-type (13%). Qualitative hydrological inferences can usually be based on the type of curve.

The geoelectrical interpretation based on curve matching reveals hydrologic resistance and depth variation (Figure 10). According to the corresponding resistivity values (ρ^1, ρ^2, ρ^3, ρ^4, and ρ^5) and thicknesses (h^1, h^2, h^3, h^4, and h^5), the geoelectric units indicate four to six sequences of lithologies, i.e., topsoil (coarse gravel and sand), alluvial layer, silty sand, clayey sand, fine sand and gravels, and clayey sand with saline water. Table 6 summarizes the VES interpretation, including the number of hydrologic layers and their corresponding resistivity values and the inferred lithology information. Appendix A presents the detailed explanation of geoelectrical stratification for all the VES surveys carried out in the Karak watershed and the resistivity variation.

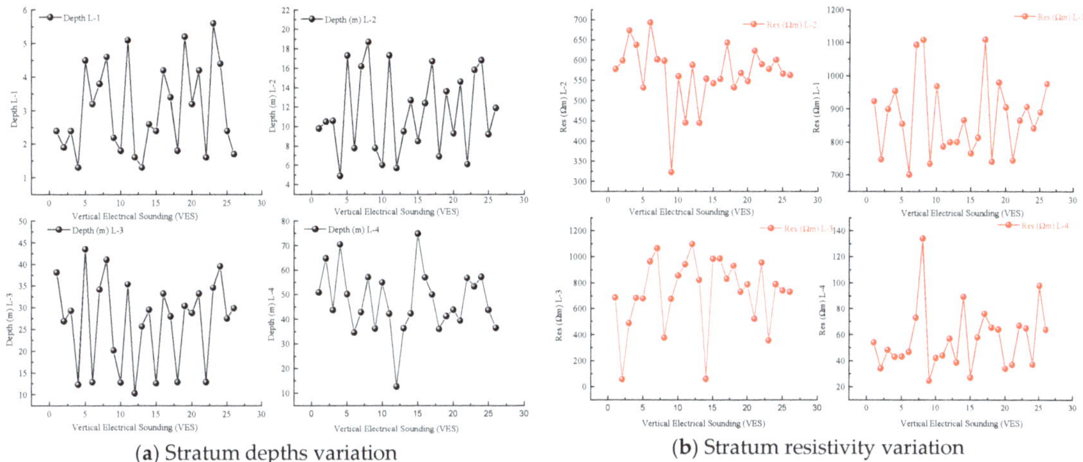

(**a**) Stratum depths variation   (**b**) Stratum resistivity variation

Figure 10. VES data interpretation result-based partial curve matching (PCM) along 26 VES stations: (**a**) hydrologic layers depth variation; (**b**) hydrologic layers resistivity variation.

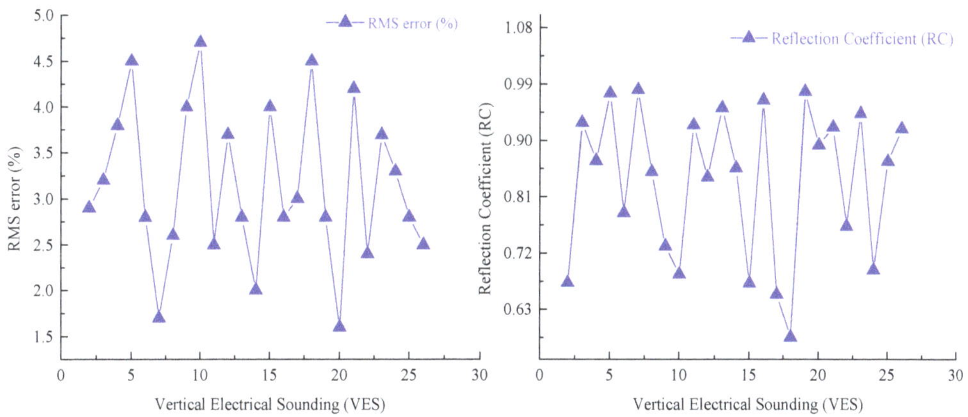

Figure 11. Root mean square (RMS) error of 26 VES stations (**left**) and the reflection coefficient variation of VES stations (**right**) in the study area.

Table 6. Average inferred hydro-stratigraphy corresponding to resistivity in the study area. The detailed VES interpretation results are shown in Appendix A.

Inferred Hydro-Stratigraphic Lithology	Inferred Resistivity (ρ) Variation	Reflection Coefficient Variation	Thickness Variation (m)
Topsoil	954–1109	0.6723–0.7889	3.3–4.5
Coarse gravel and sand	748–923.1	0.7313–0.7626	1.3–4.7
Silty sand mixed lithology	323–673	0.7841–0.8663	4.6–8.2
Clayey sand	685–1098.3	0.8911–0.9523	6.8–28.4
Fine sand and gravels	34–98.8	0.9643–0.9752	12.6–22.8
Clayey sand with saline water	26–84.2	0.5885–0.7434	10.8–32.5

5.5.2. VES Correlation with Boreholes

For better delineation of the hydro-stratigraphy, six VES results adjacent to two boreholes (BH06/BH09) were correlated with known lithological information. Performance analysis shows that VES1 yields five lithological units (coarse gravel/sand, silty sand mixed lithology, clayey sand, and fine sand/gravel) (Figure 12). The zone of interest with water saturation lies at a depth of 29.8 m. VES3 penetrates up to 48.1 m where water is predominantly saline, with freshwater saturation having a lithology of coarse gravel/sand, silty sand mixed lithology, silty sand/gravels, fine sand/gravel, and clayey sand/saline water. VES2 yields three lithological units, where the zone of interest lies at a shallow depth. Furthermore, VES9, VES17, and VES8 correlated with borehole BH09 show suitable matches, where salinity and freshwater saturation are encountered at a shallow depth (25 to 35 m) due to capillary action. However, the VES8 upper portion is mainly composed of unconsolidated alluvium, and the freshwater zone is at a shallow depth due to elevation. The main lithological characteristics of the topsoil at each VES station are predominantly alluvium. The VES and borehole log signature performance analysis show suitable matches between them (Figure 12).

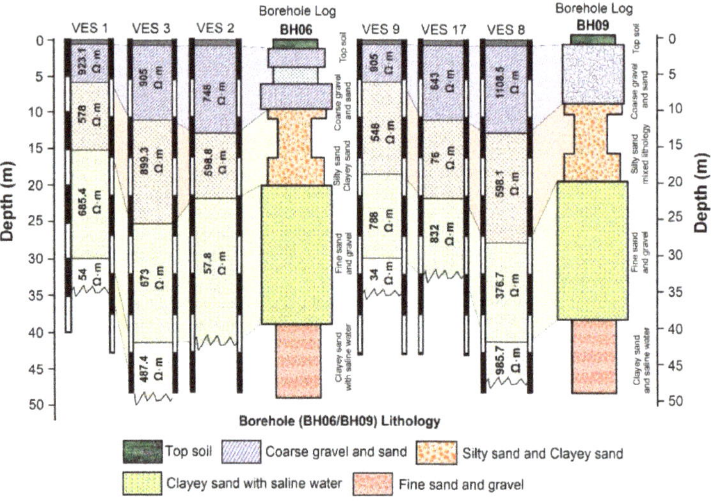

Figure 12. Correlation of VES data interpretation results with borehole lithological information.

5.5.3. GW_{pot} Based on VES

Aiming at monitoring aquifer potential, a preliminary conceptualization of geoelectrical properties governing the reflection coefficient, the aquifer's overburden thickness, and resistivity is needed during VES measurements. These basic and essential interpretative criteria are described below.

The reflection coefficient (RC) is an essential geoelectric factor, as it helps to identify the permeable hydrologic layers carrying the GW_{pot}. The RC values of the VES positions in the surveyed area were calculated using Loke's method [72]. Figure 13a shows the changes in RC values detected by each VES station. Differences in subsurface resistivity and lithology cause the RC fluctuations. The calculated RC values were contoured in Surfer 15 software, and an RC map shows a value range of 0.50–0.95 (Figure 14a). Olayinka [73] observed that the subsurface topography usually shows a good aquifer when the overburden is relatively thick and/or the reflection coefficient is low (<0.8). RC mapping has been found to be useful in investigating the hydrogeological aquifer because it reveals whether a permeable aquifer is filled with water. Therefore, an anisotropy coefficient for this parameter was considered in this study.

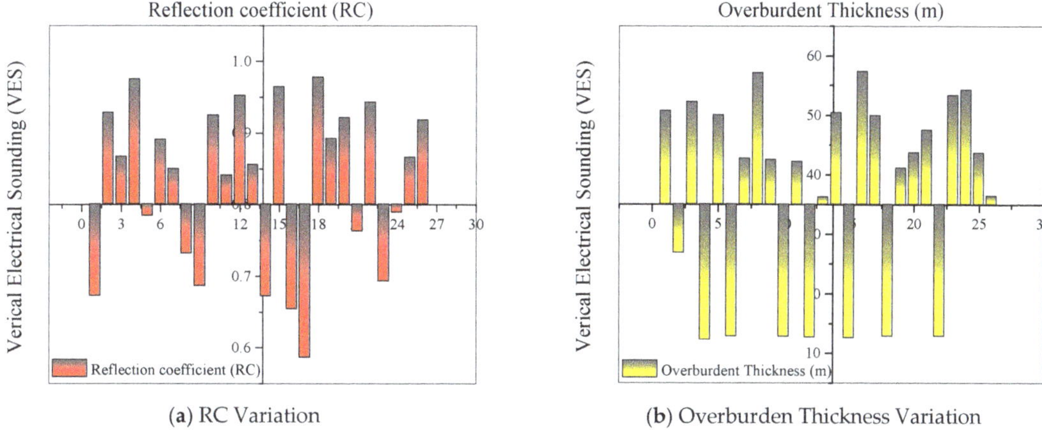

(**a**) RC Variation

(**b**) Overburden Thickness Variation

Figure 13. (**a**) Reflection coefficient (RC); (**b**) overburden thickness along 26 VES stations in the surveyed area.

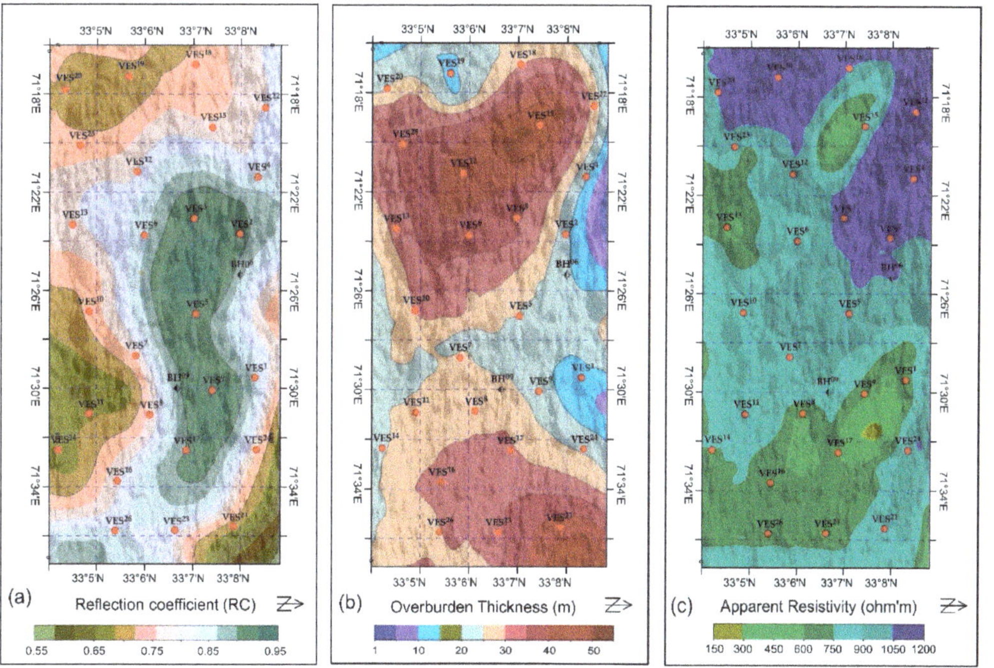

Figure 14. (**a**) Reflection coefficient map; (**b**) overburden thickness map; (**c**) apparent resistivity map based on the interpretation of VES data.

An overburden thickness/isopach map was plotted and contoured according to the interpreted depths to the sedimentary rock (Figure 14b). The isopach map illustrates the thickness variation in a hydro-stratigraphic layer, a tabular unit, or a stratum [29]. Isopach mapping is essential in the hydrogeological investigation because it shows the number of hydrogeologic layers above the aquifer, and where groundwater can be observed in areas considering the overburden thickness. The overburden thickness variation of the aquifer along VES can be seen in Figure 13b. The overburden thickness in the surveyed area varies between 6.3 and 65.6 m. The isopach map shows that the overburden thickness in the northern, eastern, and southern parts of the surveyed area ranges from 20 to 50 m (Figure 14b). In contrast, the relatively thin overburden thickness of 5–15 m is virtually around the central and western parts of the surveyed area. The overburden thickness is shallow in most probing stations, indicating that the basement is close to the surface. Therefore, groundwater in these areas is highly dependent on the occurrence of fractures [29].

The apparent resistivity values of all VES stations were contoured to produce an iso-resistivity map (Figure 14c), indicating that the apparent resistivity increases radially outward from the center of the region and the resistivity values are 10–1150 Ωm. The resistivity of the bedrock represents the resistivity of the deepest hydrological layer in the surveyed area. It has been found that the resistivity of the bedrock is of significance in many aspects of hydrogeological and hydro-geophysical measurements because it plays a vital role in assessing the potential of groundwater. After all, the resistivity of the bedrock has the potential to reveal fractured aquifers.

The lower RC and relatively high overburden thickness can increase a well's groundwater productivity [74]. In this study, the considered GW_{pot} geoelectrical factors includes reflection coefficient, overburden thickness, and iso-resistivity obtained from the interpretation of VES data. This quantification of aquifer potential indirectly verified the accuracy of the MIF model and its predictive performance. The VES stations in the surveyed area were divided into high yield, medium yield, and low yield groundwater by employing Olayinka's basic criteria [73].

(1)   High GW_{pot}: the overburden thickness is greater than 13 m with an RC less than 0.8.
(2)   Medium GW_{pot}: the overburden thickness is 13-30 m with an RC greater than or equal to 0.8.
(3)   Low GW_{pot}: the overburden thickness is less than 13 m with an RC greater than or equal to 0.8.

Considering these criteria, the RC and overburden thickness were used to produce the parameters for categorizing VES stations by the GW_{pot}, i.e., VES1, VES5, VES6, VES8, VES9, VES14, VES16, VES17, VES21, VES24, and VES26 have high yield GW_{pot} (Figure 15a), VES3, VES7, VES11, VES13, VES19, VES20, and VES25 have medium yield GW_{pot} (Figure 15b), and VES2, VES4, VES10, VES12, VES15, VES18, and VES22 have low yield GW_{pot} (Figure 15c). Based on these groundwater potentiality variations among the VES stations, a final GW_{pot} contour map of the surveyed area was generated, and it demonstrates that the northern, northeastern and eastern parts have excellent GW_{pot} for future exploitation and development, while the low and medium GW_{pot} regions are located in the western and central parts of the surveyed area (Figure 16). The VES-based groundwater potential map was compared with the groundwater potential map obtained by the RS and GIS-based MIF method. This indicated that the MIF method is accurate and consistent in predicting GW_{pot}.

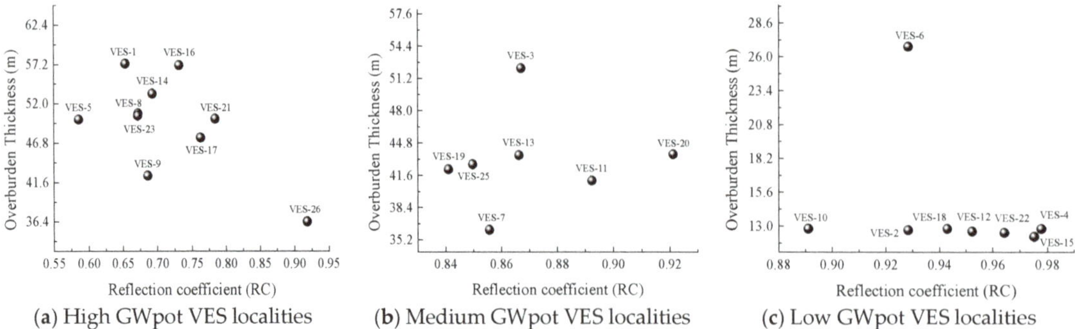

(a) High GWpot VES localities   (b) Medium GWpot VES localities   (c) Low GWpot VES localities

Figure 15. Groundwater potential VES distribution corresponding overburden thickness and reflection coefficient (RC): (**a**) high yield GW_{pot} VES stations, (**b**) medium yield GW_{pot} VES stations, and (**c**) low yield GW_{pot} VES stations.

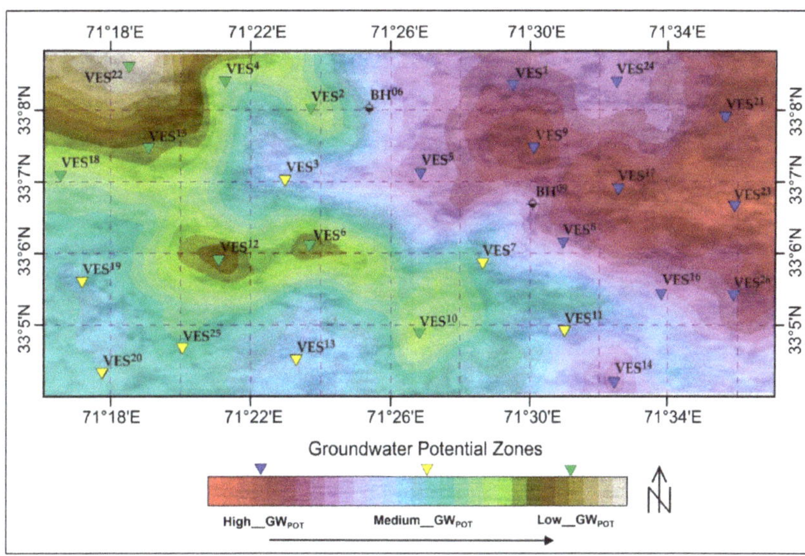

Figure 16. Groundwater potential map of the vertical electrical sounding (VES) surveyed area.

5.6. Geophysical Well Logs Interpretation

Information obtained from technical reports of SPLs and NRLs (short and long) and drilling protocols show that the slightly denser thick and deep sandstone is an effective aquifer type for groundwater exploitation in the study area (Figure 17). The geophysical well logs approach has great significance in determining the exact location (depth) of any permeable aquifers and impermeable aquitards (Table 7). In this study, NRLs (short and long) were appropriately calibrated and quantitatively interpreted. Moreover, log measurements were converted to the apparent resistivity and adjusted for mud resistivity, bed thickness, borehole diameter, mud cake, and invasion to arrive at true resistivity (Figure 17). SPL interpretation can be complex, particularly in freshwater aquifers. This complexity commences to the perversion of groundwater and misinterpretations of spontaneous potential (SP) logging. SPLs record the potential or voltage caused by contact between a shale/clay layer and an aquifer. The natural flow of current and the SP curve were offered under the salinity conditions. The NRLs (short/long), SPLs, and drilling protocol at a depth of 152.4 m showed that the major lithology's units are clay, gravel-boulders, and sandstone (Table 8). The quality of groundwater measured by TDS is fresh. The static water level depth is about 88.3 m (Figure 17). The proposed slot opening, and the estimated discharge volume, are 1/40–1/50 and 11.35–13.24 cubic meters per hour (m^3/h), respectively (Table 8).

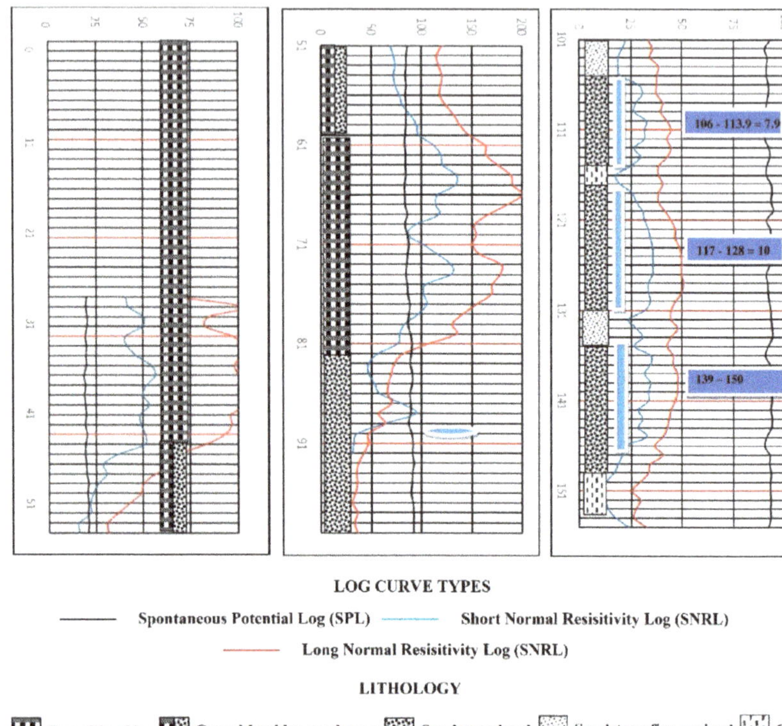

Figure 17. Spontaneous potential (SP), short normal resistivity (SNR), and long normal resistivity (LNR) log curves obtained in Well -1 of the experimental site of the Marwatan Banda, Karak.

Table 7. The following screen schedule is proposed for conversion.

No	Depth (m)	Screen (m)	Slot Size (m)
01	106.6–113.9	7.3	1/12.1–1/15.2
02	117.6–128.6	10.9	1/12.1–1/15.2
03	139.5–150.5	10.9	1/12.1–1/15.2

Table 8. Derived borehole lithology-based Normal resistivity logs (NRLs) (short/long) and spontaneous potential logs (SPLs).

No	Depth (m)	Classified Lithology	Thickness (m)
01	0–134	Gravel-boulder	134
02	134–196	Gravel-boulder-sandstone	62
03	196–269	Gravel-boulder	73
04	269–238	Sandstone hard	59
05	238–344	Sandstone fine grained	16
06	344–377	Sandstone hard	33
07	377–383	Clay	06
08	383–429	Sandstone hard	46
09	429–442	Sandstone fine grained	13
10	442–488	Sandstone hard	46
11	488–503	Clay	15

6. Discussion

The Karak watershed, located in Northern Pakistan, has experienced significant economic development associated with hydrology and groundwater exploitation. The superficial resource depletion, the irregular spatial-temporal distribution of precipitation, and the deformation of the Indian and Eurasian tectonic plate environment, which affect the occurrence and movement of groundwater, together with widespread salt in the northern mountainous catchments, which is dissolved by runoff water and polluted groundwater due to deep infiltration, have made groundwater a key resource in the study area. However, the collaboration of remote sensing observations, aquifer geoelectrical properties and accurate hydrogeological measurements, and the optimization of groundwater influential factors are major challenges. Therefore, the GW_{pot} mapping are essential for planning artificial recharge programs to mitigate groundwater decline [6]. The multi-criteria decision-making (MCDM)-based multi-influence factor (MIF) model approach can be useful for groundwater resource management (GRM) and monitoring purposes, which is an efficient bivariate statistical technique mainly used to calculate the degree to which each dependent or independent conditioning factor influences the GW_{pot}. The MIF model has become a powerful tool for delineating regional GW_{pot} and narrowing down the target areas for conducting detailed hydrogeological and hydro-geophysical surveys in the scattered areas. However, in the MIF method, weights and ranks are subjectively assigned according to expert knowledge and literatures. In a comprehensive analysis, it is important to determine the weight of each category because the output result depends on the correct weight distribution. It is used to depict groundwater prediction zones taking into account various surface and subsurface hydrological influential factors. However, several studies report that the importance and predictive power of GCFs employed in GW_{pot} assessment is usually controlled by geological, morphological, hydrological, and climatic environments [8–15,17]. According to Nampak [75], topographical features (e.g., elevation and slope) have a negative impact on GW_{pot}, while lineament density and drainage density have positive impacts. Similar research reports that topographical, soil cover, structural and hydrogeological characteristics affect precipitation runoff and permeability, thereby affecting the occurrence of GW_{pot}. Hou et al. [76] reported that lithology, altitude, and drainage density have a greater impact on the occurrence of GW_{pot}, while land use and soil type have the least impacts. In this study, a GW_{pot} map was generated based on the MIF model to identify regional GW_{pot} of the Karak watershed. Several GCFs were concluded to have significant impacts on groundwater production. For example, the high GW_{pot} zones on the final map are closely correlated to lineament density and drainage density. Usually, the lineaments indicate the areas of faults and fractures, leading to increased secondary porosity and permeability. This factor is of great significance in hydrogeology because it provides a pathway for groundwater infiltration. However, the lineament density is only an indirect indicator of the GW_{pot} in the Karak watershed, because the lineaments usually show a permeable area. In the study area, a larger slope produces a smaller recharge, because surface water will quickly flow over the steep slope during rainfall, so there is not enough time for water seeping into the ground and recharge the unsaturated zone. However, the distribution of LU/LC usually depends on the subsurface soil and geological conditions, thereby increasing the groundwater recharge on the surfaces covered by vegetation (such as agricultural plants and forests).

The hydrogeological interpretation of the 2D high-resolution resistivity tomography dataset of six traverses revealed the prospect of groundwater at different depths with variation in the resistivities in the aquifer zone. The high resistivity of the subsurface geological sediments was well delineated, which shows a large resistivity contrast within the complex geological background in the study area. This phenomenon is suggested to be caused by different degrees of weathering, fracturing and saturated weathered/fractured part of the sediments in the Karak region. In future, four to five boreholes/wells will be drilled in potential areas identified by ERT and VES to check the availability of groundwater and the performance of geoelectric surveys. The analyzed regional GW_{pot}, hydrogeological and aquifer geoelectrical information provides a beneficial prospect for the development of GRM in the study area. However, the geoelectrical exploration methods can only locally verify the result of GW_{pot} mapping, and they are too costly and time-consuming to cover the whole study area. The acquired results are expected to help practitioners to drill boreholes/wells in order to supply domestic water and irrigation in the Karak watershed of Northern Pakistan. Moreover, combined geospatial and geoelectrical methods through the MIFs model and Olayinka's basic criteria will help to assess groundwater resources in other similar areas worldwide.

7. Conclusions

This study addresses the applicability of the comprehensive MCDM-MIF model and hydro-geophysical investigation in GRM in the Karak watershed. The GIS-based MIF model facilitates the regional GW_{pot} assessment using the topographical, geological, hydrological, and land-cover GCFs, meanwhile, the geophysical exploration and data interpretation reveals the hydrogeological structure and aquifer geoelectrical characteristics. The main findings are as follows:

(1)   According to MCDM-MIF model, approximately 9.7% (72.3 km^2), 52.4% (1307.7 km^2), 31.3% (913.4 km^2), and 6.6% (44.8 km^2) areas of the total Karak watershed are classified into the low, medium, high, and very high GW_{pot}, respectively. The southern, southeastern, and the limited northeastern areas have high to medium GW_{pot} due to the distribution of Quaternary alluvial and agricultural land with high infiltration capacity. The final GW_{pot} map will help to manage sustainable groundwater resources in the study area.

(2)   The predictive performance of MCDM-MIF model is consistent with the groundwater level (GWL) data (as AUC value is 68%, confusion matrix is 68%, and Kappa (K) analysis is 65%).

(3)   The ERT approach with an optimal compromise between electrode distance and profile length highlights the complexity of hydrogeological layers and reveal that GW_{pot} is structurally controlled and adjacent to clayey sand and silicate aquifers (sandstone). The identified drilling locations on ERT traverses are of great significance for the expansion of drinking water supply and irrigation in the future. The performance analysis between ERT-predicted lithology and well-log lithology indicates suitable matches.

(4)   Hydro-stratigraphic information followed by apparent resistivity distribution at each VES station shows that the study area is mainly composed of coarse gravel and sand, followed by clayey sand with saline water. According to Olayinka's basic standards, the aquifer geoelectrical characteristics, e.g., reflection coefficient, aquifer overburden thickness and apparent resistivity distribution, were conceptualized. The interpreted potential zones based on VES show satisfactory matches with MIF-based groundwater potential. The drilling protocol and well logs data interpretation of NRLs (short/long) and SPLs reveal that deep sandstone is an effective aquifer type for the groundwater exploitation in the study area.

Author Contributions: Conceptualization, U.K. and B.Z.; methodology, U.K. and H.F.; software, Z.J. and M.W.; validation, M.Y. and B.Z.; data curation, H.F. and M.Y.; writing—original draft preparation, U.K. and B.Z.; writing—review and editing, U.K. and B.Z.; funding acquisition, B.Z. All authors have read and agreed to the published version of the manuscript.

Funding: This research was funded by the National Natural Science Foundation of China (grant number 42072326 and 41772348) and the National Key Research and Development Program of China (grant number 2019YFC1805905 and 2017YFC0601503).

Institutional Review Board Statement: Not applicable.

Informed Consent Statement: Not applicable.

Conflicts of Interest: No conflict of interest exists in the submission of this manuscript, and the manuscript was approved by all authors for publication.

Appendix A

Table A1. Summary results of VES data interpretation demonstrate the inferred lithologies corresponding to resistivity variation and hydrogeologic layers.

VES No.	RMS Error (%)	No. of Layers	Resistivity (Ohm.m)	Thickness (m)	Depth (m)	Reflection Coefficient	Inferred Hydro-Stratigraphic Lithology
1	2.9	Layer 1	923.1	2.4	2.4	0.6723	Coarse gravel and sand
		Layer 2	578	7.2	9.8		Silty sand mixed lithology
		Layer 3	685.4	28.4	38.2		Clayey sand
		Layer 4	54	12.6	50.8		Fine sand and gravels
2	3.2	Layer 1	748	1.9	1.9	0.9284	Coarse gravel and sand
		Layer 2	598.8	8.6	10.5		Silty sand and gravels
		Layer 3	57.8	16.4	26.9		Fine sands and gravels
		Layer 4	34.2	Infinite	Infinite		Clayey sand and saline water
3	3.8	Layer 1	899.3	2.4	2.4	0.8671	Coarse gravel and sand
		Layer 2	673	8.2	10.6		Silty sand mixed lithology
		Layer 3	487.4	18.7	29.3		Silty sand and gravels
		Layer 4	48.3	14.4	43.7		Fine sand and gravel
		Layer 5	26	8.6	52.3		Clayey sand and saline water
4	4.5	Layer 1	954	1.3	1.3	0.9752	Topsoil
		Layer 2	637.5	3.6	4.9		Silty sand and gravels
		Layer 3	43	7.4	12.3		Fine sand and gravels
		Layer 4	683.1	Infinite	Infinite		Clayey sands
5	2.8	Layer 1	854.7	4.5	4.5	0.7841	Coarse gravel and sand
		Layer 2	532.1	12.8	17.3		Silty sand and gravels
		Layer 3	678.2	26.2	43.5		Clayey sands
		Layer 4	43.2	6.6	50.1		Clayey sand and saline water
6	1.7	Layer 1	701	3.2	3.2	0.8911	Coarse gravel and sand
		Layer 2	693.4	4.6	7.8		Silty sand mixed lithology
		Layer 3	964.5	5.1	12.9		Clayey sands
		Layer 4	45.8	Infinite	Infinite		Clayey sand and saline water
7	2.6	Layer 1	1093.5	3.8	3.8	0.8497	Coarse gravel and sand
		Layer 2	601.5	12.4	16.2		Silty sand mixed lithology
		Layer 3	1065	18.4	34.6		Clayey sand
		Layer 4	73	8.2	42.8		Fine sand and gravels
8	4.0	Layer 1	1108.5	4.5	4.6	0.7313	Topsoil
		Layer 2	598.1	14.1	18.7		Silty sand and gravels
		Layer 3	376.7	22.4	41.1		Fine sand and gravels
		Layer 4	985.7	16.1	57.2		Clayey sands layer
9	4.7	Layer 1	735	2.2	2.2	0.6861	Alluvium
		Layer 2	323	5.6	7.8		Silty sand fine lithology
		Layer 3	675.8	12.4	20.2		Silty sand and gravels
		Layer 4	24.7	16	36.2		Fine sands and gravels
		Layer 5	74	6.4	42.6		Clayey sand and saline water
10	2.5	Layer 1	967.8	1.8	1.8	0.9248	Coarse gravel and sand
		Layer 2	560	4.2	6		Silty sand mixed lithology
		Layer 3	856.6	6.8	12.8		Clayey sand
		Layer 4	42.1	Infinite	Infinite		Fine sand and gravels
11	3.7	Layer 1	787	5.1	5.1	0.8410	Coarse gravel and sand
		Layer 2	445.8	12.2	17.3		Silty sand and gravels
		Layer 3	943.2	18.1	35.4		Clayey sands
		Layer 4	44	6.8	42.2		Clayey sand and saline water

Table A1. *Cont.*

VES No.	RMS Error (%)	No. of Layers	Resistivity (Ohm.m)	Thickness (m)	Depth (m)	Reflection Coefficient	Inferred Hydro-Stratigraphic Lithology
12	2.8	Layer 1	799.8	1.6	1.6	0.9523	Coarse gravel and sand
		Layer 2	588	4.1	5.7		Silty sand mixed lithology
		Layer 3	1098.3	4.6	10.3		Clayey sand
		Layer 4	57	2.4	12.7		Fine sand and gravels
13	2.0	Layer 1	800.3	1.3	1.3	0.8557	Coarse gravel and sand
		Layer 2	454	8.2	9.5		Silty sand and gravels
		Layer 3	822.7	16.2	25.7		Clayey sands
		Layer 4	38.6	10.6	36.3		Clayey sand and saline water
14	4.0	Layer 1	866	2.6	2.6	0.6719	Alluvium
		Layer 2	554	10.1	12.7		Silty sand fine lithology
		Layer 3	600.4	16.8	29.5		Silty sand and gravels
		Layer 4	89.3	22.8	42.3		Fine sands and gravels
		Layer 5	34.9	8.2	50.5		Clayey sand and saline water
15	2.8	Layer 1	766.1	2.4	2.4	0.9643	Coarse gravel and sand
		Layer 2	543	6.1	8.5		Silty sand and gravels
		Layer 3	985	4.1	12.6		Clayey sands
		Layer 4	27	Infinite	Infinite		Clayey sand and saline water
16	3.0	Layer 1	812.9	4.2	4.2	0.6537	Coarse gravel and sand
		Layer 2	553.4	8.2	12.4		Silty sand and gravels
		Layer 3	985.6	20.8	33.2		Clayey sands
		Layer 4	58	24.2	57.4		Clayey sand and saline water
17	4.5	Layer 1	1109	3.4	3.4	0.5855	Topsoil
		Layer 2	643	8.7	12.1		Silty sand and gravels
		Layer 3	76	15.9	28		Fine sand and gravels
		Layer 4	832	22	50		Clayey sands
18	2.8	Layer 1	741	1.8	1.8	0.9778	Coarse gravel and sand
		Layer 2	533	5.1	6.9		Silty sand and gravels
		Layer 3	932	6.0	12.9		Clayey sands
		Layer 4	65.4	Infinite	Infinite		Clayey sand and saline water
19	1.6	Layer 1	979.8	5.2	5.2	0.8923	Coarse gravel and sand
		Layer 2	568	8.4	13.6		Silty sand and gravels
		Layer 3	732.6	16.8	30.4		Clayey sands
		Layer 4	64	10.8	41.2		Clayey sand and saline water
20	4.2	Layer 1	905	3.2	3.2	0.9211	Coarse gravel and sand
		Layer 2	548	6.1	9.3		Silty sand mixed lithology
		Layer 3	788	19.5	28.8		Clayey sand
		Layer 4	34	15	43.8		Fine sand and gravels
21	2.4	Layer 1	745	4.2	4.2	0.7626	Alluvium
		Layer 2	623	10.4	14.6		Silty sand fine lithology
		Layer 3	522.3	18.6	33.2		Silty sand and gravels
		Layer 4	37	6.2	39.4		Fine sands and gravels
		Layer 5	84.2	8.2	47.6		Clayey sand and saline water
22	3.7	Layer 1	865.6	1.6	1.6	0.9429	Coarse gravel and sand
		Layer 2	590	4.5	6.1		Silty sand and gravels
		Layer 3	955.7	6.8	12.9		Clayey sands
		Layer 4	67	Infinite	Infinite		Clayey sand and saline water
23	3.3	Layer 1	906.4	5.6	5.6	0.6930	Alluvium
		Layer 2	578	10.2	15.8		Silty sand fine lithology
		Layer 3	356	18.8	34.6		Silty sand and gravels
		Layer 4	65	10.4	45		Fine sands and gravels
		Layer 5	45.6	8.4	53.4		Clayey sand and saline water

Table A1. Cont.

VES No.	RMS Error (%)	No. of Layers	Resistivity (Ohm.m)	Thickness (m)	Depth (m)	Reflection Coefficient	Inferred Hydro-Stratigraphic Lithology
24	2.8	Layer 1	842	4.4	4.4	0.7889	Coarse gravel and sand
		Layer 2	600.4	12.4	16.8		Silty sand and gravels
		Layer 3	789	22.8	39.6		Clayey sands
		Layer 4	37.2	14.7	54.3		Clayey sand and saline water
25	2.5	Layer 1	890	2.4	2.4	0.8663	Coarse gravel and sand
		Layer 2	566.3	6.8	9.2		Silty sand and gravels
		Layer 3	742	18.3	27.5		Clayey sands
		Layer 4	98	16.2	43.7		Clayey sand and saline water
26	3.0	Layer 1	975.8	1.7	1.7	0.9184	Coarse gravel and sand
		Layer 2	563	10.2	11.9		Silty sand and gravels
		Layer 3	732.1	18	29.9		Clayey sands
		Layer 4	63.9	6.6	36.5		Clayey sand and saline water

References

1. Wang, L.F.; Chen, Q.; Long, X.P.; Wu, X.B.; Sun, L. Comparative analysis of groundwater fluorine levels and other characteristics in two areas of Laizhou Bay and its explanation on fluorine enrichment. *Water Sci. Technol. Water Supply* **2015**, *15*, 384–394. [CrossRef]
2. Pinto, D.; Shrestha, S.; Babel, M.S.; Ninsawat, S. Delineation of groundwater potential zones in the Comoro watershed, Timor Leste using GIS, remote sensing and analytic hierarchy process (AHP) technique. *Appl. Water Sci.* **2017**, *7*, 503–519. [CrossRef]
3. Berhanu, K.G.; Hatiye, S.D. Identification of Groundwater Potential Zones Using Proxy Data: Case study of Megech Watershed, Ethiopia. *J. Hydrol. Reg. Stud.* **2020**, *28*, 100676. [CrossRef]
4. Chen, Q.; Jia, C.; Wei, J.; Dong, F.; Yang, W.; Hao, D.; Jia, Z.; Ji, Y. Geochemical process of groundwater fluoride evolution along global coastal plains: Evidence from the comparison in seawater intrusion area and soil salinization area. *Chem. Geol.* **2020**, *552*, 119779. [CrossRef]
5. Hao, Q.; Xiao, Y.; Chen, K.; Zhu, Y.; Li, J. Comprehensive Understanding of Groundwater Geochemistry and Suitability for Sustainable Drinking Purposes in Confined Aquifers of the Wuyi Region, Central North China Plain. *Water* **2020**, *12*, 3052. [CrossRef]
6. Goldman, M.; Kafri, U. Hydrogeophysical applications in coastal aquifers. In *Applied Hydrogeophysics*; Springer: Dordrecht, The Netherlands, 2006; pp. 233–254.
7. Hasan, M.; Shang, Y.; Jin, W.; Shao, P.; Yi, X.; Akhter, G. Geophysical Assessment of Seawater Intrusion into Coastal Aquifers of Bela Plain, Pakistan. *Water* **2020**, *12*, 3408. [CrossRef]
8. Baalousha, H. Using CRD method for quantification of groundwater recharge in the Gaza Strip, Palestine. *Environ. Geol.* **2005**, *48*, 889–900. [CrossRef]
9. Shaban, A.; Khawlie, M.; Abdallah, C. Use of remote sensing and GIS to determine recharge potential zones: The case of Occidental Lebanon. *Hydrogeol. J.* **2005**, *14*, 433–443. [CrossRef]
10. Baalousha, H. Stochastic water balance model for rainfall recharge quantification in Ruataniwha Basin, New Zealand. *Environ. Geol.* **2009**, *58*, 85. [CrossRef]
11. Zghibi, A.; Merzougui, A.; Chenini, I.; Ergaieg, K.; Zouhri, L.; Tarhouni, J. Groundwater vulnerability analysis of Tunisian coastal aquifer: An application of DRASTIC index method in GIS environment. *Groundw. Sustain. Dev.* **2016**, *2-3*, 169–181. [CrossRef]
12. Zabihi, M.; Pourghasemi, H.R.; Pourtaghi, Z.S.; Behzadfar, M. GIS-based multivariate adaptive regression spline and random forest models for groundwater potential mapping in Iran. *Environ. Earth Sci.* **2016**, *75*, 1–19. [CrossRef]
13. Arshad, A.; Zhang, Z.; Zhang, W.; Dilawar, A. Mapping favorable groundwater potential recharge zones using a GIS-based analytical hierarchical process and probability frequency ratio model: A case study from an agro-urban region of Pakistan. *Geosci. Front.* **2020**, *11*, 1805–1819. [CrossRef]
14. Rahmati, O.; Samani, A.N.; Mahdavi, M.; Pourghasemi, H.R.; Zeinivand, H. Groundwater potential mapping at Kurdistan region of Iran using analytic hierarchy process and GIS. *Arab. J. Geosci.* **2015**, *8*, 7059–7071. [CrossRef]
15. Cai, Z.; Ofterdinger, U. Analysis of groundwater-level response to rainfall and estimation of annual recharge in fractured hard rock aquifers, NW Ireland. *J. Hydrol.* **2016**, *535*, 71–84. [CrossRef]
16. Pourghasemi, H.R.; Pradhan, B.; Gokceoglu, C.; Mohammadi, M.; Moradi, H.R. Application of weights-of-evidence and certainty factor models and their comparison in landslide susceptibility mapping at Haraz watershed, Iran. *Arab. J. Geosci.* **2013**, *6*, 2351–2365. [CrossRef]
17. Baalousha, H.M. Using Monte Carlo simulation to estimate natural groundwater recharge in Qatar. *Model. Earth Syst. Environ.* **2016**, *2*, 1–7. [CrossRef]

18. Pradhan, B.; Seeni, M.I.; Kalantar, B. Performance evaluation and sensitivity analysis of expert-based, statistical, machine learning, and hybrid models for producing landslide susceptibility maps. In *Laser Scanning Applications in Landslide Assessment*; Springer: Cham, Germany, 2017; pp. 193–232.
19. Xiao, Y.; Gu, X.; Yin, S.; Pan, X.; Shao, J.; Cui, Y. Investigation of Geochemical Characteristics and Controlling Processes of Groundwater in a Typical Long-Term Reclaimed Water Use Area. *Water* **2017**, *9*, 800. [CrossRef]
20. Abrams, W.; Ghoneim, E.; Shew, R.; LaMaskin, T.; Al-Bloushi, K.; Hussein, S.; AbuBakr, M.; Al-Mulla, E.; Al-Awar, M.; El-Baz, F. Delineation of groundwater potential (GWP) in the northern United Arab Emirates and Oman using geospatial technologies in conjunction with Simple Additive Weight (SAW), Analytical Hierarchy Process (AHP), and Probabilistic Frequency Ratio (PFR) techniques. *J. Arid. Environ.* **2018**, *157*, 77–96. [CrossRef]
21. Xiao, Y.; Shao, J.; Frape, S.K.; Cui, Y.; Dang, X.; Wang, S.; Ji, Y. Groundwater origin, flow regime and geochemical evolution in arid endorheic watersheds: A case study from the Qaidam Basin, northwestern China. *Hydrol. Earth Syst. Sci.* **2018**, *22*, 4381–4400. [CrossRef]
22. Díaz-Alcaide, S.; Martínez-Santos, P. Review: Advances in groundwater potential mapping. *Hydrogeol. J.* **2019**, *27*, 2307–2324. [CrossRef]
23. Raju, R.S.; Raju, G.S.; Rajasekhar, M. Identification of groundwater potential zones in Mandavi River basin, Andhra Pradesh, India using remote sensing, GIS and MIF techniques. *HydroResearch* **2019**, *2*, 1–11. [CrossRef]
24. Das, S.; Gupta, A.; Ghosh, S. Exploring groundwater potential zones using MIF technique in semi-arid region: A case study of Hingoli district, Maharashtra. *Spat. Inf. Res.* **2017**, *25*, 749–756. [CrossRef]
25. Zygouri, V.; Verroios, S.; Anninou, A.; Koukouvelas, I. Fuzzy cognitive mapping vs. statistic approach of geomorphological analysis of active faults: Preliminary results from a case study within the Corinth Graben, Greece. *IFAC PapersOnLine* **2018**, *51*, 372–377. [CrossRef]
26. Baalousha, H.M. Development of a groundwater flow model for the highly parameterized Qatar aquifers. *Model. Earth Syst. Environ.* **2016**, *2*, 67. [CrossRef]
27. Srivastava, P.K.; Bhattacharya, A.K. Groundwater assessment through an integrated approach using remote sensing, GIS and resistivity techniques: A case study from a hard rock terrain. *Int. J. Remote Sens.* **2006**, *27*, 4599–4620. [CrossRef]
28. Farid, A.; Jadoon, K.; Akhter, G.; Iqbal, M.A. Hydrostratigraphy and hydrogeology of the western part of Maira area, Khyber Pakhtunkhwa, Pakistan: A case study by using electrical resistivity. *Environ. Monit. Assess.* **2013**, *185*, 2407–2422. [CrossRef]
29. Farid, A.; Khalid, P.; Jadoon, K.Z.; Iqbal, M.A.; Small, J. An application of variogram modelling for electrical resistivity soundings to characterize depositional system and hydrogeology of Bannu Basin, Pakistan. *Geosci. J.* **2017**, *21*, 819–839. [CrossRef]
30. Mondal, N.C.; Rao, A.V.; Singh, V.P. Efficacy of electrical resistivity and induced polarization methods for revealing fluoride contaminated groundwater in granite terrain. *Environ. Monit. Assess.* **2010**, *168*, 103–114. [CrossRef]
31. Son, Y. Assessment of concentration in contaminated soil by potentially toxic elements using electrical properties. *Environ. Monit. Assess.* **2010**, *176*, 1–11. [CrossRef]
32. Devi, S.P.; Srinivasulu, S.; Raju, K.K. Delineation of groundwater potential zones and electrical resistivity studies for groundwater exploration. *Environ. Earth Sci.* **2001**, *40*, 1252–1264. [CrossRef]
33. Anechana, R.; Noye, R.M.; Menyeh, A.; Manu, E.; Okrah, C. Electromagnetic Method and Vertical Electrical Sounding for Groundwater Potential Assessment of Kintampo North Municipality of Ghana. *J. Environ. Earth Sci.* **2015**, *5*, 9–16.
34. Gupta, G.; Patil, S.N.; Padmane, S.T.; Erram, V.C.; Mahajan, S. Geoelectric investigation to delineate groundwater potential and recharge zones in Suki river basin, north Maharashtra. *J. Earth Syst. Sci.* **2015**, *124*, 1487–1501. [CrossRef]
35. Olorunfemi, M.; Fasuyi, S. Aquifer types and the geoelectric/hydrogeologic characteristics of part of the central basement terrain of Nigeria (Niger State). *J. Afr. Earth Sci.* **1993**, *16*, 309–317. [CrossRef]
36. Frid, V.; Sharabi, I.; Frid, M.; Averbakh, A. Leachate detection via statistical analysis of electrical resistivity and induced polarization data at a waste disposal site (Northern Israel). *Environ. Earth Sci.* **2017**, *76*, 233. [CrossRef]
37. Kumar, D.; Mondal, S.; Nandan, M.J.; Harini, P.; Sekhar, B.M.V.S.; Sen, M.K. Two-dimensional electrical resistivity tomography (ERT) and time-domain-induced polarization (TDIP) study in hard rock for groundwater investigation: A case study at Choutuppal Telangana, India. *Arab. J. Geosci.* **2016**, *9*, 355. [CrossRef]
38. Hamzah, U.; Samsudin, A.R.; Malim, E.P. Groundwater investigation in Kuala Selangor using vertical electrical sounding (VES) surveys. *Environ. Earth Sci.* **2006**, *51*, 1349–1359. [CrossRef]
39. Olasehinde, P.I.; Bayewu, O.O. Evaluation of electrical resistivity anisotropy in geological mapping: A case study of Odo Ara, West Central Nigeria. *Afr. J. Environ. Sci. Technol.* **2011**, *5*, 553–566.
40. Webb, S.J.; Ngobeni, D.; Jones, M.; Abiye, T.; Devkurran, N.; Goba, R.; Lee, M.; Burrows, D.; Pellerin, L. Hydrogeophysical investigation for groundwater at the Dayspring Children's Village, South Africa. *Lead. Edge* **2011**, *30*, 434–440.
41. Manu, E.; Agyekum, W.A.; Duah, A.A.; Mainoo, P.A.; Okrah, C.; Asare, V.S. Improving Access to Potable Water Supply using Integrated Geophysical Approach in a Rural Setting of Eastern Ghana. *Elixir Environ. For.* **2016**, *95*, 40714–40719.
42. Kruseman, G.P.; Naqvi, S.A.H. Hydrogeology and groundwater resources of the North-West Frontier Province Pakistan. *WAPDA Hydrogeol. Dir. Peshawar* **1988**.
43. Gee, E.; Gee, D. Overview of the geology and structure of the Salt Range, with observations on related areas of northern Pakistan. *Geol. Soc. Am. Spec. Paper* **1989**, *232*, 95–112.

44. Ali, F.; Khan, M.I.; Ahmad, S.; Rehman, G.; Rehman, I.; Ali, T.H. Range front structural style: An example from Surghar Range, North Pakistan. *J. Himal. Earth Sci.* **2014**, *47*, 193.

45. Nazir-ur-Rehman, A.S.; Fayaz, A.; Iftikhar, A.; Amin, S. Joints/fractures analyses of Shinawah area, district Karak, Khyber Pakhtunkhwa, Pakistan. *J. Himal. Earth Sci. Vol.* **2017**, *50*, 93–113.

46. Jaswal, T.M.; Lillie, R.J.; Lawrence, R.D. Structure and evolution of the northern Potwar deformed zone, Pakistan. *AAPG Bull.* **1997**, *81*, 308–328.

47. Akhter, G.; Farid, A.; Ahmad, Z. Determining the depositional pattern by resistivity–seismic inversion for the aquifer system of Maira area, Pakistan. *Environ. Monit. Assess.* **2012**, *184*, 161–170. [CrossRef]

48. Ullah, K.; Arif, M.; Shah, M.T. Petrography of sandstones from the Kamlial and Chinji formations, southwestern Kohat plateau, NW Pakistan: Implications for source lithology and paleoclimate. *J. Himal. Earth Sci.* **2006**, *39*, 1–13.

49. Shahzad, F.; Mahmood, S.A.; Gloaguen, R. Drainage network and lineament analysis: An approach for Potwar Plateau (Northern Pakistan). *J. Mt. Sci.* **2009**, *6*, 14–24. [CrossRef]

50. Kalantar, B.; Al-Najjar, H.A.H.; Pradhan, B.; Saeidi, V.; Halin, A.A.; Ueda, N.; Naghibi, S.A.; Najjar, A.-. Ueda Optimized Conditioning Factors Using Machine Learning Techniques for Groundwater Potential Mapping. *Water* **2019**, *11*, 1909. [CrossRef]

51. Magesh, N.; Chandrasekar, N.; Soundranayagam, J.P. Delineation of groundwater potential zones in Theni district, Tamil Nadu, using remote sensing, GIS and MIF techniques. *Geosci. Front.* **2012**, *3*, 189–196. [CrossRef]

52. Sener, E.; Davraz, A.; Ozcelik, M. An integration of GIS and remote sensing in groundwater investigations: A case study in Burdur, Turkey. *Hydrogeol. J.* **2005**, *13*, 826–834. [CrossRef]

53. Edet, A.E.; Okereke, C.S.; Teme, S.C.; Esu, E.O. Application of remote-sensing data to groundwater exploration: A case study of the Cross River State, southeastern Nigeria. *Hydrogeol. J.* **1998**, *6*, 394–404. [CrossRef]

54. Pavelic, P.; Giordano, M.; Keraita, B.N.; Ramesh, V.; Rao, T. *Groundwater Availability and Use in Sub-Saharan Africa: A Review of 15 Countries*; International Water Management Institute: Colombo, Sri Lanka, 2012.

55. Ndlovu, S.; Mpofu, V.; Manatsa, D.; Muchuweni, E. Mapping groundwater aquifers using dowsing, slingram electromagnetic survey method and vertical electrical sounding jointly in the granite rock formation: A case of Matshetshe rural area in Zimbabwe. *J. Sustain. Dev. Afr.* **2010**, *12*, 199–208.

56. Steelman, C.M.; Kennedy, C.S.; Capes, D.C.; Parker, B.L. Electrical resistivity dynamics beneath a fractured sedimentary bedrock riverbed in response to temperature and groundwater–surface water exchange. *Hydrol. Earth Syst. Sci.* **2017**, *21*, 3105–3123. [CrossRef]

57. Stephenson, D.J. *Rockfill in Hydraulic Engineering*; Elsevier: Amsterdam, The Netherlands, 1979.

58. Hutti, B.; Nijagunappa, R. Identification of groundwater potential zone using geoinformatics in Ghataprabha basin, North Karnataka, India. *Int. J. Geomat. Geosci.* **2011**, *2*, 91.

59. Zghibi, A.; Mirchi, A.; Msaddek, M.H.; Merzougui, A.; Zouhri, L.; Taupin, J.-D.; Chekirbane, A.; Chenini, I.; Tarhouni, J. Using Analytical Hierarchy Process and Multi-Influencing Factors to Map Groundwater Recharge Zones in a Semi-arid Mediterranean Coastal Aquifer. *Water* **2020**, *12*, 2525. [CrossRef]

60. Malczewski, J. On the Use of Weighted Linear Combination Method in GIS: Common and Best Practice Approaches. *Trans. GIS* **2000**, *4*, 5–22. [CrossRef]

61. Chen, W.; Panahi, M.; Khosravi, K.; Pourghasemi, H.R.; Rezaie, F.; Parvinnezhad, D. Spatial prediction of groundwater potentiality using ANFIS ensembled with teaching-learning-based and biogeography-based optimization. *J. Hydrol.* **2019**, *572*, 435–448. [CrossRef]

62. Rajaveni, S.P.; Brindha, K.; Elango, L. Geological and geomorphological controls on groundwater occurrence in a hard rock region. *Appl. Water Sci.* **2017**, *7*, 1377–1389. [CrossRef]

63. Chepchumba, M.C.; Raude, J.M.; Sang, J.K. Geospatial delineation and mapping of groundwater potential in Embu County, Kenya. *Acque Sotter. Ital. J. Groundw.* **2019**, *8*. [CrossRef]

64. Jensen, J.R. *Introductory Digital Image Processing: A Remote Sensing Perspective*; Prentice-Hall Inc.: Upper Sadder River, NJ, USA, 1996.

65. Usman, M.; Liedl, R.; Shahid, M.A.; Abbas, A. Land use/land cover classification and its change detection using multi-temporal MODIS NDVI data. *J. Geogr. Sci.* **2015**, *25*, 1479–1506. [CrossRef]

66. Muchingami, I.; Hlatywayo, D.; Nel, J.; Chuma, C. Electrical resistivity survey for groundwater investigations and shallow subsurface evaluation of the basaltic-greenstone formation of the urban Bulawayo aquifer. *Phys. Chem. Earth, Parts A/B/C* **2012**, *50*, 44–51. [CrossRef]

67. Sasaki, Y. Resolution of resistivity tomography inferred from numerical simulation 1. *Geophys. Prospect.* **1992**, *40*, 453–463. [CrossRef]

68. Loke, M.H.; Barker, R.D. Rapid least-squares inversion of apparent resistivity pseudosections by a quasi-Newton method1. *Geophys. Prospect.* **1996**, *44*, 131–152. [CrossRef]

69. Ernstson, K.; Kirsch, R. Geoelectrical methods. In *Groundwater Geophysics: A Tool for Hydrogeology*; Spinger: Berlin, Germany, 2006; pp. 84–117.

70. Lenkey, L.; Hámori, Z.; Mihálffy, P. Investigating the hydrogeology of a water-supply area using direct-current vertical electrical soundings. *Geophysics* **2005**, *70*, H11–H19. [CrossRef]

71. Xiao, Y.; Yin, S.; Hao, Q.; Gu, X.; Pei, Q.; Zhang, Y. Hydrogeochemical appraisal of groundwater quality and health risk in a near-suburb area of North China. *J. Water Supply Res. Technol.* **2020**, *69*, 55–69. [CrossRef]
72. Anomohanran, O. Investigation of Groundwater Potential in some selected towns in Delta North District of Nigeria. *Int. J. Appl. Sci. Technol.* **2013**, *3*, 61–66.
73. Olayinka, A. Non uniqueness in the interpretation of bedrock resistivity from sounding curves and its hydrological implications. *Water Resour. J. NAH* **1996**, *7*, 55–60.
74. Bayewu, O.O.; Oloruntola, M.O.; Mosuro, G.O.; Laniyan, T.A.; Ariyo, S.O.; Fatoba, J.O. Geophysical evaluation of groundwater potential in part of southwestern Basement Complex terrain of Nigeria. *Appl. Water Sci.* **2017**, *7*, 4615–4632. [CrossRef]
75. Nampak, H.; Pradhan, B.; Manap, M.A. Application of GIS based data driven evidential belief function model to predict groundwater potential zonation. *J. Hydrol.* **2014**, *513*, 283–300. [CrossRef]
76. Hou, E.; Wang, J.; Chen, W. A comparative study on groundwater spring potential analysis based on statistical index, index of entropy and certainty factors models. *Geocarto Int.* **2018**, *33*, 754–769. [CrossRef]

Article

Analysis of the Spatiotemporal Annual Rainfall Variability in the Wadi Cheliff Basin (Algeria) over the Period 1970 to 2018

Mohammed Achite [1,2], **Tommaso Caloiero** [3], **Andrzej Wałęga** [4,*], **Nir Krakauer** [5] **and Tarek Hartani** [6]

1. Faculty of Nature and Life Sciences, Laboratory of Water & Environment,
   University Hassiba Benboual of Chlef, Chlef 02180, Algeria; m.achite@univ-chlef.dz
2. National Higher School of Agronomy, ENSA, Hassan Badi 16200, El Harrach, Algiers 16004, Algeria
3. National Research Council-Institute for Agricultural and Forest Systems in Mediterranean (CNR-ISAFOM),
   87036 Rende, Italy; tommaso.caloiero@cnr.it
4. Department of Sanitary Engineering and Water Management, University of Agriculture in Krakow,
   Mickiewicza 24/28 Street, 30-059 Krakow, Poland
5. Department of Civil Engineering and NOAA-CREST, The City College of New York,
   New York, NY 10031, USA; nkrakauer@ccny.cuny.edu
6. Laboratory Management and Valorization of Agricultural and Aquatic Ecosystems, Morsli Abdellah Tipaza
   University Center, Oued Merzoug, Tipaza 42000, Algeria; rik_hartani@yahoo.fr
* Correspondence: andrzej.walega@urk.edu.pl

Abstract: In the context of climate variability and hydrological extremes, especially in arid and semi-arid zones, the issue of natural risks and more particularly the risks related to rainfall is a topical subject in Algeria and worldwide. In this direction, the spatiotemporal variability of precipitation in the Wadi Cheliff basin (Algeria) has been evaluated by means of annual time series of precipitation observed on 150 rain gauges in the period 1970–2018. First, in order to identify the natural year-to-year variability of precipitation, for each series, the coefficient of variation (CV) has been evaluated and spatially distributed. Then, the precipitation trend at annual scale has been analyzed using two nonparametric tests. Finally, the presence of possible change points in the data has been investigated. The results showed an inverse spatial pattern between CV and the annual rainfall, with a spatial gradient between the southern and the northern sides of the basin. Results of the trend analysis evidenced a marked negative trend of the annual rainfall (22% of the rain gauges for a significant level equal to 95%) involving mainly the northern and the western-central area of the basin. Finally, possible change points have been identified between 1980 and 1985.

Keywords: precipitation; climate change; Sen's estimator; Mann-Kendall; Wadi Cheliff basin

Citation: Achite, M.; Caloiero, T.; Wałęga, A.; Krakauer, N.; Hartani, T. Analysis of the Spatiotemporal Annual Rainfall Variability in the Wadi Cheliff Basin (Algeria) over the Period 1970 to 2018. *Water* **2021**, *13*, 1477. https://doi.org/10.3390/w13111477

Academic Editor: Matthew Therrell

Received: 2 May 2021
Accepted: 22 May 2021
Published: 25 May 2021

Publisher's Note: MDPI stays neutral with regard to jurisdictional claims in published maps and institutional affiliations.

1. Introduction

The Mediterranean basin is climatically affected by the interaction between mid-latitude and tropical processes, being located in a transition zone between the arid climate of North Africa and the temperate and rainy climate of Europe. For this reason, it is considered a major hotspot of climate change, subject to strong warming and drying, with increasing consequences on spatial and temporal precipitation distribution [1]. Within this context, spatial and temporal precipitation analyses with different methodologies has been recently performed in the Mediterranean basin [2] and, especially, in Northern Africa [3]. The majority of these studies were principally based on non-parametric tests, which are better suited than parametric ones to deal with non-normally distributed data in hydrometeorology [4]. In particular, different results have been obtained between the eastern and western side of the region. In effect, the west-central part is characterized by a negative rainfall trend [5], albeit irregular and high variable across the decades. By contrast, the eastern side presents positive rainfall tendencies in some areas [6,7], and negative trends in others such as Israel [8,9]. In the Middle East and North Africa (MENA) region, which includes North Africa, Donat et al. [10] detected an opposite behavior in the period

1980–2010: a marked positive trend in the western side, and some consistent tendency toward dryer conditions in the eastern part. In the past years, several studies evidenced a negative rainfall trend in Northwest Africa [11,12]. For example, in the Maghreb region, Tramblay et al. [13] detected a strong negative trend in annual rainfall and number of wet days for the period 1950–2009. This trend behavior was more marked for Morocco and Western Algeria. In particular, in Algeria, average annual rainfall evidenced a decrease beginning around the second half of the 1970s [14]. This tendency has been forecast to continue over the 21st century [15,16] and to be particularly significant in semiarid areas.

Giorgi [17] demonstrated that this large decrease in average rainfall is coupled with an increase in rainfall variability, especially during the warm season. For this reason, besides rainfall trend, it is especially important to evaluate the inter-annual variability of rainfall, which has received little attention so far [18]. In fact, the inter-annual rainfall variability is a measure of the year-to-year variability in cumulative rainfall occurrences and allows us to identify years with rainfall abundance and years with rainfall scarcity. In order to evaluate the inter-annual rainfall variability, first the relative variability index has been proposed [19], but recently, the coefficient of variation (CV) has found wide application. Several authors evidenced an increase in inter-annual variability at global scale [16,20–22], but fewer studies have been performed at regional scales. For example, Gajbhiye et al. [23] analyzed CV in the Sindh river basin (India) for annual and seasonal (monsoon, post monsoon, summer and winter) rainfall events evidencing that the inter-annual variability of post monsoon rainfall is greater than that of the annual rainfall. Similar results have been obtained again in India, in the Jharkhand State [24]. He and Gautam [25] detected an increasing tendency of annual, winter and spring rainfall variabilities in California, which suggests an increasing frequency of precipitation extremes. Młyński et al. [26] detected that variability of annual extremes of precipitation in southern Poland is linked with types of cyclonic circulations. Thus, many studies evidenced that inter-annual variability of annual precipitation on many region of the World is visible and can be caused by climate change. The variability of precipitation can strongly influence water resources and thus the spatiotemporal occurrence of hydrological extremes like floods, droughts and water scarcity, and associated socioeconomic problems [27,28].

The aim of this paper was to study the spatiotemporal variability of annual rainfall in a semi-arid area by examining the annual precipitation across the Wadi Cheliff basin in Algeria. In particular, the spatial distribution of CV has been analyzed and a trend investigation on annual rainfall has been carried out using nonparametric tests. The study was performed on data recorded at 150 stations during an observation period of 49 years.

2. Methodology

CV is a statistical measure of the difference between the data points and the mean value of a series. Greater values of CV indicate larger variability and vice versa. The CV value for each series can be computed as follows:

$$CV = \frac{\sigma}{\mu} \tag{1}$$

where σ is the annual precipitation standard deviation and μ is the mean annual precipitation.

In order to analyze possible trend in annual rainfall series two non-parametric tests for trend detection have been used: the Theil-Sen (TS) estimator [29] for the evaluation of the slopes of the trends and the Mann–Kendall (MK) test [30,31] for assessment of the statistical significance. These estimators have been selected because they are not susceptible to the influence of extreme values and thus are more powerful than linear regression methods in trend slope evaluation in the presence of outliers in the series.

The first step in the calculation of the *TS* estimator is to evaluate the values of the gradient Q_i, given N pairs of data:

$$Q_i = \frac{x_j - x_k}{j - k} for \quad i = 1, \ldots, N \tag{2}$$

in which x_j and x_k are the data values at times j and k (with $j > k$), respectively.

If there is only one datum in each time period, then $N = n(n-1)/2$, where n is the number of time periods. If there are multiple observations in one or more time periods, then $N < n(n-1)/2$, where n is the total number of observations.

The *TS* estimator is then computed as the median Q_{med} of the N values of Q_i, ranked from the smallest to the largest:

$$Q_{med} = \begin{cases} Q_{[(N+1)/2]} & if\ N\ is\ odd \\ \frac{Q_{[N/2]} + Q_{[(N+2)/2]}}{2} & if\ N\ is\ even \end{cases} \tag{3}$$

The Q_{med} sign reveals the trend behaviour, while its value indicates the magnitude of the trend.

In order to evaluate the significance of the trend according to Mann-Kendall test, the statistic S must be first estimated as:

$$S = \sum_{i=1}^{d-1} \sum_{j=i+1}^{d} sgn(x_j - x_i);\ with\ \ sgn(x_j - x_i) = \begin{cases} 1\ if\ (x_j - x_i) > 0 \\ 0\ if\ (x_j - x_i) = 0 \\ -1\ if\ (x_j - x_i) < 0 \end{cases} \tag{4}$$

Here, x_j and x_i are the variable values in the years j and i (with $j > i$), respectively, and d is the dimension of the series.

Given independent and randomly ordered values, for the $d > 10$, the statistic S is distributed following a normal distribution with zero mean and variance:

$$VAR(S) = \left[d(d-1)(2d+5) - \sum_{i=1}^{m} t_i i(i-1)(2i+5) \right] / 18 \tag{5}$$

with t_i a number of i-fold ties.

Finally, the standardized statistic Z_{MK} can be computed as:

$$Z_{MK} = \begin{cases} \frac{S-1}{\sqrt{Var(S)}} for S > 0 \\ 0 for S = 0 \\ \frac{S+1}{\sqrt{Var(S)}} for S < 0 \end{cases} \tag{6}$$

By applying a two-tailed test, for a specified significance level α, the statistical significance of the trend can be evaluated. In particular, in this work, the rainfall series have been examined for three different significance levels (SL) equal to 90%, 95% and 99%.

Finally, in order to detect possible change points in the annual rainfall series, a particular form of the nonparametric Mann–Whitney (MW) test, developed by Pettitt [32], was applied.

3. Study Area and Data

The Wadi Cheliff is the longest river in the country and plays a vital role in the socioeconomic development of the main regions in Algeria. The Wadi originates from the Saharan Atlas, near Aflou in the mountains of the Jebel Amour, and is approximately 750 km long (Figure 1).

The Wadi Cheliff Basin (WCB) covers an area of 43,750 km^2 and lies between $0°7'44''$ E to $3°31'7''$ E and between $33°53'13''$ N and $36°26'34''$ N (Figure 1). The topography of the basin is complex and rugged. The altitude varies from -4 m to 1969 m. The tributaries distribute symmetrically from the south to the north along the main secondary wadis.

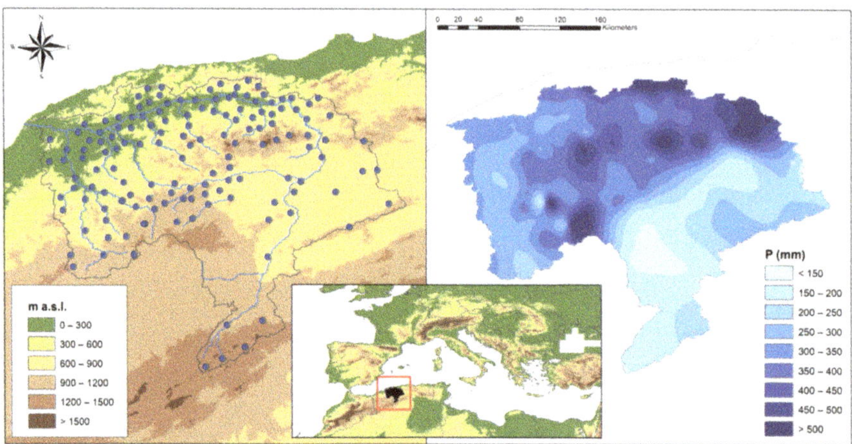

Figure 1. Location of the selected 150 rain gauges on a DEM (**left**) and spatial distribution of the average annual precipitation (**right**).

Climatically, the region is arid and semi-arid. The mean annual temperature decrease gradually from the north to south with a minimum registered at Tissimsilt region (14.20 °C) and a maximum at Cheliff region (18.7 °C) [14]. The extreme maximum and minimum temperatures occur in July and January as 42 °C and −5 °C respectively. The mean annual precipitation is from 161 mm to 662 mm (1970–2018), 80% of which falls between November and March. For this study, datasets of 150 rainfall stations (Figure 1) with long-term monthly precipitation records from 1970 to 2018 across the WCB were taken from the National Agency of the Water Resources (ANRH). However, the period of the records for these stations varies and some have missing records, and thus, to improve data quality, only the observing stations with data series accounting for 70% or more of the overall period were chosen for our study. After excluding the stations with too many missing values, the double mass curve technique was used to analyze the remaining missing data. The data was subjected to quality control and data gap filling using the linear regression method. The period of study has been chosen to be 1970–2018, which is as long as possible depending on the availability of recorded data for majority of stations in the region.

4. Results and Discussion

In this work, first the CV has been evaluated and spatially distributed. Then, the precipitation trend at annual scale has been analyzed. Finally, the presence of possible change points in the data has been investigated.

Figure 2 shows the results of the inter-annual rainfall variability analysis performed through the CV. In particular, in the boxplot, the characterization of the average CV series is represented. The CV ranges between about 16.0% (minimum CV value) and 56.5% (maximum CV value), thus evidencing high variability that is typical of the Mediterranean basin [33]. This CV range is similar to the ones obtained in past studies performed in Eastern and Southern Africa, especially for the maximum values. In fact, a high inter-annual variability has been detected in north-eastern Kenya, with CV values higher than 55% [34]. Conversely, a belt along western Uganda, Rwanda, Burundi, Tanzania and northwest Zambia with moist climate conditions evinced CV values lower than 10%.

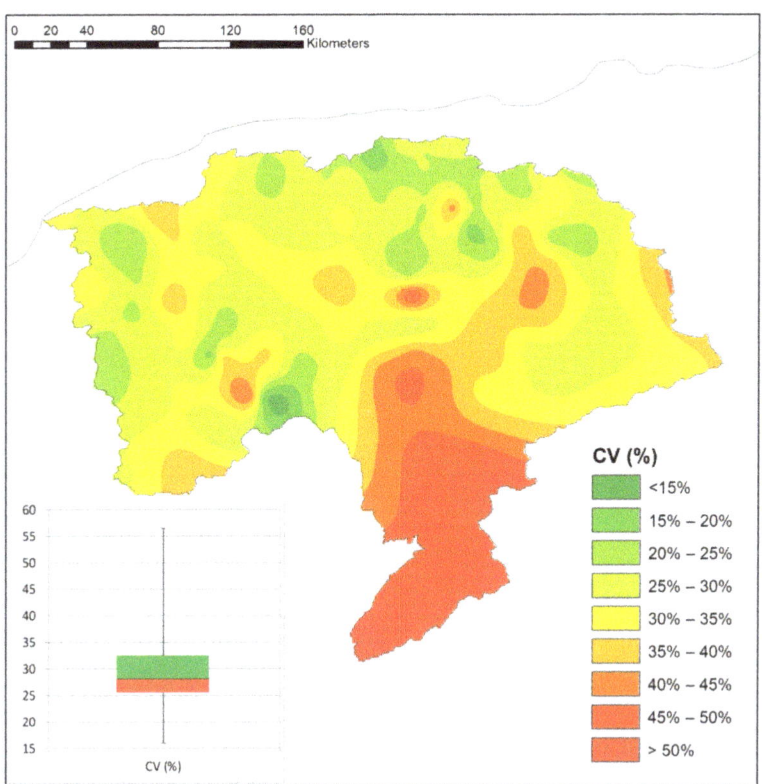

Figure 2. Spatial distribution of the CV and characterization of the CV through boxplot (Bottom and top of the box: first and third quartiles. Band inside the box: median. Ends of the whiskers: minimum and maximum values. Red color: values below the median. Green color: values above the median).

In addition to providing the characterization of the average CV values, the analysis of the spatial distribution of the CV values is paramount for understanding the risk of extreme events (Figure 2). Areas with higher inter-annual variability in rainfall are more susceptible to extreme events such as floods and droughts [35]. Results showed an inverse CV spatial pattern with respect to those observed for the annual rainfall, with a spatial gradient between the southern and the northern sides of the basin. Thus, the highest variability (CV values up to 56%) has been detected in the mountainous areas of the south side of the region, which also shows the lowest values of annual precipitation. Conversely, the northern areas of the basin, which show the highest precipitation, evidenced the lowest CV values. Both the CV and the annual rainfall spatial pattern maps were obtained with a spline algorithm.

The trend analysis has been performed at annual scale for the period 1970–2018 for three different significance levels (SL): 90%, 95% and 99% (Figure 3a). As a result, a prevalent negative rainfall trend has been evidenced. In fact, about 26.2% (SL = 90%), 16.8% (SL = 95%) and 7.4% (SL = 99%) of the rainfall series of the study area showed a negative trend. On the contrary, a positive rainfall trend has been detected in about 11.4%, 8.7% and 0.7% of the series, for a SL = 90%, 95% and 99%, respectively. This rainfall reduction at annual scale confirms the results obtained in other Mediterranean areas. e.g., [36], including in southern Italy [37–40] and in some regions of central Italy, such as Abruzzo [41] and Marche [42].

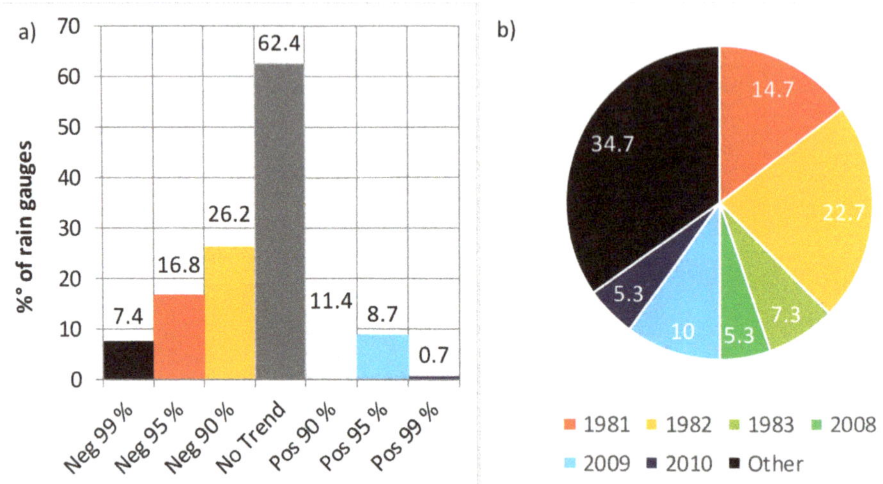

Figure 3. Percentages of annual rainfall series presenting positive or negative significant trends (**a**) and most probable change point years (**b**).

Spatially, for a SL = 95%, the negative trend mainly involved the northern and the western-central areas of the basin, with a maximum decrease in annual precipitation of more than 20 mm/10 years (Figure 4). On the contrary, a positive trend has been evidenced in the eastern side of the basin (>10 mm/10 years) and, particularly, in the southwestern area, with a maximum increase of more than 20 mm/10 years.

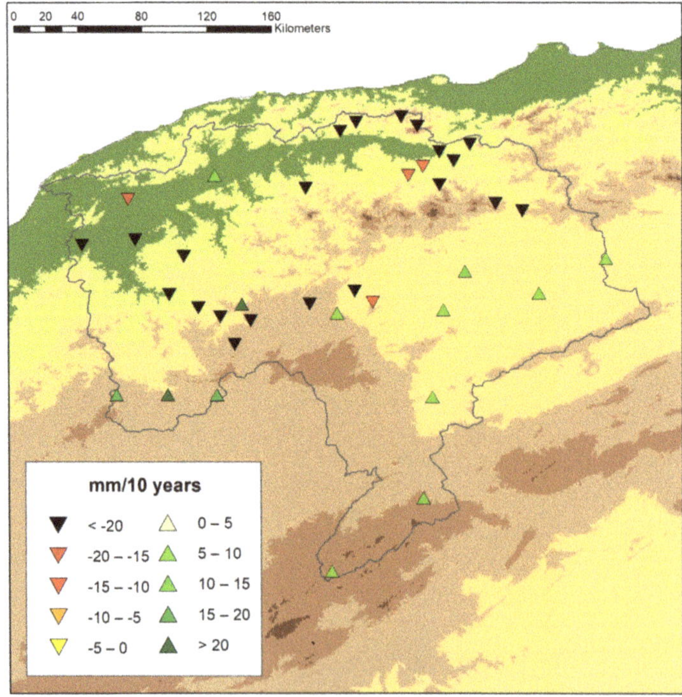

Figure 4. Spatial results of the trend analysis.

Finally, the identification of the shifts in annual precipitation observed in the Wadi Cheliff identified that the years 1982 (22% of the rain gauges), 1981 (14%) and 1983 (7%) can be considered the most probable change point years for the greatest number of stations (Figure 3b). By comparison, several studies, e.g., [43] that included earlier time periods identified change points in the decade 1960–1970.

As an example, in the Supplementary Material, Figure S1 shows the negative and the positive trend behavior, and the change point, for two of the most characteristic stations.

In order to better understand the rainfall behavior detected in this paper, it could be useful to refer to some climatic factors influencing rainfall. In fact, as evidenced by several authors, the Mediterranean rainfall regime is strongly linked to general atmospheric circulation patterns such as the El Niño Southern Oscillation (ENSO) [44], the Mediterranean Oscillation (MO) [45] and the Western Mediterranean Oscillation (WeMO) [34,46]. As regards Algeria, Meddi et al. [47] showed that the temporal variability of the annual precipitation in the west of the country is influenced by ENSO, while Tramblay et al. [13] evidenced that rainfall in North African countries such as Morocco, Algeria and Tunisia are mainly affected by the North Atlantic Oscillation (NAO). In particular, a predominant negative phase of the NAO occurred between 1940 and 1980, corresponding to a period when precipitation was above normal; it was followed by a predominant positive phase, which significantly contributed to the rainfall reduction observed from the beginning of the 1980s in the Mediterranean basin and, also, in Algeria. Similar results have been obtained by Singla et al. [48] who showed a decrease in rainfall in some regions of Morocco from the 1970s onwards and evidenced a strong relationship between rainfall and the NAO phases. In fact, a rainfall decrease is connected to a positive phase of the NAO, which occurrences increased in this century, and some studies forecasted a further increase in its occurrence in the future [49,50].

Differently from past studies analyzing rainfall trend in northern Africa, this study also focused on the identification of change points in the rainfall series, which in the past years has been mainly performed in central Africa. In particular, this study evidenced similar results with the ones obtained further south in West Africa, where studies have tended to identify change points in the 1980s, around the peak of the well-known severe Sahel drought [51–53].

5. Conclusions

With the aim to better understand the annual rainfall variability in a semi-arid area, in this paper 150 rainfall series of the Wadi Cheliff basin (Algeria) were analyzed. First, for each series, the year-to-year variability of precipitation has been studied through the coefficient of variation. Then, a trend analysis has been performed using two non-parametric tests. Finally, the presence of possible change points in the data has been investigated. The following main results were obtained:

1. the CV range between about 16.0% and 56.5%, thus evidencing high variability typical of the Mediterranean basin;
2. a spatial gradient in the CV values between the southern and the northern sides of the basin has been identified, with the highest values detected in the mountainous areas of the south side of the region and the northern areas showing the lowest CV values;
3. a general negative trend has been evidenced for the annual rainfall;
4. the negative trend mainly involved the northern and the western-central area of the basin while a positive trend has been evidenced in the eastern side of the basin;
5. the years 1981, 1982, and 1983 can be considered the most probable change point years for the largest number of stations in the basin.

Supplementary Materials: The following are available online at https://www.mdpi.com/article/10.3390/w13111477/s1: Figure S1. Example of negative and positive trends with change point (vertical orange line) for two characteristic stations.

Author Contributions: Conceptualization, M.A. and T.C.; methodology, M.A. and T.C.; software, M.A. and T.C.; validation, M.A. and T.C.; formal analysis, M.A., T.C. and A.W.; investigation, M.A., T.C., A.W., N.K. and T.H.; data curation, M.A.; writing—original draft preparation, T.C.; writing—review and editing, M.A., T.C., A.W. and N.K. and visualization, M.A. and T.C. All authors have read and agreed to the published version of the manuscript.

Funding: This research received no external funding. The APC was funded by MDPI Editor.

Institutional Review Board Statement: Not applicable.

Informed Consent Statement: Not applicable.

Data Availability Statement: The study did not report any data.

Conflicts of Interest: The authors declare no conflict of interest.

References

1. IPCC. Summary for Policymakers. In *Fifth Assessment Report of the Intergovernmental Panel on Climate Change*; Cambridge University Press: Cambridge, UK, 2013.
2. Caloiero, T.; Caloiero, P.; Frustaci, F. Long-term precipitation trend analysis in Europe and in the Mediterranean basin. *Water Environ. J.* **2018**, *32*, 433–445. [CrossRef]
3. Zittis, G. Observed rainfall trends and precipitation uncertainty in the vicinity of the Mediterranean, Middle East and North Africa. *Theor. Appl. Climatol.* **2018**, *134*, 1207–1230. [CrossRef]
4. Onyutha, C. Identification of sub–trends from hydro–meteorological series. *Stoch. Environ. Res. Risk Assess.* **2016**, *30*, 189–205. [CrossRef]
5. Longobardi, A.; Villani, P. Trend analysis of annual and seasonal rainfall time series in the Mediterranean area. *Int. J. Climatol.* **2010**, *30*, 1538–1546. [CrossRef]
6. Maheras, P.; Tolika, K.; Anagnostopoulou, C.; Vafiadis, M.; Patrikas, I.; Flocas, H. On the relationships between circulation types and changes in rainfall variability in Greece. *Int. J. Climatol.* **2001**, *24*, 1695–1712. [CrossRef]
7. Altin, T.B.; Barak, B. Changes and trends in total yearly precipitation of the Antalya District, Turkey. *Procedia Soc. Behav. Sci.* **2014**, *120*, 586–599. [CrossRef]
8. Shohami, D.; Dayan, U.; Morin, E. Warming and drying of the eastern Mediterranean: Additional evidence from trend analysis. *J. Geophys. Res.* **2011**, *116*, D22101. [CrossRef]
9. Ziv, B.; Saaroni, H.; Pargament, R.; Harpaz, T.; Alpert, P. Trends in rainfall regime over Israel, 1975–2010, and their relationship to large-scale variability. *Reg. Environ. Chang.* **2014**, *14*, 1751–1764. [CrossRef]
10. Donat, M.G.; Peterson, T.C.; Brunet, M.; King, A.D.; Almazroui, M.; Kolli, R.K.; Boucherf, D.; Al-Mulla, A.Y.; Nour, A.Y.; Aly, A.A.; et al. Changes in extreme temperature and precipitation in the Arab Region: Long-term trends and variability related to ENSO and NAO. *Int. J. Climatol.* **2014**, *34*, 581–592. [CrossRef]
11. Meddi, M.; Hubert, P. Impact de la modification du régime pluviométrique sur les ressources en eau du Nord–Ouest de l'Algérie. In *Hydrology of the Mediterranean and Semiarid Regions*; Servat Najem, E., Leduc, W., Shakeel, E.A., Eds.; IAHS Publ.: Wallingford, UK, 2003; pp. 229–235.
12. Goubanova, K.; Li, L. Extremes in temperature and precipitation around the Mediterranean basin in an ensemble of future climate scenario simulations. *Glob. Planet. Chang.* **2007**, *57*, 27–42. [CrossRef]
13. Tramblay, Y.; El Adlouni, S.; Servat, E. Trends and variability in extreme precipitation indices over Maghreb countries. *Nat. Hazards Earth Syst. Sci.* **2013**, *13*, 3235–3248. [CrossRef]
14. Elmeddahi, Y.; Issaadi, A.; Mahmoudi, H.; Tahar Abbes, M.; Goossen, M.F.A. Effect of climate change on water resources of the Algerian Middle Cheliff basin. *Desalination Water Treat.* **2014**, *52*, 2073–2081. [CrossRef]
15. Ragab, R.; Prudhomme, C. Climate change and water resources management in arid and semi–arid regions: Prospective and challenges for the 21st Century. *Biosyst. Eng.* **2002**, *81*, 3–34. [CrossRef]
16. Giorgi, F.; Lionello, P. Climate change projections for the Mediterranean region. *Glob. Planet. Chang.* **2008**, *63*, 90–104. [CrossRef]
17. Giorgi, F. Climate change hot-spots. *Geophys. Res. Lett.* **2006**, *33*, L08707. [CrossRef]
18. Longobardi, A.; Boulariah, O. Preliminary long term changes in inter–annual precipitation variability in a Mediterranean region. In *Managing Water Resources for a Sustainable Future*; Garrote, L., Tsakiris, G., Tsihrintzis, V., Vangelis, H., Tigkas, D., Eds.; EWRA Editorial Office: Athens, Greece, 2019; pp. 461–462.
19. Conrad, V. The variability of precipitation. *Mon. Weather Rev.* **1941**, *69*, 5–11. [CrossRef]
20. Giorgi, F.; Bi, X. Regional changes in surface climate interannual variability for the 21st century from ensembles of global model simulations. *Geophys. Res. Lett.* **2005**, *32*, L13701. [CrossRef]
21. Pendergrass, A.G.; Knutti, R.; Lehner, F.; Deser, C.; Sanderson, B.M. Precipitation variability increases in a warmer climate. *Sci. Rep.* **2017**, *7*, 17966. [CrossRef]
22. He, C.; Li, T. Does global warming amplify interannual climate variability? *Clim. Dyn.* **2019**, *52*, 2667–2684. [CrossRef]

23. Gajbhiye, S.; Meshram, C.; Mirabbasi, R.; Sharma, S. Trend analysis of rainfall time series for Sindh river basin in India. *Theor. Appl. Climatol.* **2016**, *125*, 593–608. [CrossRef]

24. Chandniha, S.K.; Meshram, S.G.; Adamowski, J.F.; Meshram, C. Trend analysis of precipitation in Jharkhand State, India. *Theor. Appl. Climatol.* **2017**, *130*, 261–274. [CrossRef]

25. He, M.; Gautam, M. Variability and trends in precipitation, temperature and drought indices in the State of California. *Hydrology* **2016**, *3*, 14. [CrossRef]

26. Młyński, D.; Cebulska, M.; Wałęga, A. Trends, Variability, and Seasonality of Maximum Annual Daily Precipitation in the Upper Vistula Basin, Poland. *Atmosphere* **2018**, *9*, 313. [CrossRef]

27. Tian, W.; Liu, X.M.; Wang, K.W.; Bai, P.; Liang, K.; Liu, C.M. Evaluation of six precipitation products in the Mekong River Basin. *Atmos. Res.* **2021**, *255*, 105539. [CrossRef]

28. Qi, B.Y.; Liu, H.H.; Zhao, S.F.; Liu, B.Y. Observed precipitation pattern changes and potential runoff generation capacity from 1961–2016 in the upper reaches of the Hanjiang River Basin, China. *Atmos. Res.* **2021**, *254*, 105392. [CrossRef]

29. Sen, P.K. Estimates of the regression coefficient based on Kendall's tau. *J. Am. Stat. Assoc.* **1968**, *63*, 1379–1389. [CrossRef]

30. Mann, H.B. Nonparametric tests against trend. *Econometrica* **1945**, *13*, 245–259. [CrossRef]

31. Kendall, M.G. *Rank Correlation Methods*; Hafner Publishing Company: New York, NY, USA, 1962.

32. Pettitt, A.N. A non–parametric approach to the change point problem. *Appl. Stat.* **1979**, *8*, 126–135. [CrossRef]

33. Insua-Costa, D.; Lemus-Canovas, M.; Miguez-Macho, G.; Llasat, M.C. Climatology and ranking of hazardous precipitation events in the western Mediterranean area. *Atmos. Res.* **2021**, *255*, 105521. [CrossRef]

34. Muthoni, F.K.; Odongo, V.O.; Ochieng, J.; Mugalavai, E.M.; Mourice, S.K.; Hoesche-Zeledon, I.; Mwila, M.; Bekunda, M. Long-term spatial-temporal trends and variability of rainfall over Eastern and Southern Africa. *Theor. Appl. Climatol.* **2019**, *137*, 1869–1882. [CrossRef]

35. Halifa-Martin, A.; Lorente-Plazas, R.; Pravia-Sarabia, E.; Montavez, J.P.; Jimenez-Guerrero, P. Atlantic and Mediterranean influence promoting an abrupt change in winter precipitation over the southern Iberian Peninsula. *Atmos. Res.* **2021**, *253*, 105485. [CrossRef]

36. Del Rio, S.; Herrero, L.; Fraile, R.; Penas, A.P. Spatial distribution of recent rainfall trends in Spain (1961–2006). *Int. J. Climatol.* **2011**, *31*, 656–667.

37. Coscarelli, R.; Gaudio, R.; Caloiero, T. Climatic trends: An investigation for a Calabrian basin (southern Italy). In *The Basis of Civilization—Water Science?* Rodda, J.C., Ubertini, L., Eds.; IAHS Publ.: Wallingford, UK, 2004; pp. 255–266.

38. Longobardi, A.; Buttafuoco, G.; Caloiero, T.; Coscarelli, R. Spatial and temporal distribution of precipitation in a Mediterranean area (southern Italy). *Environ. Earth Sci.* **2016**, *7*, 189. [CrossRef]

39. Caloiero, T.; Coscarelli, R.; Ferrari, E. Application of the innovative trend analysis method for the trend analysis of rainfall anomalies in southern Italy. *Water Resour. Manag.* **2018**, *32*, 4971–4983. [CrossRef]

40. Caloiero, T.; Coscarelli, R.; Gaudio, R.; Leonardo, G.P. Precipitation trend and concentration in the Sardinia region. *Theor. Appl. Climatol.* **2019**, *137*, 297–307. [CrossRef]

41. Scorzini, A.R.; Leopardi, M. Precipitation and temperature trends over central Italy (Abruzzo Region): 1951–2012. *Theor. Appl. Climatol.* **2019**, *135*, 959–977. [CrossRef]

42. Gentilucci, M.; Barbieri, M.; Lee, H.S.; Zardi, D. Analysis of Rainfall Trends and Extreme Precipitation in the Middle Adriatic Side, Marche Region (Central Italy). *Water* **2019**, *11*, 1948. [CrossRef]

43. Caloiero, T.; Coscarelli, R.; Ferrari, E.; Mancini, M. Trend detection of annual and seasonal rainfall in Calabria (southern Italy). *Int. J. Climatol.* **2011**, *31*, 44–56. [CrossRef]

44. Lloyd-Hughes, B.; Saunders, M.A. Seasonal prediction of European spring precipitation from El Niño–Southern Oscillation and local sea surfaces temperatures. *Int. J. Climatol.* **2002**, *22*, 1–14. [CrossRef]

45. Conte, M.; Giuffrida, A.; Tedesco, S. *The Mediterranean Oscillation. Impact on Precipitation and Hydrology in Italy Climate Water*; Publications of the Academy of Finland: Helsinki, Finland, 1989.

46. Martín-Vide, J.; Lopez-Bustins, J.A. The Western Mediterranean Oscillation and rainfall in the Iberian Peninsula. *Int. J. Climatol.* **2006**, *26*, 1455–1475. [CrossRef]

47. Meddi, M.; Assani, A.A.; Meddi, H. Temporal variability of annual rainfall in the Macta and Tafna catchments, northwestern Algeria. *Water Resour. Manag.* **2010**, *24*, 3817–3833. [CrossRef]

48. Singla, S.; Mahe, G.; Dieulin, C.; Driouech, F.; Milano, M.; El Guelai, F.Z.; Ardoin-Bardin, S. Évolution des relations pluie-débit sur des bassins versants du Maroc. In *Global Change: Facing Risks and Threats to Water Resources*; Servat, E., Demuth, S., Dezetter, A., Daniell, T., Ferrari, E., Ijjaali, M., Jabrane, R., van Lanen, H.A.J., Huang, Y., Eds.; IAHS Publ.: Wallingford, UK, 2010; pp. 679–687.

49. Schilling, J.; Freier, K.P.; Hertig, E.; Scheffran, J. Climate change, vulnerability and adaptation in North Africa with focus on Morocco. *Agric. Ecosyst. Environ.* **2012**, *156*, 12–26. [CrossRef]

50. Mariotti, A.; Pan, Y.; Zeng, N.; Alessandri, A. Long–term climate change in the Mediterranean region in the midst of decadal variability. *Clim. Dyn.* **2015**, *44*, 1437–1456. [CrossRef]

51. Sarr, M.A.; Zorome, M.; Seidou, O.; Bryant, C.R.; Gachon, P. Recent trends in selected extreme precipitation indices in Senegal-A changepoint approach. *J. Hydrol.* **2013**, *505*, 326–334. [CrossRef]

52. Tarhule, A.; Tarhule-Lips, R.F.A. Discontinuities in precipitation series in the West African Sahel. *Phys. Geogr.* **2001**, *22*, 430–448. [CrossRef]

53. De Longueville, F.; Hountondji, Y.C.; Kindo, I.; Gemenne, F.; Ozer, P. Long-Term Analysis of Rainfall and Temperature Data in Burkina Faso (1950–2013). *Int. J. Climatol.* **2016**, *36*, 4393–4405. [CrossRef]

Article

Bivariate Frequency of Meteorological Drought in the Upper Minjiang River Based on Copula Function

Fangling Qin [1], Tianqi Ao [1,2,*] and Ting Chen [1,3,*]

1   College of Water Resources & Hydropower, Sichuan University, Chengdu 610065, China; qinfangling123@163.com
2   State Key Laboratory of Hydraulic and Mountain River Engineering, Sichuan University, Chengdu 610065, China
3   Heavy Rain and Drought-Flood Disasters in Plateau and Basin Key Laboratory of Sichuan Province, Chengdu 610072, China
*   Correspondence: aotianqi@scu.edu.cn (T.A.); cting43@foxmail.com (T.C.); Tel.: +86-028-8540-1149 (T.A.); +86-028-6028-0028 (T.C.)

Abstract: Based on the Standardized Precipitation Index (SPI) and copula function, this study analyzed the meteorological drought in the upper Minjiang River basin. The Tyson polygon method is used to divide the research area into four regions based on four meteorological stations. The monthly precipitation data of four meteorological stations from 1966 to 2016 were used for the calculation of SPI. The change trend of SPI1, SPI3 and SPI12 showed the historical dry-wet evolution phenomenon of short-term humidification and long-term aridification in the study area. The major drought events in each region are counted based on SPI3. The results show that the drought lasted the longest in Maoxian region, the occurrence of minor drought events was more frequent than the other regions. Nine distribution functions are used to fit the marginal distribution of drought duration (D), severity (S) and peak (P) estimated based on SPI3, the best marginal distribution is obtained by chi-square test. Five copula functions are used to create a bivariate joint probability distribution, the best copula function is selected through AIC, the univariate and bivariate return periods were calculated. The results of this paper will help the study area to assess the drought risk.

Keywords: upper Minjiang River; marginal distribution; copula; bivariate joint distribution; return period

Citation: Qin, F.; Ao, T.; Chen, T. Bivariate Frequency of Meteorological Drought in the Upper Minjiang River Based on Copula Function. *Water* **2021**, *13*, 2056. https://doi.org/10.3390/w13152056

Academic Editors: Andrzej Walega and Tamara Tokarczyk

Received: 6 June 2021
Accepted: 24 July 2021
Published: 28 July 2021

Publisher's Note: MDPI stays neutral with regard to jurisdictional claims in published maps and institutional affiliations.

1. Introduction

Drought is a frequent natural disaster, which affects ecology, social economy, and agriculture to a large extent. The change of drought may be faster than the average climate change with global warming [1,2]. What is more serious is that due to the expansion of the scale of industry and agriculture, social and economic development, global warming and the rapid growth of the world's population, the demand for water has risen sharply. The shortage of water resources has increased, and the global drought trend is obvious [3].

Drought is usually divided into hydrological, meteorological, agricultural, and socio-economic drought. When the precipitation is lower than the normal level for a period of time, meteorological drought will occur [4], which may affect all other types of drought, so the evaluation of meteorological drought is important [5]. Over the past few decades, different drought indexes have been developed to assess drought conditions [6,7], including Standardized Precipitation Index (SPI) [8], Standardized Runoff Index (SRI) [9], Standardized Precipitation Evaporation Index (SPEI) [10], Standardized Hydrological Index (SHI) [11], Palmer Drought Severity Index (PDSI) [12] and so on, among which the SPI and SPEI are the most widely used [5]. According to reports, if the inter-annual temperature change in a region is not so obvious, then the results of using SPI or SPEI as research indicators will not be much different [13]. Therefore, this study chooses the SPI value as the meteorological drought assessment index.

There are many advantages of using SPI, such as simple calculations and the ability to measure drought conditions on different time scales [14]. Based on the SPI value, it is easy to extract drought characteristics, such as drought severity (S), drought duration (D) and drought peak (P) [15,16]. The analysis of drought characteristics can be univariate or multivariate. Univariate method is a traditional drought frequency analysis method [17]. However, due to the strong correlation between drought characteristics, multivariate analysis can more comprehensively characterize the drought situation. The Copula function is an excellent method for evaluating the joint probability distribution of multiple variables. Its most important advantage is that it does not need to be used on the premise that the marginal distribution of a univariate is independent [18]. At present, the copula function has been used to modeling the multivariate joint distribution of drought [19,20], flood [21,22], the joint change of precipitation and flood [23] and so on in the hydrological field.

The study area is the upper Minjiang River basin (UMR). The UMR is located in Sichuan Province, China. It is a critical water source for domestic, agricultural and industrial production in the Sichuan Basin [24]. However, the UMR has a complex geographical environment and a fragile ecological environment. Some areas have a non-zonal arid valley climate. There are large areas of arid valleys in the study area, and the foehn effect is significant [25], which makes drought become an important disaster in the area. Thus, it is a urgent need to study the drought situation in UMR.

Based on the SPI and copula function, this study analyzed the meteorological drought in UMR. The study area was partitioned into several regions based on the location of four meteorological stations using the Tyson polygon method. The monthly precipitation data of four meteorological stations from 1966 to 2016 were used to calculate the SPI values, and major drought events in various regions were counted based on SPI3 values. Drought duration, severity and peak were estimated by SPI3 value. Nine distribution functions were used to fit the marginal distributions of the three drought characteristics, and the optimal marginal distribution was obtained by chi-square test. Five common copula functions were used to create a bivariate joint probability distribution based on SPI3, and the best copula function was selected through AIC. Finally, the univariate and bivariate joint return period were calculated. The results of this study are significant to the management and distribution of water resources and the prevention of drought in UMR.

2. Data and Method

2.1. Data and Study Area

The study area in this paper is the upper Minjiang River basin (UMR). The UMR is located in Sichuan Province, China. There are many tributaries and dense river networks in the basin. It is the biggest tributary of the upper Yangtze River. The UMR is located on the southeastern edge of the Qinghai-Tibet Plateau, with high mountains and deep rivers in the area, its topography is low in the southeast and high in the northwest, which is a typical alpine valley landform [26,27]. However, due to the alternate control of the south tributary of the westerly wind, the warm Indian Ocean current, and the southeast Pacific monsoon, and under the influence of the complex and diverse geographical environment, the area has formed a unique arid valley climate feature: foehn winds in the area are strong, the atmosphere is dry all year round, and the dry and wet seasons are obvious. About 70% of annual precipitation is centralized in summer, with large annual evaporation, extreme drought in winter, and serious floods and drought disasters [25]. In addition, the UMR is located in the Longmenshan fault zone, the neotectonic movement is strong, which makes the entire mountain ecosystem fragile and changeable. In general, the UMR has a complex geographical environment and a fragile ecological environment [27]. Based on such a severe situation, the UMR was selected as the study area of this article.

The UMR basin includes all areas of Songpan, Lixian and Heishui, and parts of Wenchuan and Maoxian. There are a total of five meteorological stations. Due to the lack of precipitation data in some years in Wenchuan, this paper selects the precipitation data of other four stations as the data used in this study. Figure 1 shows that the selected four meteorological stations are evenly distributed in the UMR, which is reasonable.

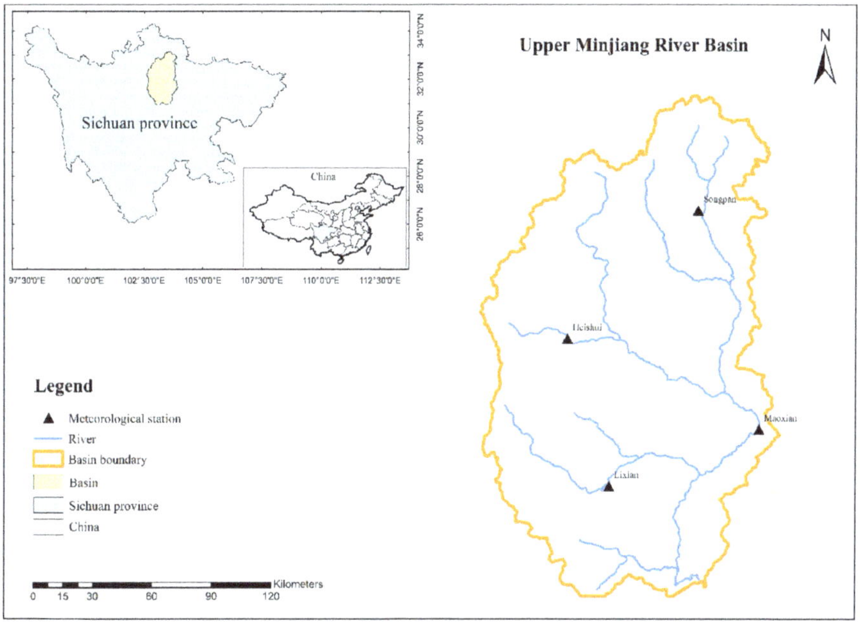

Figure 1. The location of the study area and meteorological stations.

The four meteorological stations in UMR were used to calculate the monthly precipitation data from the daily precipitation observation data from 1966 to 2016, and the monthly precipitation data were used for the calculation of SPI.

2.2. Method

2.2.1. Meteorological Drought Index Spi and Drought Characteristics

The drought index is an important variable used to assess the degree of drought and extract the drought characteristics (drought duration, drought severity, drought peak, etc.). Among them, SPI is one of the most widely used drought index s, which is recommended by the World Meteorological Organization for drought monitoring [28]. SPI was proposed by Mckee [8], its calculation is based on a multi-year monthly precipitation data series. The information of SPI response on different time scales is also different [29]. In this study, the SPIProgram downloaded from the website http://drought.unl.edu/MonitoringTools/ DownloadableSPIProgram.aspx (accessed on 15 January 2021) is only used to calculate the value of SPI on 1, 3 and 12 month time scales (SPI1, SPI3 and SPI12), the drought situation in the study area was analyzed by SPI3. Table 1 lists the SPI climate classification provided by the national standards for meteorological drought levels issued by China. According to the classification in the table, this article sets the threshold for the beginning and end of the drought time as −0.5. In addition, according to the run theory proposed by Yevjevich [30], the drought characteristics based on SPI3 is extracted. This study uses drought duration, severity, and peak to analyze drought events. The three characteristics are defined as follows:

1. Drought duration (D): The duration of $SPI \leq -0.5$;
2. Drought severity (S): The absolute value of the accumulated SPI value over the duration of the drought;

3. Drought peak (P): The absolute value of the minimum SPI value during the duration of the drought.

Table 1. Wet and drought period classification according to the SPI index.

Index Value	Class
SPI > −0.5	No drought
−0.5 ≥ SPI > −1.0	Mild drought
−1 ≥ SPI > −1.5	Moderately drought
−1.5 ≥ SPI > −2.0	Very drought
SPI ≤ −2.0	Extremely drought

Based on the preliminary identification of drought events, in order to avoid the impact of small drought events on the analysis of statistical characteristics of drought event samples, the following treatments are made for small drought events:

1. Small drought events with drought duration of only 1 month and severity less than 1 were not included in the drought event sample;
2. When the non-drought duration between two drought events is 1 unit period and the drought severity is less than −0.2, the two adjacent drought events will be merged into one drought event.

2.2.2. Mann-Kendall Test

The Mann-Kendall (MK) test is often used to test the changing trends of the meteorological and hydrological time series data. Its advantage is that the tested data series don't have to follow a certain distribution [31]. The MK test null hypothesis H_0 is that the change trend of the data sequence $X = \{X_1, X_2, \ldots \ldots, Xn\}$ is not significant. When the statistical parameter $|Z| \geq 1.96$, the null hypothesis is rejected within the 95% confidence interval, that is, the trend of the data series is significant. When Z is positive, it means the trend is up, otherwise, it indicates a decline in the trend [32]. This paper uses the MK trend test method to check the significance of the downward or upward trend of the SPI sequences within the 95% confidence interval. The specific calculation process of the z value is as follows [33]:

$$S = \sum_{i=1}^{n-1} \sum_{j=i+1}^{n} sign(x_j - x_i) \tag{1}$$

$$sign(x_j - x_i) = \begin{cases} 1 & if(x_j - x_i) > 0 \\ 0 & if(x_j - x_i) = 0 \\ -1 & if(x_j - x_i) < 0 \end{cases} \tag{2}$$

The formula for calculating the variance of S is:

$$var(S) = \frac{n(n-1)(2n+5) - \sum_{k-1}^{m} t_k(t_k - 1)(2t_k + 5)}{18} \tag{3}$$

In Equation (3), n is the number of data, k is the number of repetitions, m is the number of unique numbers (the number of groups), and t_k is the number of repetitions for each repetition. When $n > 10$, the formula for calculating the statistical parameter Z is:

$$Z = \begin{cases} \frac{S-1}{\sqrt{var(S)}} & if\ S > 0 \\ 0 & if\ S = 0 \\ \frac{S-1}{\sqrt{var(S)}} & if\ S > 0 \end{cases} \tag{4}$$

2.2.3. Marginal Distribution

In order to establish a binary probability distribution between drought duration, severity and peak, we must first define the univariate distribution of these characteristics. Several alternative probability distributions functions are taken into consideration in this study, namely: Weibull (wbl), Normal, Log-normal (logn), Gamma (gam), Exponential (exp), Logistic (log), Log-logistic, General Extreme Value (gev), and Generalized Pareto (gpa) distribution. In this paper, the parameters of the marginal distribution are evaluated using the maximum likelihood estimation (MLE) method. Spearman (ρ) and Kendall (τ) are used to examine the correlation between different drought characteristics.

2.2.4. Chi-Square Test

In order to determine the best-fitting univariate marginal distribution of each characteristics, this study uses the chi-square test to estimate the best-fitting marginal distribution. The formula for calculating the chi-square value is as follows [5,34]:

$$x^2 = \sum_{k=1}^{n} \frac{(O_k - E_k)^2}{E_k} \tag{5}$$

Among them, n is the number of the disjoint group intervals; k is the serial number of the disjoint group intervals, O_k is the number of observations in the k-th disjoint group intervals; E_k is the expected number of observations in the k-th disjoint group intervals (according to the distribution being tested). The probability distribution function with the smallest Chi-Square value is chosen as the optimal distribution function.

2.2.5. Copula Function

The copula concept comes from Sklar's theorem [35]. In the Copula function, the multivariate probability distribution and the univariate marginal distribution are connected by Sklar's theorem. Then based on the joint cumulative probability distribution of the marginal distribution $F1(x_1), F2(x_2), \ldots \ldots, Fn(x_n)$ (the $x_1, x_2, ..., x_n$ are random variables), copula function can be defined [5]. Suppose that x and y are two random variables with joint distributions $F_{X,Y}(x,y)$ and marginal distribution functions $F_X(x)$ and $F_Y(y)$, according to Sklar's theorem [36], there is a Copula function $C(x,y)$:

$$F_{X,Y}(x,y) = C(F_X(x), F_Y(y)) \tag{6}$$

If $F_X(x)$ and $F_Y(y)$ are consecutive, this Copula is unique. On the contrary, if $F_X(x)$, $F_Y(y)$ and Copula function $C(x,y)$ are given, the above formula defines the joint distribution function of $F_X(x)$ and $F_Y(y)$ [37–39].

Commonly used Copula functions are generally divided into five types, including Archimedean Copula, Metaelliptical Copula, Plackette Copula, mixed Copula, and empirical Copula. Since Archimedean Copula and Metaelliptical Copula functions are easy to construct and can capture dependent structures with several characteristics, they have become very attractive functions in bivariate hydrological frequency analysis [29,39]. In this paper, three commonly used Archimedean Copula (Clayton, Frank and Gumbol-Hougaard) and two commonly used Metaelliptical Copula (Gaussian and t Student Copula) were selected, and the inference function for margin (IFM) method [40] was used to estimate the parameters of copula functions, that is, first calculate the parameter values of the marginal distribution through the MLE method, and then use the obtained marginal distribution parameters to obtain the unknown parameters in the copula functions.

2.2.6. Function Evaluation

The fitting efficiency of the candidate Copula function is evaluated based on the Akaike Information Criterion (*AIC*). The smaller the value of *AIC*, the higher the fitting efficiency. The calculation method of *AIC* is as follows [6,41]:

$$AIC = n \cdot \log(MSE) + 2k \ and \ MSE = \left\{ \frac{1}{n-k} \sum_{i=1}^{n} (X_C(i) - X_E(i))^2 \right\}$$
$$or \ AIC = -2 \cdot \log(MLE) + 2k \tag{7}$$

Among them, k represents the number of fitting parameters, *MSE* represents the mean square error of the fitted copula function relative to the empirical copula, and X_C and X_E are the joint distribution functions based on the parameters and the empirical copula, respectively. *MLE* is the maximum likelihood of the copula function. Therefore, the copula with the smallest *AIC* value is the optimal copula.

2.2.7. Return Period

Shiau and Shen [42] proposed the return period theory of drought events. When the drought characteristic is greater than the preset value, the return period can be calculated from the expected value of the drought interval and the cumulative probability distribution corresponding to the characteristics. The calculation formula is:

$$T_D = \frac{E(L)}{1 - F_D(D)} \tag{8}$$

$$T_S = \frac{E(L)}{1 - F_S(S)} \tag{9}$$

$$T_P = \frac{E(L)}{1 - F_P(P)} \tag{10}$$

In the formula, $E(L)$ is the expected value of the drought interval. $F_D(D)$, $F_S(S)$, and $F_P(P)$ are the cumulative probability distributions of drought duration, severity, and peak, respectively. T_D, T_S, and T_P are the D, S, and P recurrence period, respectively.

According to the nature of drought, univariate analysis may cause underestimation or overestimation of drought risk [37]. Drought characteristics are related random variables, so studying the joint regression period of these characteristic quantities is more helpful to the assessment of local drought risks and the management of water resources. This article will analyze the bivariate joint probability distribution. The bivariate joint return period between drought duration, drought severity, and drought peak is divided into two situations. Here, D and S are used as examples. The combination of other characteristics is the same: (1) The return period of $D \geq d$ and $S \geq s$ is expressed by T_{DS}; (2) The return period of $D \geq d$ or $S \geq s$ is expressed by T'_{DS}. The calculation method is as follows [6,42]:

$$T_{DS} = \frac{E(L)}{P(D \geq d \ and \ S \geq s)} = \frac{E(L)}{1 - F_D(d) - F_S(S) + C(F_D(d), F_S(s))} \tag{11}$$

$$T'_{DS} = \frac{E(L)}{P(D \geq d \ or \ S \geq s)} = \frac{E(L)}{1 - C(F_D(d), F_S(s))} \tag{12}$$

3. Results and Discussion

3.1. Temporal and Spatial Trend of Drought Situation

Based on the locations of 4 meteorological stations, the ArcGIS geographic information platform was used to generate Tyson polygons, and the study area was divided into four regions. In order to explore the changes in drought trends in various regions, this paper uses the MK trend test method to calculate the Kendall trend statistics of the SPI1, SPI3, and SPI12 at each meteorological station. The results are shown in Table 2 and Figure 2.

According to the statistical distribution shown in Figure 2, the UMR can be divided into two categories, Songpan and Heishui are classified as Class I region, Maoxian and Lixian are classified as Class II region. The change trend of drought index in all time scales of Class I was significantly increasing except SPI12 in the Heishui region, which was not significantly increasing. The rising trend of SPI on the 1-month and 3-month (not cross-seasonal) timescales was more significant than the SPI on the 12-month timescales. The SPI sequence of Class II region showed a general downward trend, indicating that drought events were more likely to occur in Class II regions than before. Table 2 shows that on the time scale of 1 month and 3 months, although the drought index sequence of Maoxian region shows an upward trend, its trend rate is 0 (in fact, it is a positive number very close to 0). It can be seen that the upward trend is extremely insignificant. On a 12-month (cross-season) scale, the SPI series of Maoxian and Lixian have a significant downward trend. The statistical results show the historical dry-wet evolution phenomenon of humidification in short-term and drought in long-term in the UMR.

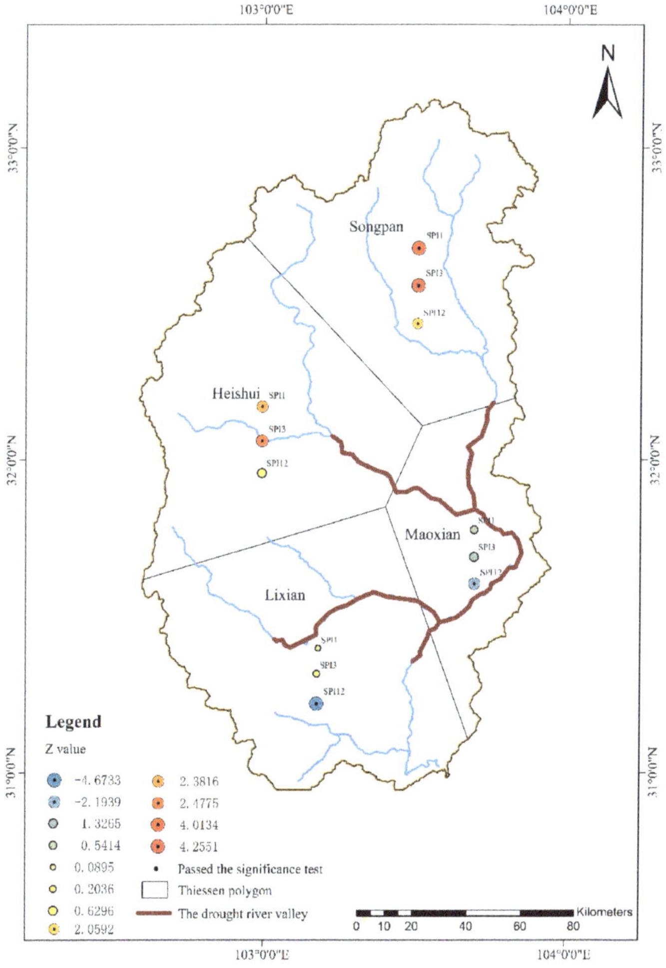

Figure 2. Spatial distribution of variation trends of SPI1, SPI3 and SPI12 (unit: month).

Table 2. Test results of change trend of drought index at different time scales.

Region	SPI	Z-Score	Slope	Change Trend
Songpan	SPI1	2.3816	0.0005	Significant upward trend
	SPI3	2.4775	0.0005	Significant upward trend
	SPI12	0.6296	0.0001	Unsignificant upward trend
Heishui	SPI1	4.0134	0.0009	Significant upward trend
	SPI3	4.2551	0.001	Significant upward trend
	SPI12	2.0592	0.0005	Significant upward trend
Maoxian	SPI1	0.0895	0	Unsignificant upward trend
	SPI3	0.2036	0	Unsignificant upward trend
	SPI12	−4.6733	−0.0011	Significant downward trend
Lixian	SPI1	−0.5414	−0.0001	Unsignificant downward trend
	SPI3	−1.3265	−0.0003	Unsignificant downward trend
	SPI12	−2.1939	−0.0005	Significant downward trend

The distribution of the dry valleys in the UMR is showed in Figure 2. It can be seen that the dry valleys are distributed in all the mainstreams of the UMR in Maoxian region, and part of the mainstreams of the UMR in Heishui and Lixian regions. The length of the dry valley in Maoxian region is the longest, followed by Lixian region. Combining the calculation results of the SPI change trend, it can be seen that, relatively speaking, the Maoxian and Lixian regions where the dry valleys are more widely distributed are more likely to become drier, that is, there is a greater risk of drought.

3.2. Meteorological Drought Assessment

SPI of different time scales reflects the different cumulative effect of drought. Figure 3 shows the SPI1, SPI3, and SPI12 sequences of the four regions. Comparing the three sequences of SPI1, SPI3, and SPI12, it can be seen that the SPI with a shorter time scale (SPI1 and SPI3) is more discrete, drought events occur more frequently, which means that the SPI with a short time scale is more capable of responding to small drought events. The long-term SPI(SPI12) treats several consecutive minor drought events as one drought event, so the long-term SPI can better reflect the long-term trend of drought, and relatively speaking, drought events last longer.

It can be found from the SPI sequences (Figure 3): in the Songpan region, from 1966 to 1972, from 1978 to 1991, and from 1996 to 2008, the SPI values were mostly negative, and the SPI values in other periods were mostly positive. This means that most of the drought events occurred in the period from 1966 to 1972, from 1978 to 1991, and from 1996 to 2008. In Heishui region, SPI was mostly negative from 1966 to 1972, from 1986 to 1987, and from 1996 to 2008, and SPI was mostly positive in other periods; in the Maoxian region, the frequency of positive and negative SPI was similar from 1966 to 1975, and the SPI was mostly negative from 1985 to 1988 and from 1991 to 2010, it can be seen that the drought lasted for a long time in the Maoxian region; in the Lixian region, SPI was mostly negative from 1966 to 1969, 1978–1980, and 1997–2012, and mostly positive in other periods. Meanwhile, Figure 3 shows that drought events occurred more frequently in Lixian during 1997–2012, and the drought was more serious.

It can be seen from the comparison of SPI sequences of different time scales, compared with SPI1, SPI3 can integrate some small drought events, which is suitable for seasonal drought and can better reflect agricultural drought scenarios. Therefore, this study is mainly based on SPI3 for drought assessment.

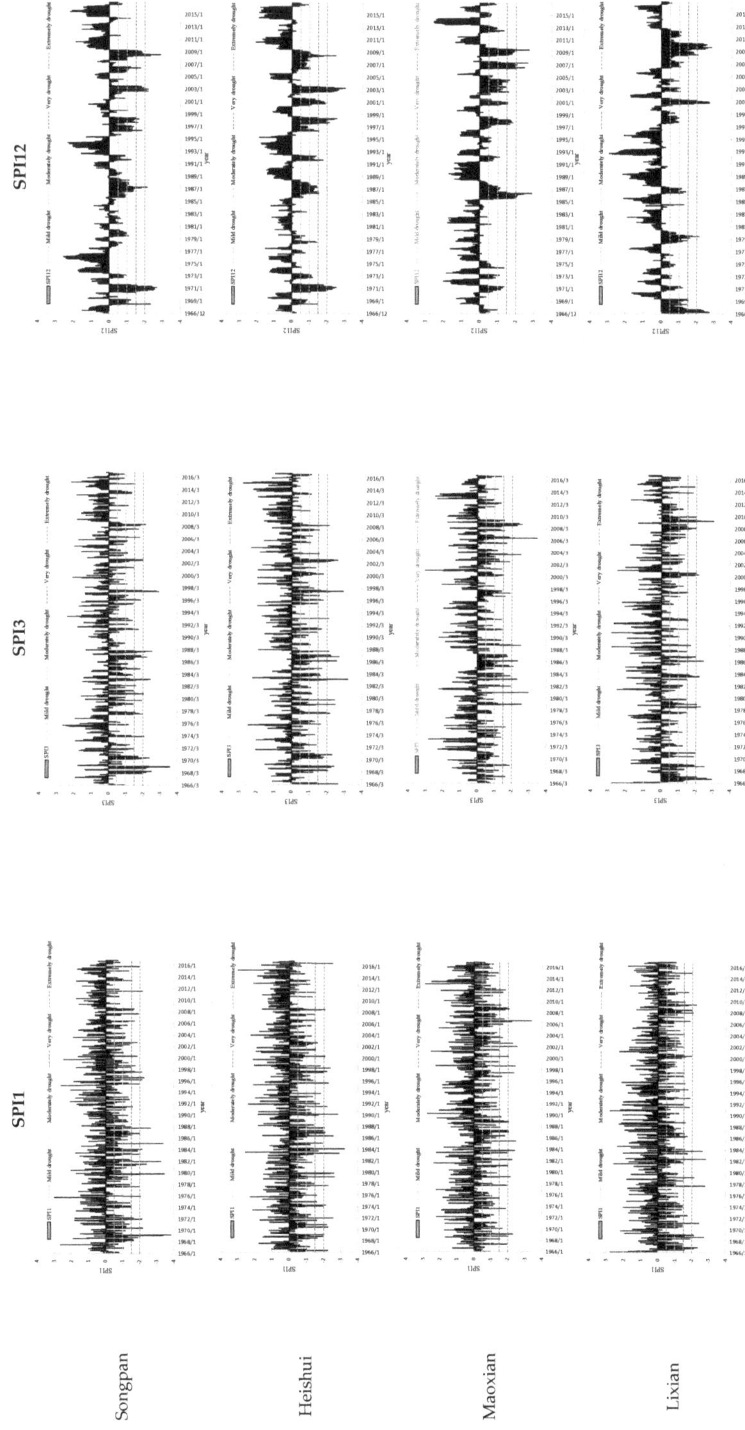

Figure 3. SPI1, SPI3, and SPI12 sequences in four regions.

According to the calculated SPI value, the drought characters of D, S, and P can be extracted to evaluate the drought. Based on the SPI3, the drought duration in Songpan, Heishui, Maoxian, and Lixian regions from 1965 to 2016 were 166, 160, 183, and 175 months, respectively. The drought duration of each region on the interdecadal scale (1960s (1966~1969), 1970s (1970~1979), 1980s (1980~1989), 1990s (1990~1999), 2000s (2000~2009), and 2010s (2010~2016)) was accounted and analyzed, the results were shown in Figure 4. Figure 4 shows that the drought duration of Songpan and Heishui in the 1980s and 2000s was longer than that of other decades. Maoxian region in the 1980s and 2000s had a longer drought duration, while Lixian region in the 1960s and 2000s had a longer drought duration. The drought duration of the four regions in 2010s was relatively short. Overall, the drought duration was relatively long in the 1980s and 2000s and was the shortest in the 2010s.

These adverse effects of drought disasters in the UMR are recorded in the above-mentioned reports.

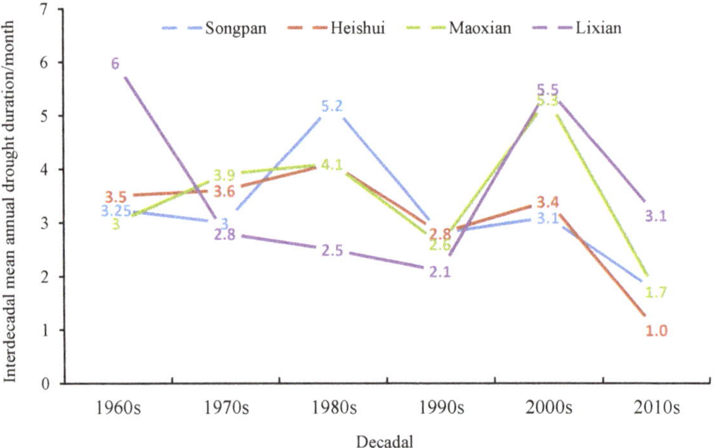

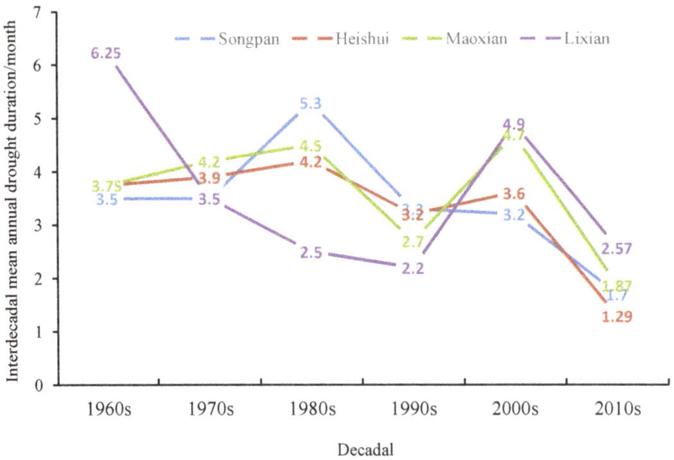

Figure 4. Variation trend of interdecadal average annual drought duration in different regions.

All historical drought events in four regions from 1966 to 2016 were analyzed as follows. According to monthly statistics, historical drought events in Songpan mostly occurred in March; historical drought events in Heishui mostly started in January, June and December; and historical drought events in Maoxian mostly started in March June and October, historical drought events in Lixian mostly started in January, February, and April. According to seasonal statistics, drought events mostly occurred in spring and winter in the Songpan region, the proportion are 29.17% and 29.17%, respectively. In the Heishui region, the proportion of drought events that occurred in winter was 36.17%. The proportion of drought events in spring and summer was 27.66% and 29.79%, respectively, in the Maoxian region. In the Lixian region, the proportion of drought events that occurred in spring and winter was 24.49% and 32.65%, respectively.

Table 3 lists some of the more serious drought events. It shows that the drought duration in Maoxian and Lixian is not only longer than that in the other two regions, but the severity of major historical drought events is also stronger than that in other regions, indicating that the drought risk in Maoxian and Lixian is relatively high.

Table 3. Statistics of severe drought events in various regions from 1966 to 2016.

Region	The Beginning and End of the Drought	Severity of Drought (S)
Songpan	February to July 1968	$S = 12.64$
	December 1968 to April 1969	$S = 11.46$
	June to December 1970	$S = 12.07$
	September 1986 to April 1987	$S = 12.92$
Heishui	May 1970 to March 1971	$S = 16.08$
	September 1986 to May 1987	$S = 11.99$
	July to November 1997	$S = 9.84$
	June to December 2002	$S = 13.5$
Maoxian	June 1985 to June 1986	$S = 16.48$
	January 1997 to January 1998	$S = 12.79$
	July 2006 to February 2007	$S = 11.4$
	July 2008 to February 2009	$S = 14.02$
Lixian	July 1966 to June 1967	$S = 19.13$
	May to December 2000	$S = 12.22$
	January to September 2006	$S = 14.65$
	April to October 2009	$S = 11.52$

According to news reports, most of the drought disasters in Sichuan in the past 20 years occurred in the 2000s, which confirms the reliability of our above analysis. On 10 April 2005, Sichuan Online-West China Metropolis Daily reported the phenomenon of "Minjiang Dehydration". The reporters found in Nanxin Town, Maoxian that the UMR had dried up and the sand was cracked, like a Gobi (http://news.sina.com.cn/o/2005-04-10/07385606826s.shtml, the accessed date is 7 July 2021). On 20 August 2006, Sichuan Online-Huaxi Metropolis Daily reported the phenomenon of "Minjiang River Drying". The snow cover of the five counties in the UMR in 2006 was lower than usual and showed a trend of decreasing year by year. The riverbed in Mianchi Township of Wenchuan dried up and cracked (http://news.sina.com.cn/c/2006-08-20/07389796123s.shtml, the accessed date is 8 July 2021). On 4 April 2007, Sichuan Online-Huaxi Metropolis Daily reported that Sichuan is facing a severe drought in spring and summer, and 5.9 million people have difficulty drinking water (http://news.sohu.com/20070404/n249186525.shtml, the accessed date is 8 July 2021). Sichuan News Net-Chengdu Business Daily reported on 27 February 2010 that since 2010, the western Sichuan Plateau has been experiencing high temperatures and low precipitation, there has been a phenomenon of droughts in autumn and winter, and the mountain snow cover was nearly 50 percent less than last year, or even at the same time for many years in February 2010.

These drought disasters have brought severe impacts on the local area in many ways:

1. Impact on humans: it has caused difficulties in drinking water for humans and animals; the dry-flow area of the Minjiang River cuts off the sources of income for residents in nearby areas who feed on and wash cars along the way; frequent "dehydration" in several sections of the Minjiang River directly affects humans when it comes to urban and rural life and industrial and agricultural production that rely on the Minjiang River for water supply;
2. Impact on agriculture: the continuous drought has caused the crops grown by local residents to turn yellow and reduce production, the supply of agricultural products is insufficient, and the price rises;
3. Impact on wild and rare animals: The construction of water conservancy projects in the UMR has changed the natural properties of the runoff and has caused a serious impact on the aquatic animals and plants of the Minjiang River. The fish species in the UMR have dropped from 40 species in the 1950s to 16 species today;
4. Impact on the environment: continuous drought has reduced the capacity of the water environment, which has aggravated the water pollution of the Minjiang River and the deterioration of the water environment. These adverse effects of drought disasters in the UMR are recorded in the above-mentioned reports.

3.3. Marginal Distribution

To explore the joint distribution of bivariate, we must first determine the marginal distribution of univariate. Calculate the value of SPI3 in each region, using Weibull (wbl), Normal, Log-normal (logn), Gamma (gam), Exponential (exp), Logistic (log), Log-logistic, General Extreme Value (gev), and Generalized Pareto (gpa) distribution functions fit the marginal distribution of D, S, and P, respectively. The chi-square goodness of fit test was used to select the optimal marginal distribution of drought duration, severity, and peak in each region under the condition of significance level $\alpha = 0.05$. The optimal marginal distribution and the corresponding parameters estimated by maximum likelihood were shown in Table 4. Table 4 illustrates that the Exponential, Log-normal, and Log-logistic distribution were selected as the best marginal distribution of the drought duration in the four regions. Log-normal and Log-logistic were chosen as the optimal marginal distributions of the drought severity characteristics in the four regions, the best marginal distributions of drought peak were Exponential, Log-normal, and Logistic. Therefore, for the characteristic of drought duration and drought severity, it is a good choice to use Log-Logistic distribution as their marginal distribution. Log-normal distribution also has good applicability for drought peak. According to the parameter values of the best marginal distribution provided in Table 4, the value of each characteristic quantity corresponding to a specific cumulative distribution probability can be easily calculated according to needs.

Table 4. Marginal distribution of drought characteristics in each region.

Region	Drought Duration	Parameter	Drought Severity	Parameter	Drought Peak	Parameter
Songpan	exp	$\mu = 3.4583$	logn	$\mu = 1.2298$ $\sigma = 0.7807$	logn	$\mu = 0.3538$ $\sigma = 0.4053$
Heishui	logn	$\mu = 1.0577$ $\sigma = 0.5916$	log-logistic	$\mu = 1.2213$ $\sigma = 0.3970$	logn	$\mu = 0.4216$ $\sigma = 0.3603$
Maoxian	log-logistic	$\mu = 1.1237$ $\sigma = 0.3614$	log-logistic	$\mu = 1.1868$ $\sigma = 0.3955$	log	$\mu = 1.4769$ $\sigma = 0.3550$
Lixian	log-logistic	$\mu = 1.1060$ $\sigma = 0.3238$	log-logistic	$\mu = 1.1455$ $\sigma = 0.3813$	logn	$\mu = 0.3749$ $\sigma = 0.3595$

In order to measure the correlation between the three characteristics of drought duration, drought severity, and drought peak, the Spearman (ρ) and Kendall (τ) correlation parameters between different drought characteristics were calculated. The closer the correlation coefficient is to 1, the stronger the correlation. Table 5 shows the calculation results. The calculation results indicate that the Spearman correlation coefficients of D and S are all higher than 0.851, reaching the maximum in Heishui area (0.886), and the Kendall correlation coefficients are all higher than 0.727, and reaching the maximum value (0.757) in Songpan and Heishui regions; the Spearman correlation coefficient of S and P are all higher than 0.721, reaching the maximum in Lixian region (0.864), Kendall correlation coefficients are all higher than 0.530, and also reaching the maximum in Lixian region (0.691), which shows that there is a significant correlation between the two pairs of characteristics. Although the correlation coefficient value of D and P are smaller than the other two pairs of characteristic combinations, the correlation coefficients of Songpan, Heishui, and Lixian regions have passed the significance test of $\alpha = 0.01$, and the correlation coefficients of Maoxian have passed the significance test of $\alpha = 0.05$, which shows that there is a significant correlation between each characteristic. Since the positive correlation between the drought characteristics and the good fitting effect of each characteristic through different distribution functions, the copula function can be used to simulate the joint probability distribution between the drought characteristics.

Table 5. Correlation coefficients among drought characteristics.

Region	D-S		D-P		S-P	
	Spearman (ρ)	Kendall (τ)	Spearman (ρ)	Kendall (τ)	Spearman (ρ)	Kendall (τ)
Songpan	0.851 **	0.757 **	0.640 **	0.458 **	0.807 **	0.614 **
Heishui	0.886 **	0.757 **	0.424 **	0.318 **	0.740 **	0.556 **
Maoxian	0.857 **	0.727 **	0.342 *	0.247 *	0.721 **	0.530 **
Lixian	0.851 **	0.732 **	0.590 **	0.463 **	0.864 **	0.691 **

** indicates that the correlation coefficient has passed the significance test of $\alpha = 0.01$. * indicates that the correlation coefficient has passed the significance test of $\alpha = 0.05$.

3.4. Joint Distribution of Drought Characteristics

This study used five common copula functions, Clayton, Frank, Gumbol-Hougaard, Gaussian, and t Student copulas, to set up the joint distribution of the drought characteristics based on SPI3, and the AIC method is used to evaluate the best copula function.

The AIC value in Table 6 indicates the appropriateness of t Student, Gaussian, Clayton, and Frank to establish the joint distribution of D-S. Gumbol-Hougaard is not applicable to establish the joint probability distribution of D-S at all regions. The five copula functions of Clayton, Frank, Gumbol-Hougaard, Gaussian, and t Student copulas are all suitable for describing the joint probability distribution of S-P, as well as D-P. The copula function with the smallest AIC value is selected as the optimal copula function of the bivariate joint probability distribution of each region. The best copula function and the corresponding parameters are shown in Table 7. Table 7 indicates that Gaussian and Frank copula functions are the best copula functions of D-S, as well as D-P. The best copula function of S-P is Gaussian copula function. It can be found from the optimal copula functions that for the entire UMR, the Gaussian Copula function is a good choice for simulating the joint distribution of D-S, D-P, and S-P.

Table 6. AIC evaluation value of each copula function.

Region	D-S		D-P		S-P	
	Copula	AIC Value	Copula	AIC Value	Copula	AIC Value
Songpan	t Student	−28.2717	t Student	−10.7198	t Student	−46.2465
	Gaussian	−27.6555	Gaussian	−13.1397	Gaussian	−48.1792
	Clayton	−15.7287	Clayton	−3.3449	Clayton	−32.6926
	Frank	−31.5535	Frank	−15.8140	Frank	−46.4643
	Gumbol	98.6545	Gumbol	−14.7983	Gumbol	−46.8826
Heishui	t Student	−69.6858	t Student	−3.2975	t Student	−30.5845
	Gaussian	−71.6859	Gaussian	−5.2975	Gaussian	−32.5845
	Clayton	−47.6935	Clayton	−1.4329	Clayton	−24.5563
	Frank	−63.0136	Frank	−4.8052	Frank	−30.1821
	Gumbol	90.2540	Gumbol	−4.9524	Gumbol	−29.6669
Maoxian	t Student	−70.0885	t Student	−4.2993	t Student	−32.0909
	Gaussian	−72.0882	Gaussian	−6.2992	Gaussian	−34.0903
	Clayton	−49.3726	Clayton	−1.2425	Clayton	−30.0014
	Frank	−61.1346	Frank	−5.1332	Frank	−33.1516
	Gumbol	73.1808	Gumbol	−5.4247	Gumbol	−26.4213
Lixian	t Student	−76.8539	t Student	−15.7463	t Student	−53.7382
	Gaussian	−78.3507	Gaussian	−17.7463	Gaussian	−55.7379
	Clayton	−56.7514	Clayton	−4.4238	Clayton	−27.8824
	Frank	−66.0320	Frank	−16.5257	Frank	−54.2777
	Gumbol	82.8494	Gumbol	−22.8013	Gumbol	−57.3134

Table 7. The optimal copula function of the bivariate joint distribution of each region.

Region	D-S		D-P		S-P	
	Copula	Parameter	Copula	Parameter	Copula	Parameter
Songpan	Frank	6.9027	Frank	4.7848	Gaussian	0.8053
Heishui	Gaussian	0.8898	Gaussian	0.3792	Gaussian	0.7219
Maoxian	Gaussian	0.8907	Gaussian	0.4016	Gaussian	0.7320
Lixian	Gaussian	0.8987	Gaussian	0.5760	Gaussian	0.8324

3.5. Frequency Analysis of Drought Characteristics in Univariate and Bivariate

According to the best marginal distribution of each characteristic selected in Section 3.3, this paper gives the univariate cumulative probability distribution diagram of each characteristic through calculation. As shown in Figure 5, the cumulative probability value corresponding to the specific value of the characteristic can be read. For example, when the cumulative probability $P(X \leq x)$ in Maoxian area is 0.8, the corresponding drought duration (D) is 5.01 months, the drought severity (S) is 5.78, and the drought peak (P) is 1.95.

According to the optimal copula function of the bivariate joint distribution of drought characteristics selected in Chapter 3.4, the bivariate joint probability distribution based on SPI3 (including $P(D \leq d, S \leq s)$, $P(D \leq d, P \leq p)$ and $P(S \leq s, P \leq p)$), the joint probability distribution of the bivariate drought characteristics can be read from Figure 6. For example, when the cumulative probability of D, S and P in Maoxian area is 0.8, the bivariate joint probability $P(D \leq 5.01, S \leq 5.78)$ is 0.7562, and $P(D \leq 5.01, P \leq 1.95)$ is 0.6764, and $P(S \leq 5.78, P \leq 1.95)$ value is 0.7177. These results will help quantify the frequency of drought events of different degrees.

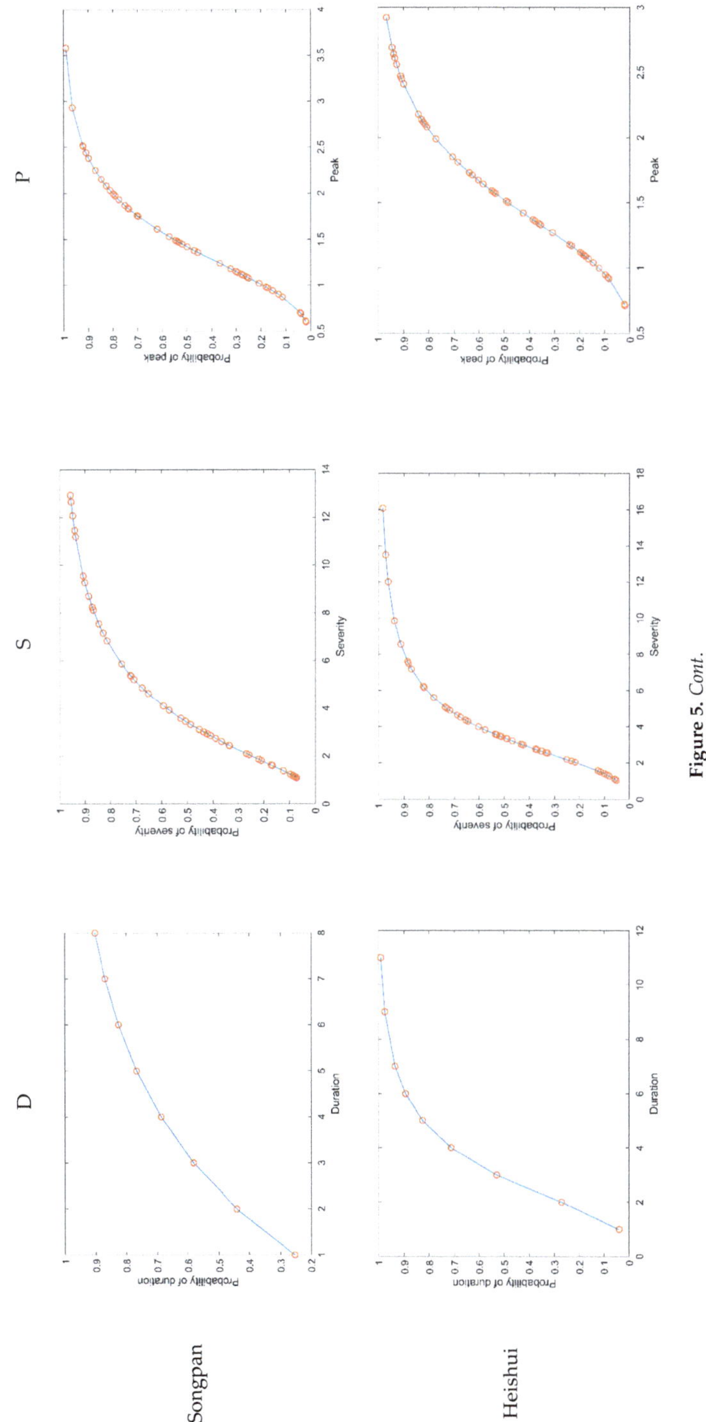

Figure 5. *Cont.*

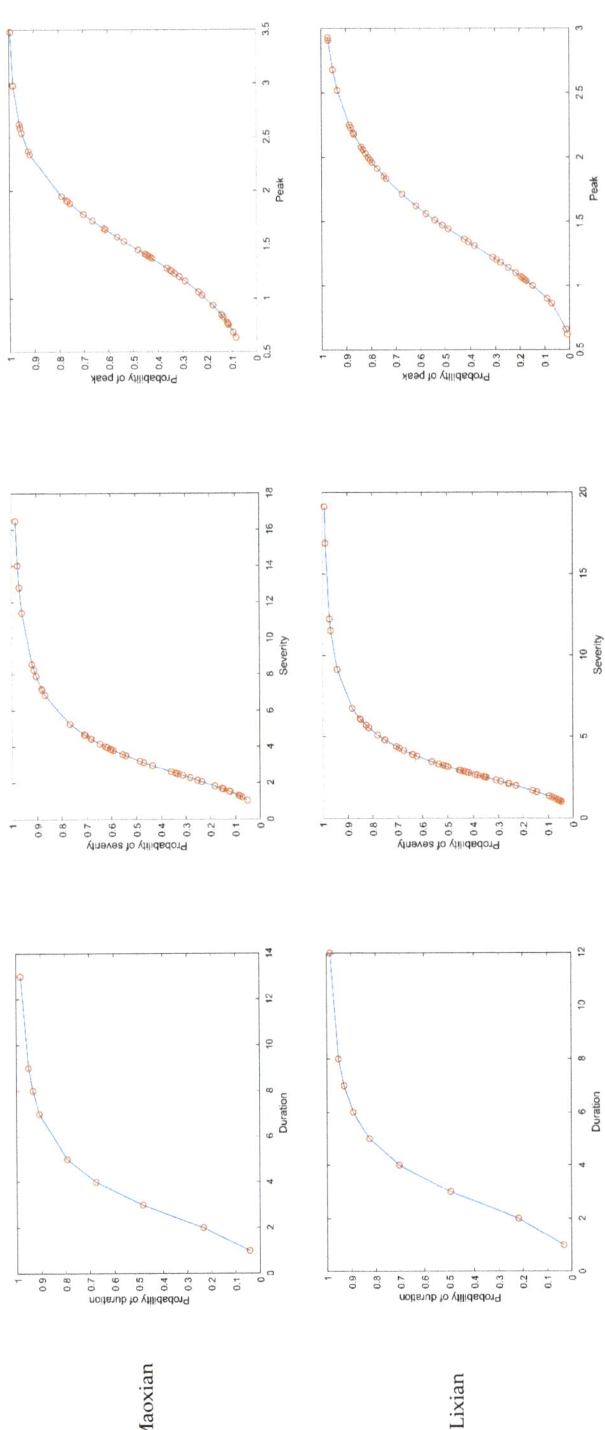

Figure 5. Univariate cumulative probability distribution of drought characteristics.

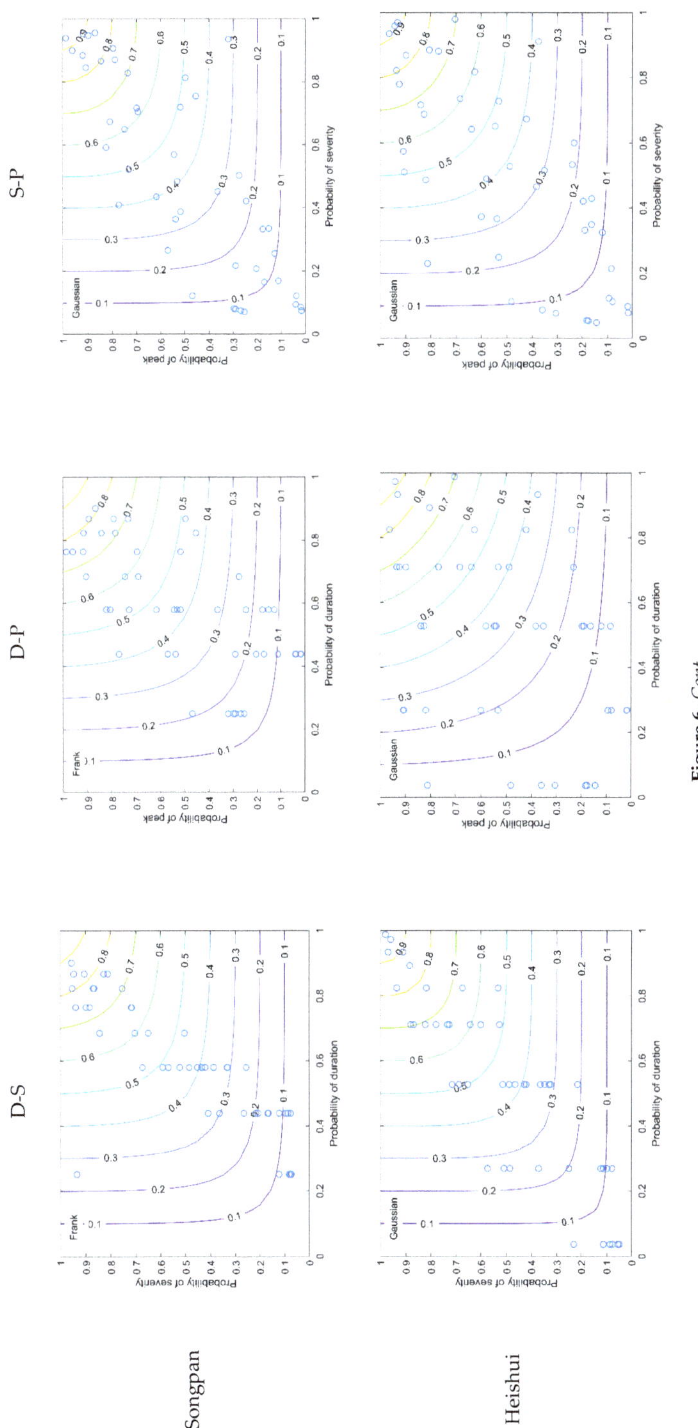

Figure 6. *Cont.*

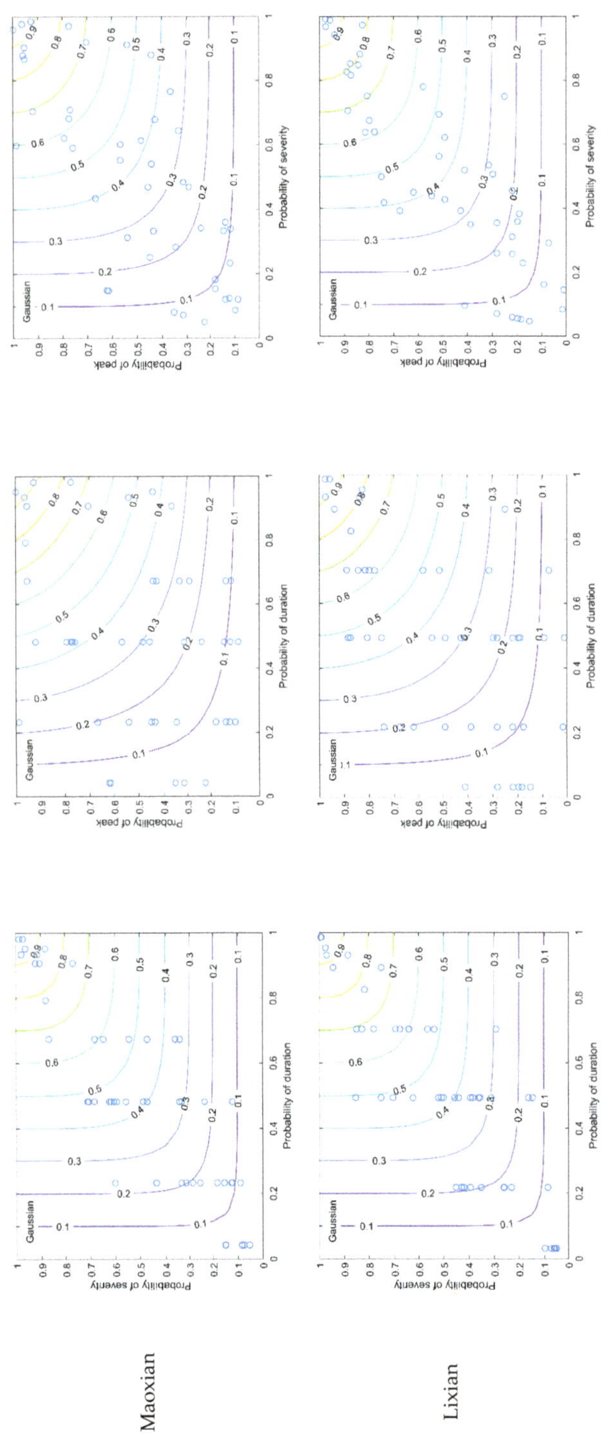

Figure 6. Data space and contour plots of joint probabilities of drought duration, severity and peak.

3.6. Return Period Analysis

Return period analysis is an important part of drought assessment. For the determination of the drought return period, the expected value of the drought interval must first be determined. According to the analysis of drought characteristic variables, the expected drought interval $E(L)$ for the four regions of Songpan, Heishui, Maoxian, and Lixian are 12.71, 12.97, 12.98, and 12.27 months, namely 1.0590, 1.0807, 1.0816, and 1.0226 years, which are similar to 1 year. The meteorological drought interval in southwest China is reported to be mainly affected by the superposition of monsoon and drought disturbances. The drought disturbances are mainly related to the ENSO circulation, which is closely related to the interannual planetary westerly disturbance and the interannual SST disturbance at the equator [43].

First the univariate return period (T) levels are taken to be 2, 5, 10, 20, and 50a; the corresponding values of D, S, and P are calculated, respectively, and the corresponding two-dimensional Copula function values are calculated by the optimal Copula functions of different characteristic variables. According to Equations (11) and (12), the corresponding bivariate joint return periods at a given univariate return period level are calculated. The computed values are showed in Table 8.

Table 8. Return periods of joint distribution of drought characteristics.

Region	T/(a)	D/(m)	S	P	D-S		D-P		S-P	
					T/a	T'/a	T/a	T'/a	T/a	T'/a
Songpan	2	2.2395	3.2300	1.3825	2.4418	1.6936	2.6271	1.6146	2.467	1.6804
	5	4.4017	6.4254	1.9702	8.1839	3.5996	9.7694	3.3598	7.6023	3.7249
	10	6.7901	9.0771	2.3645	23.3260	6.3642	29.1736	6.0342	17.4178	7.0132
	20	8.7848	12.0887	2.7801	74.0559	11.5611	98.0556	11.1356	39.5149	13.3881
	50	12.4807	15.7218	3.3386	−398.1203	26.2751	540.3061	26.2129	115.6114	31.8976
Heishui	2	2.7375	3.1812	1.4701	2.3226	1.7561	3.0589	1.4857	2.5768	1.6342
	5	4.6431	5.6618	2.0244	7.0912	3.9849	14.4479	3.0983	9.0435	3.5538
	10	5.9894	7.8725	2.3855	14.7355	7.5679	37.7339	5.7637	20.1849	6.6464
	20	7.6368	10.7294	2.7262	32.0018	14.5451	108.3952	11.0163	47.6709	12.6546
	50	9.7106	15.5482	3.1969	88.3791	34.8613	427.4921	26.5528	147.4754	30.1031
Maoxian	2	2.9060	3.0720	1.4188	2.3205	1.7573	3.0238	1.4941	2.5618	1.6403
	5	4.9202	5.5221	1.9341	6.6997	3.9882	12.5709	3.1206	8.3238	3.5732
	10	6.7365	7.5708	2.2566	14.6717	7.5849	35.9097	5.8088	19.8023	6.6889
	20	8.6969	10.46945	2.4997	31.9433	14.5572	101.4634	11.0933	46.5004	12.7397
	50	12.5496	14.9461	2.8703	88.1931	34.8903	391.3169	26.7062	142.9931	30.2969
Lixian	2	2.9797	3.0913	1.4401	2.3289	1.7525	2.8477	1.5412	2.4470	1.6911
	5	4.7525	5.2891	1.9535	6.6645	4.0008	10.3043	3.3008	7.3706	3.7832
	10	6.1348	7.3876	2.3063	14.5214	7.6257	26.5472	6.1602	16.6493	7.1461
	20	7.8213	9.7371	2.6239	31.4066	14.6714	67.9017	11.7271	37.2396	13.6711
	50	11.2152	14.3690	3.0104	86.6317	35.1409	232.1980	28.0164	106.4765	32.6709

Table 8 illustrates that the univariate return period is between the joint return period T ('and' event) and T' ('or' event). The bivariate joint return period T is always bigger than T', because the calculation of the return period of the 'and' event is more restrictive than that of the 'or' event. Taking the Maoxian region as an example, the 50-year return periods of univariate of D and S are both between the $T_{DS} = 88.1931$a and the $T'_{DS} = 34.8903$a. In addition, under the same univariate return period level, the duration of drought in Maoxian was greater than that of the other three regions, and the severity and peak of drought in Songpan were greater than those of the other three regions, which indicates that the drought duration of Maoxian lasted longer than other three regions, and the severity and peak of the drought in the Songpan is more severe than other regions. Since the optimal marginal distribution function and optimal copula function have been obtained above, and the corresponding parameters have been calculated, in addition to the return periods corresponding to the drought events of different degrees that have been calculated in Table 8, the return period corresponding to the value of a particular drought duration, drought severity, or drought peak can also be determined according to need.

4. Conclusions and Suggestions

4.1. Conclusions

Drought assessment is critical to water resources planning and management. This article aims to comprehensively analyze the meteorological drought in the UMR.

In this paper, the change trends of SPI in different time scales in four regions were analyzed. The results show that the SPI sequence on a short time scale is more discrete and more able to reflect small drought events. The long-term SPI can better reflect the long-term trend of drought. The UMR showed the historical dry-wet evolution of humidification in short-term and drought in long-term. By analyzing the trend of SPI at various time scales, it is found that Maoxian and Lixian regions where the dry valleys are more widely distributed are more likely to become more arid.

Based on SPI3, the duration, severity, and peak of meteorological drought were estimated, and the drought events in each region were calculated. The results showed that the drought lasted the longest in Maoxian from 1966 to 2016, which was 183 months, the droughts in Songpan, Heishui, and Lixian lasted 166, 160, and 175 months, respectively. According to the decadal statistics of the drought duration in each region, the results show that the drought duration in the study area was relatively long in the 1980s and 2000s, and the drought duration was the shortest in 2010s. Drought events in the study area mostly started in winter and spring. Compared with the statistics of notable drought events in different regions, Maoxian not only has a longer drought duration, but also has a higher severity of historical drought events. Lixian has the highest severity of drought events in history.

According to the results of the chi-square test, this study determines the optimal marginal distribution of drought characteristics from Weibull (wbl), Normal, Log-normal (logn), Gamma (gam), Exponential (exp), Logistic (log), Log-logistic, General Extreme Value (gev), and Generalized Pareto (gpa) distribution functions. For drought duration, it is a good choice to use Log-logistic distribution as its marginal distribution. Log-normal distribution also has good applicability for drought peaks. The drought severity in different regions has different optimal marginal distributions, including Exponential, Log-normal, Logistic, and Log-logistic distributions.

Due to the dependence of the drought characteristics, this study uses Clayton, Frank, Gumbol-Hougaard, Gaussian and t Student five copula functions to fit the bivariate joint distribution to present a more realistic joint distribution result. According to the AIC value, the joint distribution of drought characteristics that is most suitable to describe each region is determined. The results show that due to differences in the correlation between drought characteristics in different regions, the applicable copula functions may also be different. For example, the optimal copula functions for D-S and D-P in different regions include Gaussian and Frank copula functions. As far as the entire study area is concerned, the Gaussian copula function is a good choice for the simulation of the joint probability distribution of the D-S, D-P and S-P.

In addition, based on the optimal marginal distribution and the optimal copula function, this paper calculates the univariate return period and the bivariate joint return period of drought characteristics to reflect the frequency of drought events of different degrees.

In general, Maoxian and Lixian have a higher risk of drought than Songpan and Heishui. According to the drought indices at different scales, almost all the SPI sequences at different scales in Songpan and Heishui showed an obvious increasing trend, while the SPI12 in Maoxian and Lixian showed an obvious trend of becoming drier. Maoxian has the longest drought duration among the four historical drought events. From the perspective of drought severity, the historical drought events in Lixian were more serious than those in the other three regions. However, this does not mean that the drought disaster in Songpan and Heishui is not serious, because except for Maoxian, the drought lasted for 183 months, the historical drought duration in the other three regions is more than 160 months, and serious drought events have occurred in all regions.

In short, the results of this paper can supply effective information for the study area to assess drought risk, so as to optimize the allocation of water resources and reduce the impact of drought on the UMR in the future.

4.2. Suggestions

Due to the frequent occurrence of drought disasters in the UMR, this article puts forward some suggestions for drought disaster management.

First of all, a good ecological environment is a strong barrier against drought disasters. Aiming at the fragile ecological environment in the UMR, new drought-resistant tree species can be cultivated, and various types of plants such as arbor, shrubs, grass, and cane can be planted to build a multi-level structure of the forest system to strengthen ecological barriers. Secondly, local residents can choose to plant crops with strong drought resistance to avoid the residents' diet from being greatly affected when drought disasters occur. Based on the concept of water conservation, relevant departments of the Sichuan government can re-allocate the limited water resource input in terms of urban and rural life, industrial and agricultural production, encourage and promote residents and factories in Sichuan to take concrete water-saving measures, and improve people's water-saving awareness.

Author Contributions: T.A., T.C. and F.Q. designed this research; T.C. and F.Q. collected the data; F.Q. analyzed the data and wrote the draft. T.A. and T.C. revised the manuscript. All authors have read and agreed to the published version of the manuscript.

Funding: This research was funded by the Regional Innovation Cooperation Program from Science &Technology Department of Sichuan Province (2020YFQ0013), the China Scholarship Council (201806240035), and National Natural Science Foundation of China (50979062).

Institutional Review Board Statement: Not applicable.

Informed Consent Statement: Not applicable.

Data Availability Statement: Data is contained within the article.

Acknowledgments: The authors are grateful to the editors and the anonymous reviewers for their constructive comments and suggested revisions.

Conflicts of Interest: The authors declare no conflict of interest.

References

1. Ali, Z.; Almanjahie, I.M.; Hussain, I.; Ismail, M.; Faisal, M. A novel generalized combinative procedure for Multi-Scalar standardized drought Indices-The long average weighted joint aggregative criterion. *Tellus Ser. A Dyn. Meteorol. Oceanogr.* **2020**, *72*, 1–23. [CrossRef]
2. Dai, A.; Trenberth, K.E.; Qian, T.T. A global dataset of Palmer Drought Severity Index for 1870–2002: Relationship with soil moisture and effects of surface warming. *J. Hydrometeorol.* **2004**, *5*, 1117–1130. [CrossRef]
3. Yang, J.; Chang, J.; Wang, Y.; Li, Y.; Hu, H.; Chen, Y.; Huang, Q.; Yao, J. Comprehensive drought characteristics analysis based on a nonlinear multivariate drought index. *J. Hydrol.* **2018**, *557*, 651–667. [CrossRef]
4. Duan, K.; Mei, Y. Comparison of Meteorological, Hydrological and Agricultural Drought Responses to Climate Change and Uncertainty Assessment. *Water Resour. Manag.* **2014**, *28*, 5039–5054. [CrossRef]
5. Nabaei, S.; Sharafati, A.; Yaseen, Z.M.; Shahid, S. Copula based assessment of meteorological drought characteristics: Regional investigation of Iran. *Agric. For. Meteorol.* **2019**, *276*, 107611. [CrossRef]
6. Tosunoglu, F.; Can, I. Application of copulas for regional bivariate frequency analysis of meteorological droughts in Turkey. *Nat. Hazards* **2016**, *82*, 1457–1477. [CrossRef]
7. Pei, Z.; Fang, S.; Wang, L.; Yang, W. Comparative Analysis of Drought Indicated by the SPI and SPEI at Various Timescales in Inner Mongolia, China. *Water* **2020**, *12*, 1925. [CrossRef]
8. McKee, T.; Doesken, J.; Kleist, J. The Relationship of Drought Frequency and Duration to Time Scales. In Proceedings of the 8th Conference on Applied Climatology, Anaheim, CA, USA, 17–22 January 1993; pp. 179–184.

9. Shukla, S.; Wood, A.W. Use of a standardized runoff index for characterizing hydrologic drought. *Geophys. Res. Lett.* **2008**, *35*, L02405. [CrossRef]
10. Vicente-Serrano, S.M.; Begueria, S.; Lopez-Moreno, J.I. A Multiscalar Drought Index Sensitive to Global Warming: The Standardized Precipitation Evapotranspiration Index. *J. Clim.* **2010**, *23*, 1696–1718. [CrossRef]
11. Sharma, T.C.; Panu, U.S. Analytical procedures for weekly hydrological droughts: A case of Canadian rivers. *Hydrol. Sci. J.* **2010**, *55*, 79–92. [CrossRef]
12. Park, S.; Im, J.; Jang, E.; Rhee, J. Drought assessment and monitoring through blending of multi-sensor indices using machine learning approaches for different climate regions. *Agric. For. Meteorol.* **2016**, *216*, 157–169. [CrossRef]
13. Gurrapu, S.; Chipanshi, A.; Sauchyn, D.J.; Howard, A. Comparison of the SPI and SPEI on predicting drought conditions and streamflow in the Canadian Prairies. In Proceedings of the 28th Conference on Hydrology, Atlanta, GA, USA, 2–6 February 2014.
14. Das, J.; Jha, S.; Goyal, M.K. Non-stationary and copula-based approach to assess the drought characteristics encompassing climate indices over the Himalayan states in India. *J. Hydrol.* **2020**, *580*, 124356. [CrossRef]
15. Bhunia, P.; Das, P.; Maiti, R. Meteorological Drought Study Through SPI in Three Drought Prone Districts of West Bengal, India. *Earth Syst. Environ.* **2020**, *4*, 43–55. [CrossRef]
16. Khadr, M.; Morgenschweis, G.; Schlenkhoff, A. Analysis of Meteorological Drought in the Ruhr Basin by Using the Standardized Precipitation Index. *World Acad. Sci. Eng. Technol.* **2009**, *57*, 607–616.
17. Pontes Filho, J.D.; Souza Filho, F.d.A.; Passos Rodrigues Martins, E.S.; de Carvalho Studart, T.M. Copula-Based Multivariate Frequency Analysis of the 2012-2018 Drought in Northeast Brazil. *Water* **2020**, *12*, 834. [CrossRef]
18. Zhang, L.; Singh, V.P. Gumbel-Hougaard copula for trivariate rainfall frequency analysis. *J. Hydrol. Eng.* **2007**, *12*, 409–419. [CrossRef]
19. Weng, B.; Zhang, P.; Li, S. Drought risk assessment in China with different spatial scales. *Arab. J. Geosci.* **2015**, *8*, 10193–10202. [CrossRef]
20. Ganguli, P.; Ganguly, A.R. Space-time trends in U.S. meteorological droughts. *J. Hydrol. Reg. Stud.* **2016**, *8*, 235–259. [CrossRef]
21. Requena, A.I.; Flores, I.; Mediero, L.; Garrote, L. Extension of observed flood series by combining a distributed hydro-meteorological model and a copula-based model. *Stoch. Environ. Res. Risk Assess.* **2016**, *30*, 1363–1378. [CrossRef]
22. Wahl, T.; Plant, N.G.; Long, J.W. Probabilistic assessment of erosion and flooding risk in the northern Gulf of Mexico. *J. Geophys. Res. Ocean.* **2016**, *121*, 3029–3043. [CrossRef]
23. Zhu, Y.; Liu, Y.; Wang, W.; Singh, V.P.; Ma, X.; Yu, Z. Three dimensional characterization of meteorological and hydrological droughts and their probabilistic links. *J. Hydrol.* **2019**, *578*, 124016. [CrossRef]
24. Xu, J.; Lv, C.; Yao, L.; Hou, S. Intergenerational equity based optimal water allocation for sustainable development: A case study on the upper reaches of Minjiang River, China. *J. Hydrol.* **2019**, *568*, 835–848. [CrossRef]
25. Zhou, H. Study on Mineral Elements and Drought Resistance of Dominant Shrubs in Dry Valleys of the Upper Minjiang River. Master's Thesis, Sichuan Normal University, Chengdu, China, 2019.
26. Hou, J.; Ye, A.; You, J.; Ma, F.; Duan, Q. An estimate of human and natural contributions to changes in water resources in the upper reaches of the Minjiang River. *Sci. Total Environ.* **2018**, *635*, 901–912. [CrossRef] [PubMed]
27. Ma, K.; Huang, X.; Guo, B.; Wang, Y.; Gao, L. Land Use/Land Cover Changes and Its Response to Hydrological Characteristics in the Upper Reaches of Minjiang River. In Proceedings of the International Association of Hydrological Sciences 379, Beijing, China, 6–9 November 2018.
28. Vicente-Serrano, S.M.; Dominguez-Castro, F.; Murphy, C.; Hannaford, J.; Reig, F.; Pena-Angulo, D.; Tramblay, Y.; Trigo, R.M.; Mac Donald, N.; Luna, M.Y.; et al. Long-term variability and trends in meteorological droughts in Western Europe (1851–2018). *Int. J. Climatol.* **2021**, *41*, E690–E717. [CrossRef]
29. Yang, X.; Li, Y.P.; Liu, Y.R.; Gao, P.P. A MCMC-based maximum entropy copula method for bivariate drought risk analysis of the Amu Darya River Basin. *J. Hydrol.* **2020**, *590*, 125502. [CrossRef]
30. Yevjevich, V. An Objective Approach to Definitions and Investigations of Continental Hydrologic Droughts. Ph.D. Thesis, Colorado State University, Fort Collins, CO, USA, 1967.
31. Kumar, S.; Merwade, V.; Kam, J.; Thurner, K. Streamflow trends in Indiana: Effects of long term persistence, precipitation and subsurface drains. *J. Hydrol.* **2009**, *374*, 171–183. [CrossRef]
32. Ye, X.; Li, X.; Xu, C.-Y.; Zhang, Q. Similarity, difference and correlation of meteorological and hydrological drought indices in a humid climate region–The Poyang Lake catchment in China. *Hydrol. Res.* **2016**, *47*, 1211–1223. [CrossRef]
33. De Luca, D.L.; Petroselli, A.; Galasso, L. A Transient Stochastic Rainfall Generator for Climate Changes Analysis at Hydrological Scales in Central Italy. *Atmosphere* **2020**, *11*, 1292. [CrossRef]
34. Can, I.; Tosunoglu, F. Estimating T-year flood confidence intervals of rivers in Coruh basin, Turkey. *J. Flood Risk Manag.* **2013**, *6*, 186–196. [CrossRef]
35. Sklar, A. Fonctions de Repartition a n Dimensions et Leurs Marges. *Publ. Inst. Statist. Univ. Paris* **1959**, *8*, 229–231.
36. Nelsen, R.B. Copulas and quasi-copulas: An introduction to their properties and applications. In *Logical, Algebraic, Analytic and Probabilistic Aspects of Triangular Norms*; Elsevier Science BV: Amsterdam, The Netherlands, 2005; pp. 391–413.
37. Saghafian, B.; Mehdikhani, H. Drought characterization using a new copula-based trivariate approach. *Nat. Hazards* **2014**, *72*, 1391–1407. [CrossRef]

38. Rauf, U.F.A.; Zeephongsekul, P. Copula based analysis of rainfall severity and duration: A case study. *Theor. Appl. Climatol.* **2014**, *115*, 153–166. [CrossRef]
39. Mirabbasi, R.; Fakheri-Fard, A.; Dinpashoh, Y. Bivariate drought frequency analysis using the copula method. *Theor. Appl. Climatol.* **2012**, *108*, 191–206. [CrossRef]
40. Joe, H. *Multivariate Models and Multivariate Dependence Concepts*; CRC Press: Boca Raton, FL, USA, 1997.
41. Hsu, K.L.; Gupta, H.V.; Sorooshian, S. Artificial Neural Network Modeling of the Rainfall-Runoff Process. *Water Resour. Res.* **1995**, *31*, 2517–2530. [CrossRef]
42. Shiau, J.T. Fitting drought duration and severity with two-dimensional copulas. *Water Resour. Manag.* **2006**, *20*, 795–815. [CrossRef]
43. Qian, W.H.; Zhang, Z.J. Planetary scale and regional scale anomaly signals for persistent drought events over Southwest China. *Chin. J. Geophys.* **2012**, *55*, 1462–1471.

 *water*

MDPI

Article

Simulating Rainfall Interception by Caatinga Vegetation Using the Gash Model Parametrized on Daily and Seasonal Bases

Daniela C. Lopes [1,*], Antonio José Steidle Neto [1], Thieres G. F. Silva [2], Luciana S. B. Souza [2], Sérgio Zolnier [3] and Carlos A. A. Souza [2]

[1] Campus Sete Lagoas, Federal University of São João del-Rei, Rodovia MG 424, Km 47, Sete Lagoas 35701-970, Minas Gerais, Brazil; antonio@ufsj.edu.br

[2] Academic Unit of Serra Talhada, Federal Rural University of Pernambuco, Serra Talhada 56900-000, Pernambuco, Brazil; thieres.silva@ufrpe.br (T.G.F.S.); luciana.sandra@ufrpe.br (L.S.B.S.); carlosandre08_@msn.com (C.A.A.S.)

[3] Department of Agricultural Engineering, Federal University of Viçosa, Av. Peter Henry Rolfs, s/n, Viçosa 36570-000, Minas Gerais, Brazil; zolnier@ufv.br

* Correspondence: danielalopes@ufsj.edu.br; Tel.: +55-31-3775-5521

Abstract: Rainfall partitioning by trees is an important hydrological process in the contexts of water resource management and climate change. It becomes even more complex where vegetation is sparse and in vulnerable natural systems, such as the Caatinga domain. Rainfall interception modelling allows extrapolating experimental results both in time and space, helping to better understand this hydrological process and contributing as a prediction tool for forest managers. In this work, the Gash model was applied in two ways of parameterization. One was the parameterization on a daily basis and another on a seasonal basis. They were validated, improving the description of rainfall partitioning by tree species of Caatinga dry tropical forest already reported in the scientific literature and allowing a detailed evaluation of the influence of rainfall depth and event intensity on rainfall partitioning associated with these species. Very small (0.0–5.0 mm) and low-intensity (0–2.5 mm h^{-1}) events were significantly more frequent during the dry season. Both model approaches resulted in good predictions, with absence of constant and systematic errors during simulations. The sparse Gash model parametrized on a daily basis performed slightly better, reaching maximum cumulative mean error of 9.8%, while, for the seasonal parametrization, this value was 11.5%. Seasonal model predictions were also the most sensitive to canopy and climatic parameters.

Keywords: rainfall partitioning; dry tropical forest; gash model; interception modelling

Citation: Lopes, D.C.; Steidle Neto, A.J.; Silva, T.G.F.; Souza, L.S.B.; Zolnier, S.; Souza, C.A.A. Simulating Rainfall Interception by Caatinga Vegetation Using the Gash Model Parametrized on Daily and Seasonal Bases. *Water* **2021**, *13*, 2494. https://doi.org/10.3390/w13182494

Academic Editors: Tamara Tokarczyk and Andrzej Walega

Received: 16 August 2021
Accepted: 9 September 2021
Published: 11 September 2021

Publisher's Note: MDPI stays neutral with regard to jurisdictional claims in published maps and institutional affiliations.

1. Introduction

Water availability is limited in arid and semiarid regions, with rainfall interception playing an important role on site and catchment water balances, as well as in the context of climate change [1]. Rainfall partitioning by trees is an intricate process, mainly affected by canopy and weather factors, such as the characteristics of rainfall events, becoming even more complex where vegetation is sparse [2]. Thus, rainfall interception modelling appears as an important tool for extrapolating experimental results both in time and space, helping to better understand this hydrological process, as well as to implement effective water resource management and land use planning.

Many mathematical models have been developed, validated and successfully applied to simulate rainfall partitioning in different forest types, including coniferous and hardwood stands [3–5], rainforests [6,7], deciduous and sparse canopies [8–10], mixed stands [11] and crops [12]. However, there are few studies about simulating or evaluating the rainfall interception in Caatinga vegetation [1,13–16]. This domain corresponds to an area of tropical dry forest with deciduous tree-shrubs, which covers close to one million km^2 in the Northeast of Brazil, occupying around 50% of this region [17]. Caatinga is

87

a fragile ecosystem due to the scarce water resources and the anthropogenic pressures, mainly the intensive exploitation of the region by agriculture and livestock [18]. There is a high temporal and spatial variability of the rainfall regime in this domain, both considering annual and individual events. Specifically, rainfall that occurs in the semi-arid Northeast of Brazil is concentrated over a short period, with the dry season lasting from five to nine months and resulting in uncertainties about the water regime [1]. Caatinga ecosystem is highly dynamic and its vegetation responds quickly to climatic conditions due to morphological and physiological adaptations to aridity by many species of plants. The Caatinga species comprise a whole range of deciduousness, including plants that retain their leaves throughout all the year and other that are leafless during seven months each year [15,17]. The main factor that controls the structure and distribution of vegetation is the precipitation, but photoperiod and nutrients also affect the Caatinga species [18].

More studies about rainfall interception in the Caatinga domain are important to increase the spatial and temporal accuracy in rainfall partitioning simulations and to better understand this process in dry tropical forests. Additionally, these studies may also benefit watershed and forest managers of other similar arid and semiarid ecosystems, since dry tropical forests are recognized as one of the world's major biomes and are found in a wide area extending from the Amazon basin in South America towards northern Mexico and the Caribbean [19].

The analytical Gash models [9,20] are most often used when predicting rainfall interception due to their ease of use, low parameter requirement and low programming complexity [21]. These models are capable of estimating rainfall partitioning by using a series of parameters based on canopy structure, evaporation rate and rainfall regime [22]. The original analytical Gash model [20] represents rainfall input as series of discrete storms, each comprising a wetting up period, a saturation period and a drying out period [7]. The sparse version [9] encompasses the case of forest stands with significant open spaces between tree canopies, also introducing some minor corrections [21]. The main difference between these two versions is that the sparse model is based on evaporation and canopy storage per unit area of canopy cover rather than per unit of ground area. This overcame a limitation in the description of sparse forests by the original model, which can prevent the simulated canopy from wetting up [9,11].

Both original and sparse Gash models [9,20] are typically applied using mean annual or seasonal rainfall intensity and evaporation rates, which are considered as constant parameters in all events during the simulated period [2,8,13]. The same occurs with the canopy storage capacity, canopy cover fraction and threshold value required to saturate the canopy [5,23,24]. The sparse Gash model was already parametrized on a daily basis, considering a linear relationship between leaf area index and canopy storage capacity during the plant cycle [12,22]. These model adaptations, based on estimates of parameters for individual storm events, were also compared to other methodologies [3], reinforcing that the daily changes, observed in canopy structures, especially for deciduous vegetation, tend to reduce systemic simulation inaccuracies.

Although a number of works have been focused on applying and evaluating the Gash model with different parametrizations and for distinct forest types, its parametrizations on daily and seasonal bases were not studied for the Caatinga domain. These procedures tend to better represent the effects of changes in canopy cover on the rainfall interception process, mainly in this deciduous ecosystem, where canopy structure often changes gradually, but relatively rapidly. Furthermore, the adjustments and modifications required to perform such simulations allow a more detailed and accurate evaluation of the influence of rainfall depth and event intensity on rainfall partitioning. Therefore, the objective of this study was to parametrize the sparse Gash model on daily and seasonal bases, validating these approaches for simulating rainfall interception from five Caatinga species (*Spondias tuberosa, Commiphora leptophloeos, Cnidoscolus quercifolius, Aspidosperma pyrifolium* and *Cenostigma pyramidale*), improving the description of this hydrological process already reported in the scientific literature for dry tropical forests by enhancing the temporal and spatial accuracy of the estimates performed by the model. Additionally, the influence of rainfall depth and event intensity on rainfall partitioning associated with these species was evaluated.

2. Materials and Methods

2.1. Meteorological and Rainfall Measurements

This study was conducted within a private property with Caatinga vegetation area of 81,000 m^2 and density of 930 trees ha^{-1}, located in the Floresta municipality, Pernambuco State, Brazil (08°18′31″ S, 38°31′37″ W, 378 m a.s.l.). Vegetation in the experimental plot is mainly composed by the native species *Spondias tuberosa*, *Commiphora leptophloeos*, *Cnidoscolus quercifolius*, *Aspidosperma pyrifolium* and *Cenostigma pyramidale* (Figure 1), which are randomly distributed over the study site and are representative of the Caatinga domain [17,25–27]. Table 1 presents the main characteristics of these tree species [19], which are common to other species found in the Caatinga domain [14,16].

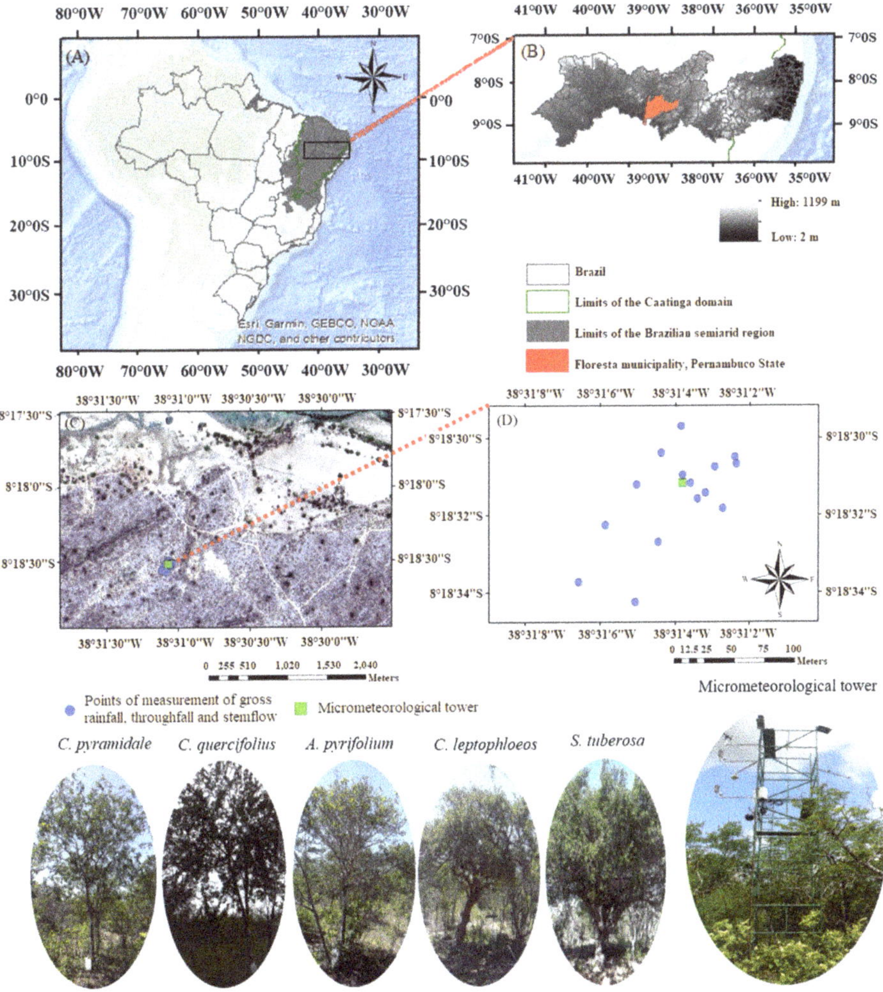

Figure 1. Study area, monitored trees and micrometeorological tower.

Table 1. Main characteristics of the tree species used in the rainfall interception measurement: number of individuals (N), average diameter at breast height (DBH), average number of stems (NB), average tree height (H) and average tree crown projected area (CPA).

Scientific Name	DBH (m)	NS (-)	H (m)	CPA (m^2)	N (-)
C. pyramidale	0.08	3	4.9	19.4	415
C. quercifolius	0.13	2	6.5	33.7	35
A. pyrifolium	0.07	3	4.1	12.5	280
C. leptophloeos	0.15	1	5.5	64.7	10
S. tuberosa	0.21	5	4.9	99.9	10

The climate of the region according to the Köppen classification is BSh, corresponding to a tropical semiarid hot type [28], and the Thornthwaite Aridity Index is 0.48 [29], confirming the region as semiarid. Average annual precipitation, wind speed and net solar radiation are 489.3 mm, 2.3 m s^{-1} and 22.2 MJ m^{-2} day^{-1}, respectively. Mean annual air temperature is around 26.1 °C, with mean monthly temperatures ranging from 23.3 °C in July to 28.3 °C in November. Mean annual relative humidity is 61.9%, with mean monthly values between 50.3% in October and 70.4% in April.

Micrometeorological measurements were performed by electronic sensors installed on a galvanized iron tower at 8 m above the ground (Figure 1). Data were registered and stored in a datalogger (CR10X, Campbell Scientific, Logan, UT, USA) and measurements were conducted continuously from 1 March 2016 to 30 September 2017. The wind speed and direction were measured by an anemometer (03002 Wind Sentry, R. M. Young Company, Traverse, MI, USA). A quantum sensor (SQ-321, Apogee Instruments, Logan, UT, USA) measured the photosynthetically active radiation (PAR), while radiation balance was measured by a net radiometer (NR-Lite, Kipp & Zonen, Delft, The Netherlands) and the global solar radiation was obtained by a pyranometer (SP-230, Apogee Instruments, Logan, UT, USA). An automatic rain gauge (CS700-L, Hydrological Services Pty, Sydney, Australia) registered the gross rainfall.

The photosynthetically active radiation transmitted through the canopies was measured at two below-canopy positions, previously defined in representative trees of the predominant Caatinga species in the study area, by hand-moving two linear quantum sensors (SQ-321, Apogee Instruments, Logan, UT, USA) from one tree to another. Additionally, three aspirated psychrometers, made of T-type thermocouples (copper-constantan), were used for obtaining the dry and wet bulb temperatures at 0.5, 1.5 and 2.5 m above the mean canopy level. Two soil heat flux plates (HFT3-REBS, Hukseflux, Delft, The Netherlands) were also installed in the top soil layer at a depth of approximately 0.05 m.

A ceptometer (LP-80, Decagon Devices, Pullman, DC, USA) was used to measure the fractional interception of photosynthetically active radiation. The incident radiation measurements were performed in open areas without physical obstacles, not including cloudy days or at dusk. The transmitted radiation measurements occurred under the tree canopies. One incident and four transmitted radiation measurements in different directions (north, east, south and west) were executed for each sample of predominant Caatinga species, during 14 campaigns with 135 readings each. Based on the fractional interceptions of photosynthetically active radiation, the integrated microprocessor of the ceptometer estimated the leaf area index based on a simplified version of the Norman–Jarvis radiation transmission and scattering model [30,31]. Polynomial equations were then fitted to measure fractional interception of photosynthetically active radiations obtained by the ceptometer and registered in the datalogger for estimating daily leaf area indices for each studied species.

Throughfall was measured by 15 manual collection gauges, placed randomly underneath the vegetation canopy, comprising three gauges per predominant Caatinga species. Measurements were performed after each rainfall event and the gauges were installed at 1.0 m above ground level, presenting orifices of 0.07 m^2. Gauges were installed at a half-way distance between canopy edge projection and stem in order to minimize the effects of spatial variability on the magnitude of average throughfall [13]. The area under each tree was divided by three diagonals considering the crown projected limits, totaling six sampling points equally spaced at angles of 60°. The gauges were distributed in three sampling points, being representative of a 120° circular sector bisected by the gauge longitudinal axis and centered on the tree position. Each gauge was relocated after every three rainfall events to a new position correspondent to the empty sampling point located beside it and following the area clockwise. This procedure minimizes errors originating from spatial variability and improves long-term sampling [32,33]. Furthermore, it allows to derive reliable mean throughfall per tree, even with a limited number of gauges [2,10,23,34].

Stemflow was measured after each rainfall event by installing twelve zinc gutters of 0.15 m in height, attached to the tree stems at 1.3 m above ground level and connected to individual plastic containers. A hose was fed into each plastic container and measurements were performed with a graduated test tube, with the purpose of reducing evaporation. Due to the tortuous trunks and rough bark of *S. tuberosa* and *C. leptophloeos*, the stemflow monitoring was restricted to the other three species, which presented a projected crown radius greater than 0.2 m and were sub-divided into two classes of diameter at breast height [35], that were 0.05 m $\leq$ DBH $<$ 0.10 m and 0.10 m $\leq$ DBH $\leq$ 0.20 m. This sampling plan assured reliable stemflow measurements, since the two unmonitored trees have canopy structures similar to other species with low stemflow [36]. *Spondias tuberosa* presents low inclined branches with many flow path obstructions that create drip points, enhancing throughfall production, while *Commiphora leptophloeos* has horizontal leaves and only one stem.

The effective rainfall interception was obtained for each rain event by subtracting the gross rainfall by the sum of measured throughfall and stemflow [34,37]. Each rainfall event was defined as a period when cumulative gross rainfall exceeded 0.2 mm, provided that there was a minimum of 6 h without rainfall between events [2,7,24,38]. A consistency analysis was performed on the rainfall and micrometeorological data with electronic spreadsheet functions to remove all inconsistent values and outliers. Visual analysis of graphs relating the variables to time complemented the data evaluation.

2.2. Sparse Gash Model Parametrized on Daily and Seasonal Bases

A daily parametrization for the sparse Gash model was proposed in this study for simulating the rainfall interception of Caatinga species [3,12,22] and was compared with a seasonal parametrization (Table 2), which was based on mean constant parameters for rainy and dry seasons.

In both approaches, the net rainfall interception was estimated as [9]

$$IN = IC + IW + IS + IA, \tag{1}$$

where IN is the net rainfall interception (mm), IC is the interception insufficient to saturate the canopy (mm), IW is the rainfall interception during canopy wetting (mm), IS is the rainfall interception during saturated canopy conditions (mm) and IA is the evaporation after rain ceased (mm).

The original and sparse Gash models also include a formulation for stemflow and evaporation of water stored on wetted trunks. However, the cumulative stemflow accounted for only 0.7% of total rainfall for the five studied Caatinga species and was considered negligible during simulations in this work. Thus, an equivalent interception was calculated, corresponding to the difference between gross rainfall and throughfall [7,8,34].

Table 2. Equations describing the components of rainfall interception in the seasonal and daily parametrizations of the sparse Gash model proposed in this paper.

Interception Component	Seasonal Basis	Daily Basis
For m storms insufficient to saturate the canopy ($P_G \leq P_S$)		
Evaporation from the whole canopy (IC)	$\sum\limits_{i=1}^{m} c_y\, P_{Gi}$	$\sum\limits_{i=1}^{m} c_i\, P_{Gi}$
For *n* storms sufficient to saturate the canopy ($P_G > P_S$)		
Wetting up of canopy (IW)	$n\, c_y\, (P_{Sy} - S_{cy})$	$\sum\limits_{i=1}^{n} c_i\, (P_{Si} - S_{ci})$
Wet canopy evaporation during storms (IS)	$\frac{c_y E_{cy}}{R_y} \sum\limits_{i=1}^{n} (P_{Gi} - P_{Sy})$	$\frac{c_i E_{ci}}{R_i} \sum\limits_{i=1}^{n} (P_{Gi} - P_{Si})$
Evaporation after storms cease (IA)	$n\, c_y\, S_{cy}$	$\sum\limits_{i=1}^{n} c_i\, S_{ci}$

M, number of storms insufficient to saturate the canopy (dimensionless); i, mean value for a rainfall event (dimensionless); c, canopy cover fraction (dimensionless); y, mean value for rainy or dry season (dimensionless); P_G, gross rainfall (mm); n, number of storms sufficient to saturate the canopy (dimensionless); P_S, threshold value required to saturate the canopy (mm); S_c, canopy storage capacity per unit area of cover (mm); E_c, evaporation rate from wet canopy per unit area of cover (mm h^{-1}); R, rainfall rate or rainfall intensity (mm h^{-1}).

2.3. Estimation of Meteorological and Canopy Parameters

The threshold value required to saturate the canopy (P_S, mm) was obtained on a seasonal or daily basis, depending on the model approach, as [9,12,22]

$$P_{Sy} = -\left(\frac{R_y S_{Cy}}{E_{Cy}}\right) \ln\left(1 - \frac{E_{Cy}}{R_y}\right), \tag{2}$$

$$P_{Si} = -\left(\frac{R_i S_{Ci}}{E_{Ci}}\right) \ln\left(1 - \frac{E_{Ci}}{R_i}\right), \tag{3}$$

When using the parametrization on a daily basis, the rainfall rate (R_i, mm h^{-1}) was the average rainfall intensity during all hours in each storm event [22]. For the seasonal sparse Gash model, mean rainfall rates (R_y, mm h^{-1}) were calculated separately [24], for rainy (December–May) and dry (June–November) seasons and then applied in a generalized form to all individual rainfall events. Rainy and dry periods were determined according to the rainfall pattern observed in the studied region, as well as the phenological and leaf area index data of the five studied Caatinga species [1,16,19].

The evaporation rate from wet canopy (E_m, mm h^{-1}), which represents the evaporation from the canopy during the storms, was also calculated for each storm event for the parametrization on a daily basis, while mean values were obtained for rainy and dry seasons when using the seasonal approach. This parameter was estimated hourly based on the Penman–Monteith equation [39], excluding the non-storm periods, with the canopy resistance set to zero [40] and using the momentum method for estimating the aerodynamic resistance [5,39]. The estimated evaporation rates from wet canopy were then divided by the canopy cover fractions before being applied in the models (E_c, mm h^{-1}), adjusting the original values for a complete canopy in proportion to the canopy cover [9].

The canopy cover fraction (c, dimensionless) described the vegetation density. Daily canopy cover fractions were calculated according to the Beer–Lambert equation [41]:

$$c = 1 - PB/PA = 1 - e^{-kL}, \tag{4}$$

where PA is the incoming photosynthetically active radiation on canopy (μmol m^{-2} s^{-1}), PB is the transmitted photosynthetically active radiation through the canopy (μmol m^{-2} s^{-1}), k is the extinction coefficient (dimensionless) and L is the leaf area index (m^2 m^{-2}).

The Beer–Lambert model describes the radiation transmittance through crop canopies as an exponential-type attenuation process, which can be also associated with the fractional photosynthetically active radiation, as well as with the leaf area index [12]. Daily canopy cover fractions were directly applied in the sparse Gash model parametrized on a daily basis, while average values for rainy and dry seasons were obtained when performing simulations with the seasonal parametrization.

The canopy storage capacity (S, mm) corresponded to the amount of water remaining in the canopy after rainfall and throughfall cease, considering evaporation equals to zero [42,43]. This parameter depends on the intensity and duration of the storm, as well as the spatial and temporal variability of trees [44]. In this study, S was assumed to have a linear relationship with the leaf area index [12]. Furthermore, the canopy storage capacity was adjusted per unit area of cover (S_c, mm), by dividing the original S value by the canopy cover fraction before applying it in the models [9].

The mean method was used for estimating a specific S value [3,45], representing the depth of water retained by leaves of each studied species. For this, scatter plots of measured rainfall interception versus gross rainfall were plotted for a number of rain events large enough to saturate the canopy of each Caatinga species and the specific canopy storage capacities were derived from the intercepts of the regression lines fitted to these data. That is, two regression lines were created, relating rainfall interception to gross rainfall for storms that are either insufficient or sufficient to saturate the canopy. The slope of each regression line was determined by an iterative least square fitting procedure. The difference between gross rainfall and throughfall at the intersection point of these two regression lines provided the estimate of S. The use of rainfall interception when plotting the regression lines yields the least stochastic errors, mainly when rainfall outside the canopy is measured without observational scatter and rainfall inside the canopy is observed with scatter [13,45]. For the simulations with daily parametrization, specific S values were multiplied by the daily leaf area index of each species, resulting in different daily S values. When applying the seasonal parametrization, daily S values were averaged considering rainy and dry periods.

2.4. Validation Analysis

The predictive capacity of the adjusted model was evaluated by the statistical parameters cumulative mean relative error, mean bias error, index of agreement and Nash–Sutcliffe efficiency [23,46,47]:

$$CMRE = 100\frac{|C_I - C_S|}{C_S}, \tag{5}$$

$$MBE = \frac{\sum (P_j - O_j)}{w}, \tag{6}$$

$$d = 1 - \frac{\sum (P_j - O_j)^2}{\sum \left(|P_j - O_m| + |O_j - O_m|\right)^2}, \tag{7}$$

$$E = 1 - \frac{\sum (P_j - O_j)^2}{\sum (O_j - O_m)^2}, \tag{8}$$

where CMRE is the cumulative mean relative error (%), C_I is the real cumulative rainfall interception (mm), C_S is the simulated cumulative rainfall interception (mm), MBE is the mean bias error (mm), O_j is the measured rainfall interception (mm), P_j is the predicted rainfall interception (mm), w is the number of testing data (dimensionless), d is the index of agreement (dimensionless), O_m is the average experimental rainfall interception (mm) and E is the Nash–Sutcliffe efficiency (dimensionless).

Values of MBE close to zero indicate that the model is useful for prediction, with negative and positive values suggesting underestimates and overestimates, respectively [46]. This indicator is related with the unit in which the evaluated property is expressed, as well as with the dataset range of values, which, in this study, represents the rainfall interception, in mm.

The CMRE, d and E values are standardized measures in which cross-comparisons for a variety of models, regardless of units, are possible. That is, these indicators are not measures of correlation or association in the formal sense, but rather measures of the degree to which the predictions obtained from a model are error-free [47,48]. E and d vary from 0 to 1, with the maximum value representing a perfect agreement between observed and predicted data. CMRE ranges between 0 and 100%. Based on approximately 111 scientific research studies about rainfall interception modelling [21], the cumulative mean error was classified in five qualitative groups: bad (>30%), applicable (10–30%), good (5–10%), very good (1–5%) and extremely good (<1%).

Additionally, validation graphs of the measured rainfall interceptions against the predicted ones were plotted. Aiming at verifying the model accuracy, the t-test was applied to the intercept (b) of each linear regression to check whether it was significantly different from 0 and to the line angular coefficient (a) to confirm whether it was significantly different from 1, at the level of 1% probability.

2.5. Statistical and Sensitivity Analyses

Sensitivity analyses were performed to identify the relative importance of canopy and climatic parameters (S, c, E_m and R) in both daily and seasonal parametrizations of sparse Gash model. For this, the values of each parameter were increased and decreased by up to 50% of their original values and the simulated results were compared to measured data [5,11,24].

Measured rainfall interceptions and estimated model parameters were also submitted to variance analysis and averages were compared by the F and Scott–Knott tests ($p < 0.05$) through the SISVAR software (Federal University of Lavras-UFLA, Lavras, Minas Gerais, Brazil) [49]. Statistical differences between Caatinga species, storm classes and simulation periods were evaluated.

3. Results

3.1. Rainfall Partitioning

The total measured rainfall between 1 March 2016 and 30 September 2017 was 429.5 mm, generated by 66 discrete rainfall events. From these, 343.7 mm (80.0%) occurred during the rainy season, while 85.8 mm (20.0%) occurred during the dry season. The average, maximum and minimum event-based rainfall amounts were 6.5 ($\pm$9.3), 36.4 and 0.2 mm, respectively. Rainfall intensities varied from 1.2 to 19.2 mm h^{-1}, with an average of 3.2 ($\pm$2.9) mm h^{-1}.

Statistical analyses indicated that the frequency distributions of the event size (Table 3) and intensity (Figure 2) among annual analysis, rainy and dry seasons did not differ significantly. The very small events (0.0–5.0 mm) were significantly more frequent than the other four classes. However, as shown in Table 3, they contributed with the lowest percentages to total gross rainfall during the rainy season and for the annual analysis. When evaluating the dry season, events from 20.1 to 40.0 mm were not observed and small events (5.1–10.0 mm) were responsible for the lowest percentages of total gross rainfall. The highest percentages of total gross rainfall were verified for the very large events (30.1–40.0 mm) during the rainy season and for middle events (10.1–20.0 mm) during dry and annual periods.

Table 3. Number of events (NE), cumulative gross rainfall (CGR) and percentage of gross rainfall (PGR) in different rainfall classes during the study period.

	Classes (mm)				
	0.0–5.0	5.1–10.0	10.1–20.0	20.1–30.0	30.1–40.0
		Annual Analysis			
NE (-)	43 (65.2%)	7 (10.6%)	10 (15.2%)	2 (3.0%)	4 (6.1%)
CGR (mm)	46.0	53.9	144.9	52.2	132.5
PGR (%)	10.7	12.6	33.7	12.1	30.9
		Rainy Season			
NE (-)	18 (50.0%)	6 (16.7%)	6 (16.7%)	2 (5.6%)	4 (11.1%)
CGR (mm)	27.3	45.0	86.8	52.2	132.5
PGR (%)	7.9	13.1	25.3	15.2	38.6
		Dry Season			
NE (-)	25 (83.3%)	1 (3.3%)	4 (13.3%)	0 (0.0%)	0 (0.0%)
CGR (mm)	18.7	9.0	58.1	0.0	0.0
PGR (%)	21.8	10.5	67.7	0.0	0.0

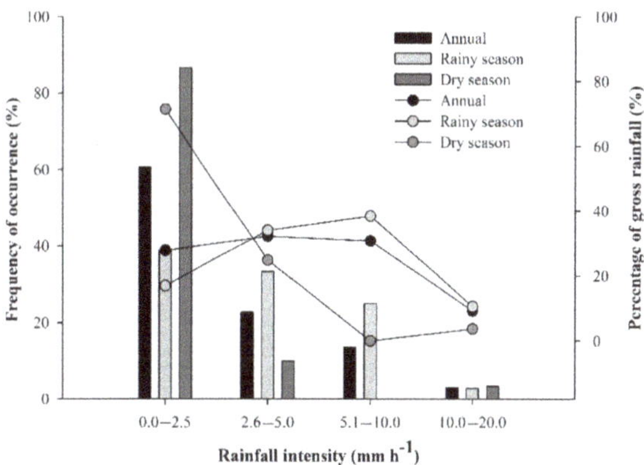

Figure 2. Frequency distributions (bars) and percentages of gross rainfall of the intensity of rainfall events (lines) at the Caatinga domain in Brazil during the measurement period (annual, rainy and dry seasons) from 1 March 2016 to 30 September 2017.

When evaluating rainfall intensity, events from 0 to 2.5 mm h^{-1} presented significantly higher frequency of occurrence. Figure 2 shows that these low intensity rainfall events contributed to a higher percentage of gross rainfall only during the dry season, while middle and large intensity rainfall events (2.6–10 mm h^{-1}) were responsible for the greatest percentages of gross rainfall both for annual and rainy season conditions.

As shown in Figure 3, the annual cumulative interception values were between 90.5 and 169.7 mm, resulting in average proportions of gross rainfall into interception of 27.9% ($\pm$6.5). When considering the rainy and dry seasons, the cumulative interceptions ranged from 43.5 to 114.7 mm and from 35.6 to 55.0 mm, with proportions of gross rainfall to interception of 17.2% ($\pm$5.7) and 10.7% ($\pm$1.8), respectively. Additionally, the annual average proportions of gross rainfall to throughfall and stemflow were 71.6% ($\pm$7.5) and 0.7% ($\pm$0.2), respectively.

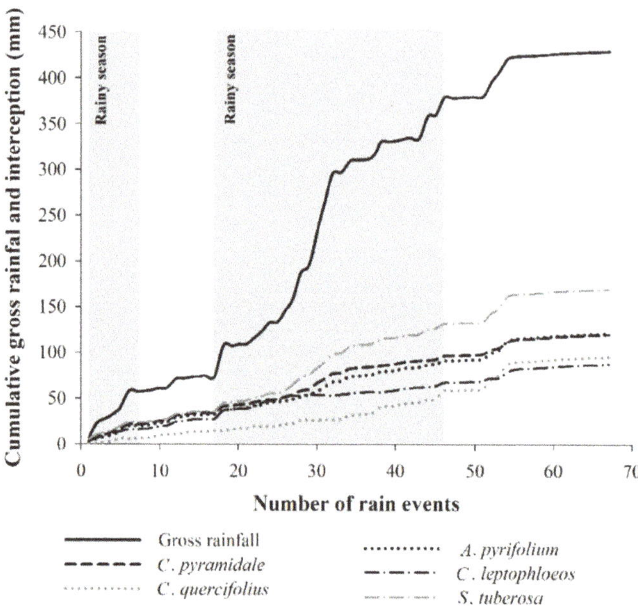

Figure 3. Cumulative measured gross rainfall and interception values from March 2016 to September 2017 at the Caatinga domain in Brazil.

Observing the different storm size classes (Table 4), the proportions of gross rainfall into interception of very small storms (<5 mm) were significantly higher for all studied Caatinga species. Statistical analysis also showed that there was no significant difference between large (20.1–30.0 mm) and very large (30.1–40.0 mm) storms, as well as between small (5.1–10.0 mm) and middle (10.1–20.0 mm) storms. Filtering by significantly equal rainfall classes (<5, 5.1–20.0, 20.1–40 mm), the proportions of gross rainfall to interception tended to decrease as gross rainfall increased. For very large events (30.1–40 mm), *C. quercifolius* presented a significantly lower percentage of rainfall interceptions (Figure 3 and Table 4) and, for the other storm classes, there was also a trend of smaller values when compared with the other studied species.

3.2. Model Parameters

The leaf area indices, as well as the parameters for the adapted Gash model applied to the five studied Caatinga species, are shown in Table 5. The values of each parameter and leaf area index did not significantly differ among Caatinga species, but *S. tuberosa* presented a larger leaf area index, S_c and c, which also helps to explain the significantly higher rainfall interception observed during the experimental trial. Parameters and leaf area index were statistically equal between seasons and annual analysis, but their numerical differences contributed to improve the simulation results, as discussed below.

Table 4. Proportions of gross rainfall portioned into interception (I:GR) considering five rainfall classes.

	Classes (mm)				
	0.0–5.0	5.1–10.0	10.1–20.0	20.1–30.0	30.1–40.0
Annual Analysis (I:GR—%)					
C. pyramidale	82.1	34.4	25.0	8.6	17.3
C. quercifolius	78.1	11.3	32.7	12.0	1.1
A. pyrifolium	80.5	37.0	28.2	9.7	13.7
C. leptophloeos	80.2	35.7	17.0	1.3	6.9
S. tuberosa	87.0	46.2	42.3	24.7	23.1
Rainy Season (I:GR—%)					
C. pyramidale	41.4	25.3	14.1	8.6	17.3
C. quercifolius	37.4	10.4	9.0	12.0	1.1
A. pyrifolium	39.8	27.7	13.2	9.7	13.7
C. leptophloeos	39.8	28.4	8.1	1.3	6.9
S. tuberosa	46.3	37.4	20.5	24.7	23.1
Dry Season (I:GR—%)					
C. pyramidale	40.7	9.1	10.9	0.0	0.0
C. quercifolius	40.7	1.0	23.7	0.0	0.0
A. pyrifolium	40.7	9.3	15.1	0.0	0.0
C. leptophloeos	40.7	7.3	8.9	0.0	0.0
S. tuberosa	40.7	8.8	21.8	0.0	0.0

3.3. Sensitivity Analyses

The sensitivity analysis of the canopy and climatic parameters to rainfall interception is presented in Figure 4. Variations observed for the five studied species were averaged for each model, since they presented very similar patterns.

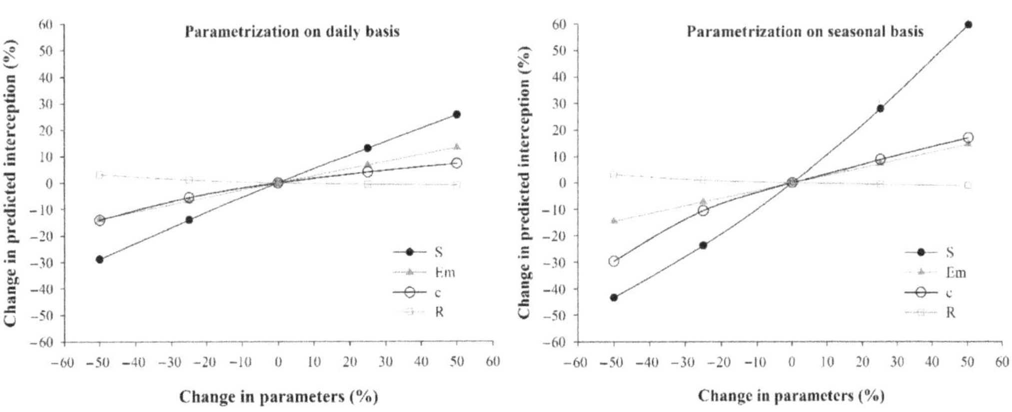

Figure 4. Sensitivity analysis of the Gash model parametrized on daily and seasonal bases applied to the Caatinga domain in Brazil.

Table 5. Measured leaf area indices (L) and estimated canopy storage capacity per unit area of cover (S_c), relative evaporation rate per unit area of cover (E_c/R), evaporation rate from wet canopy (mm h^{-1}), canopy cover fraction (c) and threshold value required to saturate the canopy (P_S) for the sparse Gash model parametrized on daily and seasonal bases applied to each studied Caatinga species. Values in brackets represent standard deviations.

Vegetation	S_c (mm)	L (mm^2 mm^{-2})	E_c/R (-)	E_m (mm h^{-1})	c (-)	P_S (mm)
			Daily Basis			
C. pyramidale	2.3–3.4	0.4–1.6	0.04–0.83	0.20–0.85	0.29–0.72	2.8–5.4
C. quercifolius	1.7–4.4	0.5–4.0	0.04–0.79	0.21–0.94	0.27–0.94	2.3–4.9
A. pyrifolium	2.8–4.0	0.9–2.0	0.03–0.79	0.19–0.80	0.52–0.80	3.3–4.6
C. leptophloeos	1.9–4.4	0.9–4.0	0.03–0.86	0.20–0.89	0.48–0.95	2.2–4.8
S. tuberosa	1.8–4.8	0.8–7.0	0.03–0.83	0.20–0.85	0.46–0.97	2.1–5.0
			Seasonal Basis (Rainy Season)			
C. pyramidale	2.30 (±0.2)	1.04 (±0.3)	0.16 (±0.2)	0.40 (±0.11)	0.67 (±0.1)	2.50
C. quercifolius	2.85 (±0.6)	1.82 (±0.7)	0.17 (±0.2)	0.44 (±0.12)	0.68 (±0.2)	3.13
A. pyrifolium	2.58 (±0.3)	1.44 (±0.2)	0.15 (±0.2)	0.38 (±0.10)	0.69 (±0.1)	2.79
C. leptophloeos	2.89 (±0.5)	2.02 (±0.6)	0.14 (±0.1)	0.42 (±0.12)	0.76 (±0.10)	3.12
S. tuberosa	2.97 (±0.6)	2.22 (±0.8)	0.14 (±0.2)	0.40 (±0.12)	0.77 (±0.12)	3.19
			Seasonal Basis (Dry Season)			
C. pyramidale	2.10 (±0.2)	0.85 (±0.3)	0.27 (±0.2)	0.38 (±0.16)	0.60 (±0.1)	2.45
C. quercifolius	2.49 (±0.4)	1.30 (±0.6)	0.31 (±0.2)	0.42 (±0.18)	0.56 (±0.2)	2.98
A. pyrifolium	2.45 (±0.3)	1.28 (±0.2)	0.24 (±0.2)	0.36 (±0.15)	0.64 (±0.1)	2.78
C. leptophloeos	2.55 (±0.4)	1.59 (±0.5)	0.25 (±0.1)	0.40 (±0.17)	0.68 (±0.1)	2.93
S. tuberosa	2.56 (±0.4)	1.69 (±0.6)	0.24 (±0.2)	0.38 (±0.15)	0.68 (±0.1)	2.92

Canopy storage capacity (S), canopy cover fraction (c) and evaporation rate from wet canopy (E_m) presented positive relationships with interception, whereas mean rainfall rate (R) resulted in a negative relationship with interception. All parameters were sensitive to interception, but S had larger effects on the Caatinga rainfall partitioning, followed by E_m, c and R, respectively. Seasonal model predictions were most sensitive to canopy and climatic parameters, which were more pronounced for the S variable.

The sensitivity analysis showed that E_m changes could lead from -13.8 (±1.5) to 13.4 (±1.3)% errors in rainfall interception when applying the sparse Gash model parametrized on a daily basis, while, for the seasonal approach, errors could vary from -14.6 (±1.6) to 14.5 (±1.6)%. Considering the R parameter, a -50% change in daily and seasonal parametrizations caused an average 3.2 (±1.5)% difference in the simulated rainfall interception.

For the daily parametrization, a decrease of 50% in c resulted in an average decrease of 14.1% (±0.8) in simulated interception, while an increase of 50% resulted in an average rise of 7.3% (±0.5). When applying the seasonal parametrization, a decrease of 50% in c could lead to an average decrease of 29.7% (±3.4) in simulated interception, while an increase of 50% could lead to an average rise of 17.0% (±0.9).

If the value of S increased by 50%, simulated rainfall interception tended to rise by 25.7% (±1.2) and 59.6 (±3.2) on average by applying the daily and seasonal parametrizations, respectively. On the other hand, the decrease of 50% resulted in average reductions of 28.8% (±1.4) and 43.3% (±2.3) for daily and seasonal parametrizations, respectively.

3.4. Rainfall Interception Simulations

Figure 5 shows that the daily and seasonal parametrizations performed very similarly, when simulating cumulative rainfall interception, with the daily parametrization performing slightly better, except for *S. tuberosa*. The average annual proportions of gross rainfall into interception were simulated as 27.1% (±5.1), when applying the daily parametrization, and as 26.5% (±5.3), when using the seasonal parametrization. Compared with the measured data, the differences were 0.8 and 1.4%, respectively. When considering the tree species, simulations for *A. pyrifolium* and *C. pyramidale* resulted in more accurate estimates. These species reached cumulative relative mean errors from 1.23 to 3.69%, considered very good [21]. The cumulative relative mean errors for the other species varied from 5.00 to 8.98% and were considered good [21].

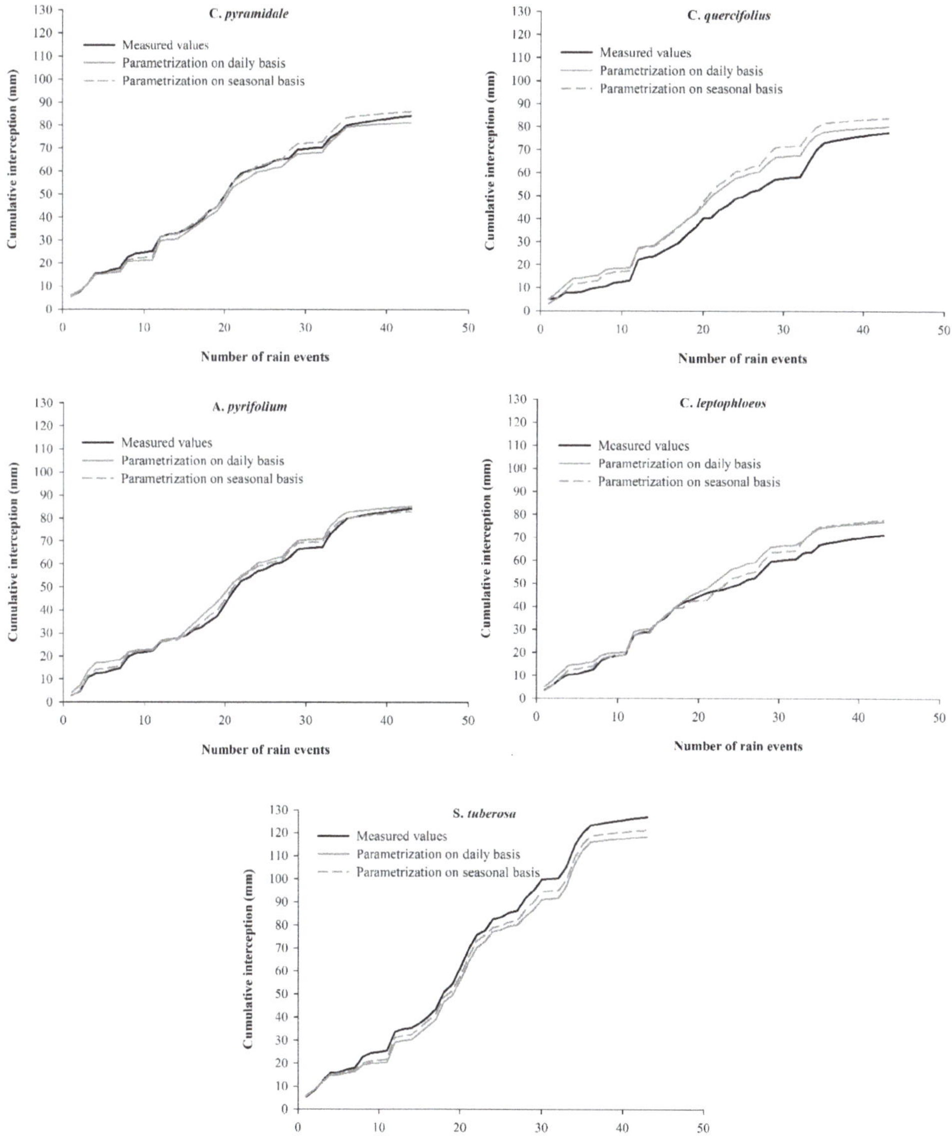

Figure 5. Cumulative measured interception values compared with those estimated by the sparse Gash model parametrized on daily and seasonal bases for each Caatinga species in Brazil.

On a per storm analysis (Figure 6), when estimates were evaluated for individual rainfall events, both simulation approaches also performed similarly, with the daily parametrization presenting slightly less scatter. Prediction errors were higher mainly for small, middle and large storms (5.1–30 mm), with interceptions between 0.2 and 6 mm. Among the tree species, best results were obtained with *S. tuberosa*.

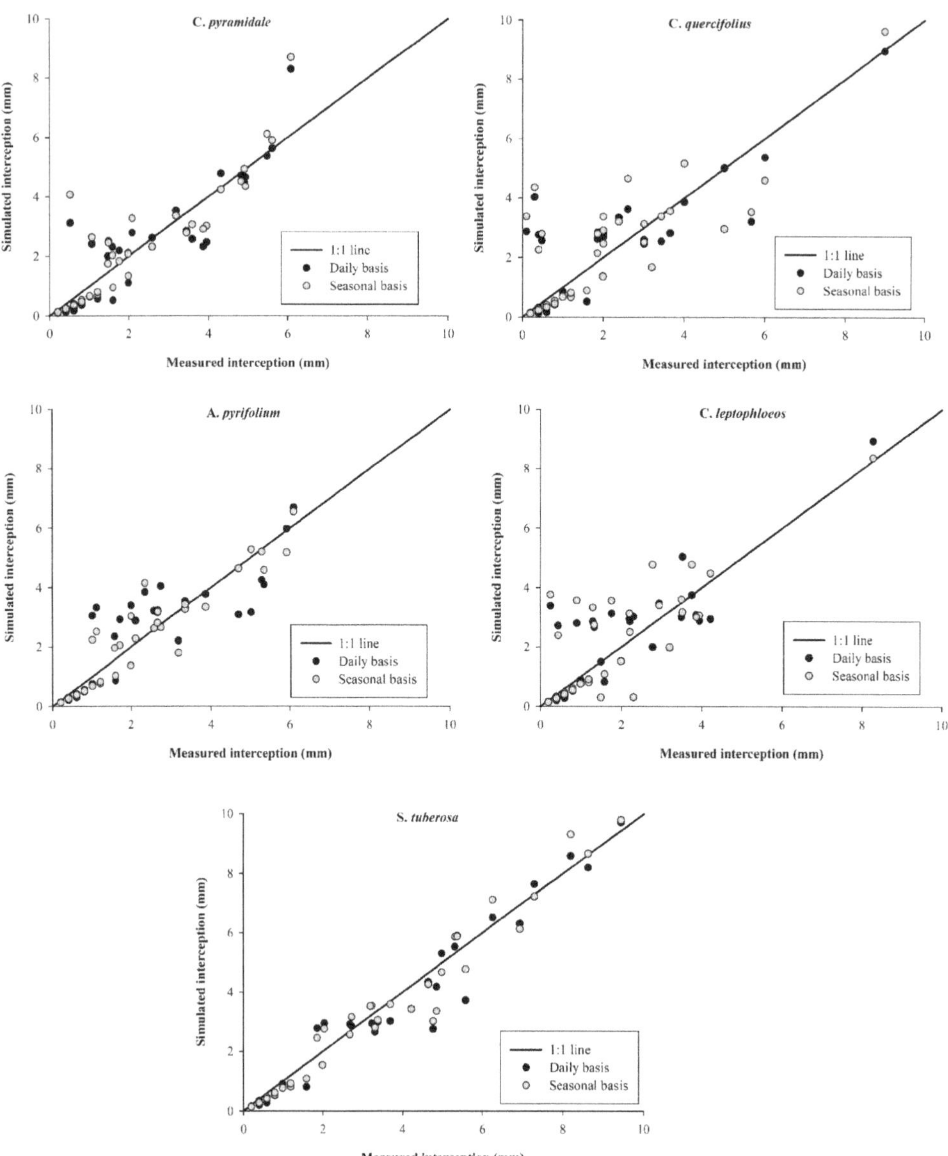

Figure 6. Validation graphs of experimental interception values against those predicted by the sparse Gash model parametrized on daily and seasonal bases applied to Caatinga species.

The reliability of the proposed models for Caatinga species was also proven by the statistical indicators (Table 6). MBE varied from −0.20 to 0.15 mm for daily and seasonal parametrizations, indicating predicted values close to measured ones and confirming the model trends of slightly underestimating the *S. tuberosa* interception for both tested approaches, as well as for *A. pyrifolium* with seasonal parametrization and *C. pyramidale* with daily parametrization. The other cases showed trends of slightly overestimating. The d and E averages were 0.94 ($\pm$0.04) and 0.75 ($\pm$0.80), respectively, when applying seasonal simulations. For the daily parametrization, these indices reached 0.94 ($\pm$0.03) and 0.76 ($\pm$0.12) on average, respectively. The d and E results proved the high accuracy and the very good agreement between measured and predicted interceptions for all Caatinga species, as well as the slightly better performance of the daily parametrization.

Table 6. Summary statistics of interception values predicted by the sparse Gash model parametrized on daily and seasonal bases applied to Caatinga species on storm-based rainfall analysis.

Vegetation	a (-)	b (-)	R^2 (-)	MBE (mm)	d (-)	E (-)
			Daily Basis			
S. tuberosa	0.99	−0.19	0.96	−0.20	0.99	0.95
C. leptophloeos	0.93	0.26	0.72	0.13	0.92	0.66
A. pyrifolium	0.90	0.22	0.77	0.02	0.94	0.75
C. quercifolius	0.82	0.39	0.69	0.06	0.91	0.66
C. pyramidale	1.01	−0.10	0.83	−0.07	0.97	0.79
			Seasonal Basis			
S. tuberosa	1.02	−0.20	0.96	−0.13	0.99	0.96
C. leptophloeos	0.93	0.26	0.66	0.15	0.89	0.53
A. pyrifolium	0.98	0.01	0.90	−0.03	0.97	0.89
C. quercifolius	0.82	0.46	0.65	0.15	0.90	0.59
C. pyramidale	1.03	−0.02	0.82	0.04	0.95	0.77

4. Discussion

4.1. Rainfall Partitioning

The partitioning pattern observed in this study for the five Caatinga species agreed with other studies of semiarid regions [1,2,16,24], where rainfall is concentrated over a short period. Highly variable rainfall depths per event (0.2–40 mm) were also reported when applying the Gash model to deciduous shrubs in the semiarid Qinghai–Tibet Plateau [22]. Furthermore, mean values of gross rainfall per event (7.2 and 5.1 mm) close to the average value found in this work were verified for semiarid regions of Spain and Iran, respectively [2,23]. On the other hand, average Caatinga rainfall intensities (3.2 mm h^{-1}) were higher than those observed in other semiarid regions. For example, a mean rainfall intensity of 1.7 mm h^{-1} was found when modelling the interception in central–western Spain [2], while rainfall intensities equal to 1.8 and 2.3 mm h^{-1} were reported for semiarid regions of China and Kenya, respectively [24,50]. The frequency distributions of the event size and intensity were in agreement with other studies about rainfall partitioning in semiarid regions [16,23,24].

Average rainfall interception was significantly higher during the rainy season, which is justified by the largest number of rainfall events and greatest rainfall amounts observed during this period, but also by the reduction in leaf amounts, as well as canopy cover, during the dry season. Rainfall interception of *S. tuberosa* was significantly higher than those verified for the other studied Caatinga species. This difference can likely be related with canopy characteristics, mainly the number of stems, diameter at breast height and tree crown projected area (Table 1), on which further studies are required. Additionally, the leafless periods of *S. tuberosa* corresponds to around 3 months [26], while the other species are leafless from approximately 4 to 7 months [25,27,51] and the emergence of leaves strongly affects the interception process, modifying the redistribution by the tree and the profile of rainwater.

The trend of decreasing the proportions of gross rainfall to interception as gross rainfall increases was also verified in other studies [23]. Additionally, the lower percent rainfall interceptions of *C. quercifolius* can be explained since the peak of its leaf fall lasts around 5 months [25] and this tree produces small leaves during the dry season that only attain their maximum size during the rainy season. For this, it presents a less dense canopy, which tends to facilitate water flow in throughfall [15].

Results obtained in this study confirm the findings of other scientific research studies [1,8,16,23], which suggested that changes in the proportions of gross rainfall to interception are mainly associated with the size of gross rainfall and can be explained since most of the gross rainfall is stored in the canopy during the very small and small rainfall storms. On the other hand, canopies tend to saturate during large and very large rainfall events, increasing throughfall, while the remaining rainfall is stored in the canopies and lost as evaporation during the storm event [9].

4.2. Model Parameters

Average leaf area indices observed in the Caatinga domain were similar to those verified for species of other semiarid regions, such as those for *R. pseudoacacia* in the Shaanxi province, China [52], as well as for *Q. ilex* and *Q. pyrenaica* in Sardon stands, Spain [2]. However, *S. tuberosa* stood out in this study, maintaining most of its leaves during the dry season, as well as reaching higher individual leaf area indices, when compared with other Caatinga species and native trees of different semiarid regions [22,23]. The average leaf area indices for rainy and dry seasons (Table 5) reflected the seasonal variations of this parameter, which increased linearly from the beginning to the end of the rainy season and decreased linearly during the dry season for all studied species. Indeed, the peak of leaf flush for the Caatinga species tends to coincide with the rainy season, but this process is also affected by the photoperiod [25].

Despite rainfall interception is known to be closely related to the leaf area index [53], other factors influence this process. The increase in canopy density causes leaves to touch, hindering the fully saturation of the entire canopy [12]. Furthermore, wind may reduce canopy storage capacity, as well as branch shape, and leaf inclination and canopy thickness may turn the leaves less wettable.

As expected and as shown in Table 5, both c and S_c followed the pattern observed in the leaf area index for all studied species, with average values of the rainy season larger than those of the dry season. The same behavior was verified for the P_S values, since all of these parameters derived from the leaf area index in the parametrizations performed in this study. The c values were similar to other simulations of rainfall interception for deciduous forests in semiarid regions [8,14,22], but S_c and P_S were larger. These results are consistent with the larger leaf area indices verified mainly during the rainy season and are probably also associated with the height of the studied Caatinga species (Table 1). The S_c and P_S values obtained in this study can be also justified by the distinct forest structures among semiarid regions, since the canopy morphology and physiology interfere in the parametrization, including the leaf amounts and canopy cover during leafed and leafless periods [17]. Additionally, S_c and P_S are closely related to Caatinga weather conditions, mainly rainfall distribution and intensity [38]. The estimate of S_c also depended on the specific canopy storage capacity, which was based on the mean method, relating the observed gross rainfall, interception and throughfall that were either insufficient or sufficient to saturate the canopy. On the other hand, the P_S also depended on the evaporation rate from wet canopy, which was estimated by the Penman–Monteith equation, considering meteorological (air temperature, net radiation, vapor pressure deficit and wind speed) and vegetation characteristics (crop height).

Canopy coverage is an important parameter in both daily and seasonal parametrizations, since it is a structural parameter, directly related to the free throughfall coefficient and the leaf area index [5]. The free throughfall coefficient, which is assumed to be one minus canopy cover, affects soil water content and nutrient cycling, since it reflects the fraction of rainfall passing through the canopy without contacting the tree surface or removing dry leaves and twigs in the canopy [24]. Indeed, Table 5 shows a trend of species with smaller leaf area indices and greater leafless periods (*C. pyramidale* and *C. quercifolius*) presenting lower canopy coverages, with consequent larger free throughfall coefficients. The obtained c values can also explain the lower rainfall interceptions verified for *C. quercifolius*, as well as the significantly high interceptions from *S. tuberosa*. These species presented respectively the lowest (0.27) and the largest (0.97) canopy coverages when applying the daily parametrization, with *C. quercifolius* resulting in the greatest range of c values during the study and the smallest canopy coverage during the dry season (Table 5). Additionally, *S. tuberosa* was the species with the largest c values both in dry and rainy seasons. Thus, there is a trend of *S.tuberosa* species protect the Caatinga floor from raindrop splash erosion, also delaying the peaks in storm runoffs, as was verified in semiarid Northeast of China [24] and Brazil [54]. On the other hand, these trees could contribute to the enhancement of soil water scarcity in the Caatinga domain, since the high c values lead to less throughfall reaching the forest floor. However, detailed comprehension about these effects merit further studies.

Considering the daily parametrization, E_m was highly changeable during the studied events, varying between 0.19 and 0.94 mm h^{-1}, with an average of 0.40 ($\pm$0.14) mm h^{-1}. The E_m values observed during rainy and dry seasons reached 0.41 ($\pm$0.02) and 0.39 ($\pm$0.02) mm h^{-1} on average, respectively. Scaled to canopy cover, E_c values ranged from 0.24 to 2.50 mm h^{-1}, with an average of 0.67 ($\pm$0.36) mm h^{-1} for the daily parametrization, while the seasonal approach resulted in E_c values of 0.62 ($\pm$0.08) and 0.57 ($\pm$0.05) mm h^{-1} on average during rainy and dry seasons, respectively (Table 5). Maximum in-storm evaporation rates from 1.83 to 3.98 mm h^{-1} were also observed in tropical dry and semiarid regions of Mexico [47], but the average evaporation rates found in this study were larger than those of semiarid regions of Spain, Iran and China [2,23,24]. These results are related with the semiarid climate type of Caatinga, which is mainly characterized by high temperatures and solar radiation, tending to increase evaporation rates when compared with other semiarid and dry tropical stands [1]. The higher evaporation rates and the distinct rainfall intensities affected the E_c/R ratios, which were lower during the rainy season (0.15 $\pm$0.01), reaching average values of 0.26 ($\pm$0.03) during the dry season. When applying the daily parametrization, E_c/R presented minimum values between 0.03 and 0.04, while maximum values varied from 0.79 to 0.86 for the five studied species.

The differences observed between seasonal and daily parametrizations, regarding the sensitivity analysis, are expected, since the seasonal model uses two sets of constant parameters, while the daily parametrization considers the daily changes of canopy structure and weather conditions, tending to better represent the associated components and processes of rainfall interception. The sensitivity analysis agreed with other studies [53,55], which found that canopy storage capacity is among the most influential parameters on simulated rainfall interception in deciduous vegetation. Other factors that affect the sensitivity of model predictions are the rainfall and climate characteristics, such as raindrop size distribution, rainfall intensity and wind speed, though those factors are not included in the Gash model [24]. Additionally, canopy basal area and height interfere in the parametrization process, as well as the woody light-blocking elements from the canopy with respect to diameter growth, represented by the wood area index [56].

4.3. Rainfall Interception Simulations

The good results observed when simulating cumulative rainfall interception may be consequence of using leaf area indices during estimates of c and S, as well as the simulations based on daily or seasonal parameter variations, which allowed the proposed models to describe the rainfall interception patterns better than other approaches [13,14].

When considering the per-storm simulations (Figure 6), very large rain events (30.1–40 mm) were less frequent (Table 3), representing 6.1% of total gross rainfall, and were not observed during the dry season. This probably led to less events with interceptions greater than 6 mm and, consequently, less scatter was verified for these situations. There were more outliers for smaller interceptions, with simulated values moving further from the 1:1 line in both parametrizations and showing that model estimation was less accurate in some individual events. For these cases, the daily parametrization was also subtly better than the seasonal one. These discrepancies certainly affected the statistical indicators (Table 6). However, the scatter patterns agreed with other works [8,23] and, despite the observed discrepancies when comparing measured and simulated individual interceptions, both daily and seasonal parametrizations resulted in a good fit, with all intercept and angular coefficients not significantly differing from 0 and 1, respectively (Table 6). Additionally, the determination coefficients between measured and predicted values varied from 65.0 to 96.0%. These results indicate the absence of constant and systematic errors, confirming the good reproducibility of the estimates from the proposed models when applied to the Caatinga vegetation.

Results of this study indicate that parametrization on daily and seasonal bases improved estimates for rain events where interception is less dependent on S_c than on E_c/R, that is, for larger and more intense events. During heavy storms the canopy tends to rapidly saturate, decreasing the influence of canopy storage capacity and increasing the control of E_c/R [8]. This behavior also explains the minor error propagation, observed when cumulative interception was simulated (Figure 5). However, for other ecosystems, this trend should be better investigated.

The sparse Gash model parametrized on a daily basis is indicated for vegetation densities that change gradually, but relatively rapidly, as well as for vegetation that changes more slowly, but is subject to infrequent rainfall [12]. It requires leaf area index monitoring and a more complex implementation, but represents important conceptual improvements in the rainfall interception simulations, giving accurate estimates from low to high intensity storms, as well as for events with different amounts. However, when it is not possible to use the more expensive instrumentation required for parametrizing this approach, or the greater data processing during simulations, the seasonal sparse Gash model is capable of considering the variability of Caatinga species regarding foliation and defoliation, which is reflected by the canopy and climate parameters associated with rainy and dry periods. It does not require leaf area monitoring and equations are simplified, resulting in a less complex simulation with reliable approximations.

4.4. Limitations and Constraints

The methodological challenges in measurement and data processing when modelling and validating rainfall interception are associated with the complex and expensive micrometeorological instrumentation, as well as the long data acquisition periods required in this process. Additionally, throughfall is highly spatially heterogeneous at small scales, while rainfall interception and stemflow are variable across species, requiring a measurement scheme capable of sufficiently take into account these differences, but also coherent with the financial and logistical constraints. For this, it is important to attempt to correctly locate the gauges and divide the area under each tree, also systematically performing the relocation of gauges by applying well defined methods and minimizing errors originating from spatial variability. These precautions were followed in this study, agreeing with other works [2,10,13,16,33] that had proved the possibility of monitoring throughfall, stemflow and rainfall interception by considering fewer sample trees per studied species in a credible way.

Among the Gash model parameters, canopy storage capacity and evaporation rate from wet canopy are the most difficult to obtain individually [21,47]. The quantification of canopy storage capacity can be determined experimentally for a particular species using laboratory methods [12]. However, the indirect methods are relatively low-cost and require no complex instrumentation, having been preferred in most of studies for Gash model parametrization [3,7,8,23], despite the long measurement period required. In this work, a regression-based method was applied, but future efforts for obtaining reliable measurements of this parameter or decomposing it into easily measurable physical components are encouraged [55]. The measurement of the evaporation rate from wet canopy also involves high costs and technical difficulties and this parameter is frequently estimated by means of the Penman–Monteith method [47,52]. This method was used in this study, performing very well, despite requiring data of many micrometeorological variables. Other formulations and methods that overcome the high data input requirements of the Penman–Monteith equation have been proposed [2,24,34], but the best method for this purpose remains controversial and deserves further attention [47].

Finally, climate change is altering the water cycle and world ecosystems, with precipitation and phenological responses of plants to habitat being directly affected. Considering the Brazilian semi-arid region, the global climate change scenarios indicate that the aridity of the region will tend to increase in the next century, evidencing its vulnerability [18]. Therefore, a good comprehension of the hydrological processes and the study of their behaviors considering the climate change projections is essential for the evaluation of water availability and the anthropogenic effects in Caatinga. Thus, rainfall interception modelling based on these projections and the use of validated models for studies of land management are required.

5. Conclusions

Observed cumulative rainfall interceptions varied from 10.1 to 26.7% of gross rainfall during the rainy season and from 8.3 to 12.8% during the dry season for the five studied Caatinga species, indicating that significantly lower throughfall and stemflow reached the soil as available water input during the dry periods of 2016 and 2017. The frequency distributions of the event size and intensity did not differ significantly between dry and rainy seasons, with the very small (0.0–5.0 mm) and low-intensity (0–2.5 mm h^{-1}) events being significantly more frequent. However, the highest percentages of total gross rainfall were verified for the very large events (30.1–40.0 mm) during the rainy season and for middle events (10.1–20.0 mm) when considering the dry season. The low intensity storms contributed to a higher percentage of gross rainfall only during the dry season, while middle and large intensity rainfall events (2.6–10 mm h^{-1}) were responsible for the greatest percentages of gross rainfall during the rainy season. The sparse Gash model parametrized on a daily basis performed slightly better than the seasonal one, but both approaches resulted in very good or good agreement between the modelled and estimated interception, with cumulative mean relative errors between 1.23 and 8.98%. Seasonal model predictions were the most sensitive to canopy and climatic parameters, with canopy storage capacity presenting larger effects on the Caatinga rainfall partitioning, followed by E_m, c and R, respectively, for both simulation approaches. Future works should focus on finding reliable methods for measuring canopy storage capacity, as well as on formulations capable of accurately estimating the evaporation rate from wet canopy requiring smaller input variables. Furthermore, the Gash model should be parametrized for other Caatinga species, with the validated approaches providing a basis for studies of land management, including the evaluation of degraded areas and effects of climate changes.

Author Contributions: Conceptualization, D.C.L. and A.J.S.N.; data curation, L.S.B.S. and C.A.A.S.; methodology, D.C.L., A.J.S.N., T.G.F.S. and S.Z.; resources, T.G.F.S. and S.Z.; validation, D.C.L. and A.J.S.N.; writing—original draft, D.C.L.; writing—review and editing, A.J.S.N. All authors have read and agreed to the published version of the manuscript.

Funding: This research was funded by the Research Support Foundation of the Pernambuco State (FACEPE-APQ-0215-5.01/10 and FACEPE-APQ-1159-1.07/14), the National Council for Scientific and Technological Development (CNPq-475279/2010-7, 476372/2012-7, 305286/2015-3, 309421/2018-7 and 152251/2018-9) and the Coordination for the Improvement of Higher Education Personnel (CAPES-Finance Code 001).

Institutional Review Board Statement: Not applicable.

Informed Consent Statement: Not applicable.

Conflicts of Interest: The authors declare no conflict of interest.

References

1. Brasil, J.B.; Andrade, E.M.; Queiroz Palácio, H.A.; Medeiros, P.H.A.; Santos, J.C.N. Characteristics of precipitation and the process of interception in a seasonally dry tropical forest. *J. Hydrol.* **2018**, *19*, 307–317. [CrossRef]
2. Hassan, S.M.T.; Ghimire, C.P.; Lubczynski, M.W. Remote sensing upscaling of interception loss from isolated oaks: Sardon catchment case study, Spain. *J. Hydrol.* **2017**, *555*, 489–505. [CrossRef]
3. Pypker, T.G.; Tarasoff, C.S.; Koh, H.S. Assessing the efficacy of two in direct methods for quantifying canopy variables associated with interception loss of rainfall in temperate hardwood forests. *Open J. Mod. Hydrol.* **2012**, *2*, 29–40. [CrossRef]
4. Rutter, A.J.; Morton, A.J.; Robins, P.C. A predictive model of rainfall interception in forests. II. Generalization of the model and comparison with observations in some coniferous and hardwoods stands. *J. Appl. Ecol.* **1975**, *12*, 367–380. [CrossRef]
5. Shi, Z.; Wang, Y.; Xu, L.; Xiong, W.; Yu, P.; Gao, J.; Zhang, L. Fraction of incident rainfall within the canopy of a pure stand of *Pinus armandii* with revised Gash model in the Liupan Mountains of China. *J. Hydrol.* **2010**, *385*, 44–50. [CrossRef]
6. Calder, I.R. A stochastic model of rainfall interception. *J. Hydrol.* **1986**, *89*, 65–71. [CrossRef]
7. Link, T.E.; Unsworth, M.; Marks, D. The dynamics of rainfall interception by a seasonal temperate rainforest. *Agric. For. Meteorol.* **2004**, *124*, 171–191. [CrossRef]
8. Fathizadeh, O.; Hosseini, S.M.; Keim, R.F.; Boloorani, A.D. A seasonal evaluation of the reformulated Gash interception model for semiarid deciduous oak forest stands. *For. Ecol. Manag.* **2018**, *409*, 601–613. [CrossRef]
9. Gash, J.H.C.; Lloyd, C.R.; Lachaud, G. Estimating sparse forest rainfall interception with an analytical model. *J. Hydrol.* **1995**, *170*, 79–86. [CrossRef]
10. Steidle Neto, A.J.; Ribeiro, A.; Lopes, D.C.; Sacramento Neto, O.B.; Souza, W.G.; Santana, M.O. Simulation of rainfall interception of canopy and litter in Eucalyptus plantation in tropical climate. *For. Sci.* **2012**, *58*, 54–60. [CrossRef]
11. Su, L.; Zhao, C.; Xu, W.; Xie, Z. Modelling interception loss using the revised Gash model: A case study in a mixed evergreen and deciduous broadleaved forest in China. *Ecohydrology* **2016**, *9*, 1580–1589. [CrossRef]
12. Van Dijk, A.I.J.M.; Bruijnzeel, L.A. Modelling rainfall interception by vegetation of variable density using an adapted analytical model. Part 1. Model description. *J. Hydrol.* **2001**, *247*, 230–238. [CrossRef]
13. Lopes, D.C.; Steidle Neto, A.J.; Queiroz, M.G.; Souza, L.S.B.; Zolnier, S.; Silva, T.G.F. Sparse Gash model applied to seasonal dry tropical forest. *J. Hydrol.* **2020**, *590*, 125497. [CrossRef]
14. Medeiros, P.H.A.; Araujo, J.C.; Bronstert, A. Interception measurements and assessment of Gash model performance for a tropical semiarid region. *Rev. Ciênc. Agron.* **2009**, *40*, 165–174. [CrossRef]
15. Silva, A.C.F.; Souto, J.S.; Santana, J.A.S.; Souto, P.C.; Nascimento, J.A.M. Distribution of rainwater by species of caatinga vegetation. *Afr. J. Agric. Res.* **2018**, *13*, 2239–2248. [CrossRef]
16. Queiroz, M.G.; Silva, T.G.F.; Zolnier, S.; Souza, C.A.A.; Souza, L.S.B.; Araújo, G.N.; Jardim, A.M.R.F.; Moura, M.S.B. Partitioning of rainfall in a seasonal dry tropical forest. *Ecohydrol. Hydrobiol.* **2020**, *20*, 230–242. [CrossRef]
17. Machado, I.C.S.; Barros, L.M.; Sampaio, E.V.S.B. Phenology of Caatinga species at Serra Talhada, PE, Northeastern Brazil. *Biotropica* **1997**, *29*, 57–68. [CrossRef]
18. Vieira, R.D.S.P.; Tomasella, J.; Alvalá, R.C.S.; Sestini, M.F.; Affonso, A.G.; Rodriguez, D.A.; Barbosa, A.A.; Cunha, A.P.M.A.; Valles, G.F.; Crepani, E.; et al. Identifying areas susceptible to desertification in the Brazilian northeast. *Solid Earth* **2015**, *6*, 347–360. [CrossRef]
19. Queiroz, M.G.; Silva, T.G.F.; Zolnier, S.; Souza, C.A.A.; Souza, L.S.B.; Steidle Neto, A.J.; Araújo, G.G.L.; Ferreira, W.P.M. Seasonal patterns of deposition litterfall in a seasonal dry tropical forest. *Agric. For. Meteorol.* **2019**, *279*, 107712. [CrossRef]
20. Gash, J.H.C. An analytical model of rainfall interception by forests. *Q. J. R. Meteorol. Soc.* **1979**, *105*, 43–55. [CrossRef]
21. Muzylo, A.; Llorens, P.; Valente, F.; Keizer, J.J.; Domingo, F.; Gash, J.H.C. A review of rainfall interception modelling. *J. Hydrol.* **2009**, *350*, 191–206. [CrossRef]
22. Zhang, S.Y.; Li, X.Y.; Jiang, Z.Y.; Li, D.Q.; Lin, H. Modelling of rainfall partitioning by a deciduous shrub using a variable parameters Gash model. *Ecohydrology* **2018**, *11*, e2011. [CrossRef]
23. Sadeghi, S.M.M.; Attarod, P.; Van Stan II, J.T.; Pypker, T.G.; Dunkerley, D. Efficiency of the reformulated Gash's interception model in semiarid afforestations. *Agric. For. Meteorol.* **2015**, *201*, 76–85. [CrossRef]
24. Ma, C.; Li, X.; Luo, Y.; Shao, M.; Jia, X. The modelling of rainfall interception in growing and dormant seasons for a pine plantation and a black locust plantation in semi-arid Northwest China. *J. Hydrol.* **2019**, *577*, 123849. [CrossRef]
25. Lima, A.L.A.; Rodal, M.J.N. Phenology and wood density of plants growing in the semi-arid region of northeastern Brazil. *J. Arid Environ.* **2010**, *74*, 1363–1373. [CrossRef]
26. Lins Neto, E.M.F.; Almeida, A.L.S.; Peroni, N.; Castro, C.C.; Albuquerque, U.P. Phenology of *Spondias tuberosa* Arruda (Anacardiaceae) under different landscape management regimes and a proposal for a rapid phenological diagnosis using local knowledge. *J. Ethnobiol. Ethnomed.* **2013**, *9*, 10. [CrossRef]
27. Moura, V.G.M. Sazonalidade Fenológica e Aspectos Funcionais de Espécies Lenhosas da Caatinga: Acompanhamento com Camera Hemisférica e in Loco. Master's Thesis, Universidade Federal Rural de Pernambuco, Recife, Brazil, 2018.
28. Alvares, C.A.; Stape, J.L.; Sentelhas, P.C.; Gonçalves, J.L.M.; Sparovek, G. Köppen's climate classification map for Brazil. *Meteorol. Z.* **2013**, *22*, 711–728. [CrossRef]
29. Thornthwaite, C.W. An approach toward a rational classification of climate. *Geogr. Rev.* **1948**, *38*, 55–94. [CrossRef]

30. Norman, J.M.; Jarvis, P.G. Photosynthesis in Sitka Spruce (*Picea Sitchensis* (Bong.) Carr.). III. Measurements of Canopy Structure and Interception of Radiation. *J. Appl. Ecol.* **1974**, *11*, 375–398. [CrossRef]
31. Pokovai, K.; Fodor, N. Adjusting Ceptometer Data to Improve Leaf Area Index Measurements. *Agronomy* **2019**, *9*, 866. [CrossRef]
32. Bryant, M.L.; Bhat, S.; Jacobs, J.M. Measurements and modeling of throughfall variability for five forest communities in the southeastern US. *J. Hydrol.* **2005**, *312*, 95–108. [CrossRef]
33. Lloyd, C.R.; Gash, J.H.C.; Shuttleworth, W.J. The measurement and modelling of rainfall interception by Amazonian rain forest. *Agric. For. Meteorol.* **1998**, *43*, 277–294. [CrossRef]
34. Pereira, F.L.; Gash, J.H.C.; David, J.S.; David, T.S.; Monteiro, P.R.; Valente, F. Modelling interception loss from evergreen oak Mediterranean savannas: Application of a tree-based modelling approach. *Agric. For. Meteorol.* **2009**, *149*, 680–688. [CrossRef]
35. Bäse, F.; Elsenbeer, H.; Neill, C.; Krusche, A.V. Differences in throughfall and net precipitation between soybean and transitional tropical forest in the southern Amazon, Brazil. *Agric. Ecosyst. Environ.* **2012**, *159*, 19–28. [CrossRef]
36. Levia, D.F.; Frost, E.E. A review and evaluation of stemflow literature in the hydrologic and biogeochemical cycles of forested and agricultural ecosystems. *J. Hydrol.* **2003**, *274*, 1–29. [CrossRef]
37. He, Z.-B.; Yang, J.J.; Du, J.; Zhao, W.-Z.; Liu, H.; Chang, X.X. Spatial variability of canopy interception in a spruce forest of the semiarid mountain regions of China. *Agric. For. Meteorol.* **2014**, *188*, 58–63. [CrossRef]
38. Murakami, S. Application of three canopy interception models to a young stand of Japanese cypress and interpretation in terms of interception mechanism. *J. Hydrol.* **2007**, *342*, 305–319. [CrossRef]
39. Allen, R.G.; Pereira, L.S.; Raes, D.; Smith, M. *Crop Evapotranspiration: Guidelines for Computing Crop Water Requirements*; Irrigation and Drainage Paper 56; Food and Agriculture Organization of the United Nations (FAO): Rome, Italy, 1998.
40. Valente, F.; David, J.S.; Gash, J.H.C. Modelling interception loss for two sparse eucalypt and pine forests in central Portugal using reformulated Rutter and Gash analytical models. *J. Hydrol.* **1997**, *190*, 141–162. [CrossRef]
41. Oyarzún, R.; Stöckle, C.; Wu, J.; Whiting, M. In field assessment on the relationship between photosynthetic active radiation (PAR) and global solar radiation transmittance through discontinuous canopies. *Chil. J. Agric. Res.* **2011**, *71*, 122–131. [CrossRef]
42. Cuartas, L.A.; Tomasella, J.; Nobre, A.D.; Hodnett, M.G.; Waterloo, M.J.; Múnera, J.C. Interception water-partitioning dynamics for a pristine rainforest in Central Amazonia: Marked differences between normal and dry years. *Agric. For. Meteorol.* **2007**, *145*, 69–83. [CrossRef]
43. Motahari, M.; Attarod, P.; Pypker, T.G.; Etemad, V.; Shirvany, A. Rainfall interception in a Pinus eldarica plantation in a semiarid climate zone: An application of the Gash model. *J. Agric. Sci. Technol.* **2013**, *15*, 981–994.
44. Véliz-Chávez, C.; Mastachi-Loza, C.A.; González-Sosa, E.; Becerril-Piña, R.; Ramos-Salinas, N.M. Canopy storage implications on interception loss modelling. *Am. J. Plant Sci.* **2014**, *5*, 3032–3048. [CrossRef]
45. Klaasen, W.; Bosveld, F.; Water, E. Water storage and evaporation as constituents of rainfall interception. *J. Hydrol.* **1998**, *212–213*, 36–50. [CrossRef]
46. Lopes, D.C.; Steidle Neto, A.J.; Vasco Júnior, R. Comparison of equilibrium models for grain aeration. *J. Stored Prod. Res.* **2015**, *60*, 11–18. [CrossRef]
47. Návar, J. Fitting rainfall interception models to forest ecosystems of Mexico. *J. Hydrol.* **2017**, *548*, 458–470. [CrossRef]
48. Willmott, C.J. On the validation of models. *Phys. Geogr.* **1981**, *2*, 184–194. [CrossRef]
49. Ferreira, D.F. Sisvar: A computer statistical analysis system. *Ciên. Agrotecnol.* **2011**, *35*, 1039–1042. [CrossRef]
50. Jackson, N.A. Measured and modelled rainfall interception loss from an agroforestry system in Kenya. *Agric. For. Meteorol.* **2000**, *100*, 323–336. [CrossRef]
51. Amorim, I.L.D.; Sampaio, E.V.D.S.B.; Araújo, E.D.L. Phenology of woody species in the Caatinga of Seridó, RN, Brazil. *Rev. Árvore* **2009**, *33*, 491–499. [CrossRef]
52. Jiao, L.; Lu, N.; Sun, G.; Ward, E.J.; Fu, B. Biophysical controls on canopy transpiration in a black locust (*Robinia pseudoacacia*) plantation on the semi-arid Loess Plateau, China. *Ecohydrology* **2016**, *9*, 1068–1081. [CrossRef]
53. Liu, Z.; Wang, Y.; Tian, A.; Liu, Y.; Ashley, A.W.; Wang, Y.; Zuo, H.; Pengtao, Y.; Xiong, W.; Xu, L. Characteristics of canopy interception and its simulation with a revised Gash model for a larch plantation in the Liupan Mountains, China. *J. For. Res.* **2018**, *29*, 187–198. [CrossRef]
54. Santos, J.C.N.; Andrade, E.M.; Medeiros, P.H.A.; Guerreiro, M.J.S.; Palácio, H.A.Q. Effect of Rainfall Characteristics on Runoff and Water Erosion for Different Land Uses in a Tropical Semiarid Region. *Water Resour. Manag.* **2017**, *31*, 173–185. [CrossRef]
55. Linhoss, A.C.; Siegert, C.M. A comparison of five forest interception models using global sensitivity and uncertainty analysis. *J. Hydrol.* **2016**, *538*, 109–116. [CrossRef]
56. Sun, X.; Onda, Y.; Kato, H. Incident rainfall partitioning and canopy interception modelling for an abandoned Japanese cypress stand. *J. For. Res.* **2014**, *19*, 317–328. [CrossRef]

Article

Study of the Overflow Transport of the Nordic Sea

Wenqi Shi [1,*], Ning Li [2,*] and Xianqing Lv [3,4]

[1] National Marine Environmental Monitoring Center, Dalian 116023, China
[2] School of Science, Dalian Jiaotong University, Dalian 116028, China
[3] Physical Oceanography Laboratory, CIMST, Ocean University of China, Qingdao 266100, China; xqinglv@ouc.edu.cn
[4] Qingdao National Laboratory for Marine Science and Technology, Qingdao 266100, China
* Correspondence: swqouc@163.com (W.S.); lndjtu@163.com (N.L.); Tel.: +86-1347-842-3671 (W.S.); +86-1584-261-1706 (N.L.)

Abstract: Changes in the climate system over recent decades have had profound impacts on the mean state and variability of ocean circulation, while the Nordic Sea overflow has remained stable in volume transport during the last two decades. The changes of the overflow flux depend on the pressure difference at the depth of the overflow outlet on both sides of the Greenland-Scotland Ridge (GSR). Combining satellite altimeter data and the reanalysis hydrological data, the analysis found that the barotropic pressure difference and baroclinic pressure difference on both sides of the GSR had a good negative correlation from 1993 to 2015. Both are caused by changes in the properties of the upper water, and the total pressure difference has no trend change. The weakening of deep convection can only change the temperature and salt structure of the Nordic Sea, but cannot reduce the mass of the water column. Therefore, the stable pressure difference drives a stable overflow. The overflow water storage in the Nordic Sea is decreasing, which may be caused by the reduction of local overflow water production and the constant overflow flux. When the upper interface of the overflow water body in the Nordic Sea is close to or below the outlet depth, the overflow is likely to greatly slow down or even experience a hiatus in the future, which deserves more attention.

Keywords: Nordic Sea; overflow flux; barotropic pressure; baroclinic pressure

Citation: Shi, W.; Li, N.; Lv, X. Study of the Overflow Transport of the Nordic Sea. *Water* **2021**, *13*, 2675. https://doi.org/10.3390/w13192675

Academic Editors: Andrzej Walega and Tamara Tokarczyk

Received: 21 July 2021
Accepted: 22 September 2021
Published: 27 September 2021

Publisher's Note: MDPI stays neutral with regard to jurisdictional claims in published maps and institutional affiliations.

1. Introduction

As an important driver of thermohaline circulation, the Nordic Sea overflow has a profound impact on environmental changes in the Arctic and even the world. In the Nordic Sea, high-density water bodies with a geopotential density (σ_θ) greater than 27.8 kg/m^3 and shallower than the Greenland-Scotland Ridge (GSR) depth can overflow. There are three overflow channels on the GSR. From west to east, they are the Denmark Strait (DS), the Iceland-Faroe Ridge (IFR), and the Faroe-Shetland Channel (FSC) (Figure 1a,b). The overflow of dense water between Greenland and Shetland consists of the Faroe Bank Channel (FBC) overflow and Wyville Thomson Ridge (WTR) overflow, and FBC is the main channel for FSC overflow. The high-density water overflowing from these channels forms the North Atlantic Deep Water (NADW), which affects the nature of the deep-water mass and the deep circulation in the North Atlantic [1–4].

Theoretical analysis and field measurement results show that the Nordic Sea overflow is hydraulically controlled. In hydraulic control theory, changes of the overflow flux through a strait depend only on the total pressure difference at the depth of the sill on both sides of the GSR [1–4]. The total pressure difference is equal to the barotropic pressure difference plus the baroclinic pressure difference, depending on the local and remote physical processes, such as convection, mixing, and circulation, which further determine the overflow flux of the Nordic Sea [4].

109

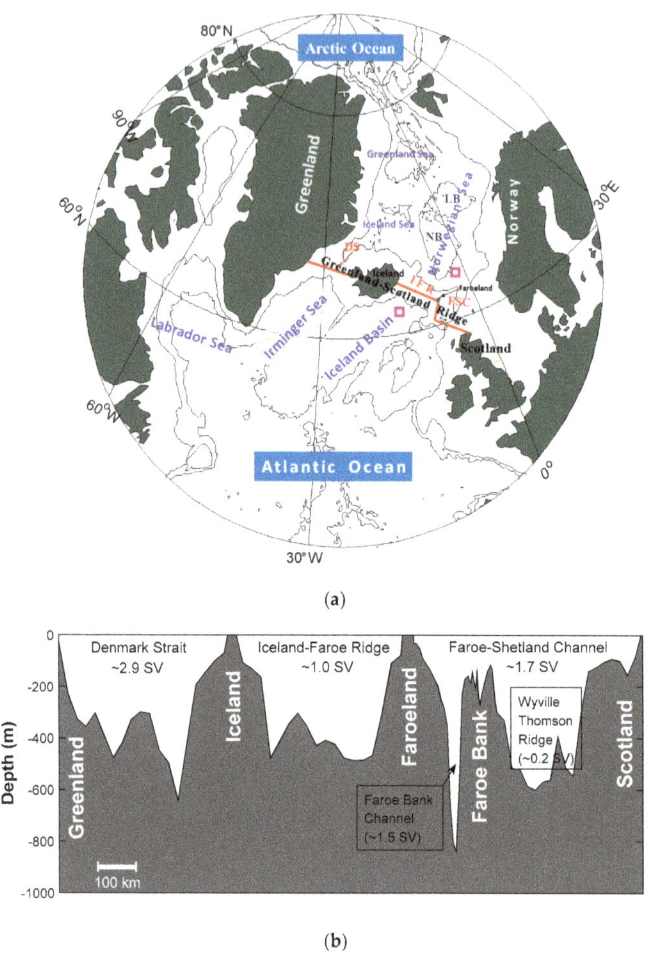

(**a**)

(**b**)

Figure 1. (**a**) Topographic map of the Nordic Sea and (**b**) bottom depth map of the GSR. Note: The Nordic Sea is composed of the Norwegian Sea, the Icelandic Sea, and the Greenland Sea. In the Norwegian Sea, LB stands for the Lofoten Basin and NB stands for the Norwegian Basin; the DS, IFR, and FSC stand for the three main overflow channels in Greenland-Scotland Ridge, namely Denmark Strait, Iceland-Faroe Ridge, and Faroe-Shetland Channel, respectively; the purple square represents the selected point for calculating the pressure difference between the two sides of the GSR; the red line is the selected section for the bottom depth map. Bottom depth of the oceanic part of the GSR and overflow flux are based on Hansen and Østerhus [2].

Affected by climate change, the deep convection in the Nordic Sea has been weakened significantly from the 1960s to the beginning of the 21st century; by about 2006, the depth of deep convection in Greenland was less than 1000 m [5,6]. Recent studies showed that although there is a decreasing trend in atmospheric forcing from 1993 to 2016, the depth of convection in the Greenland Sea in winter has a deepening tendency. This is due to the increase in the salinity of seawater in the upper 1500 m, which results in the weakening of stratification inside the Greenland Sea circulation [7,8]. Modern climate models have found that the overflow flux of the Nordic Sea has a good consistency with the Greenland Sea convection, showing a weakening trend [9–12]. However, this weakening is not reflected by the measured data. The field measurement found that the overflow flux of the Nordic

Sea remained strong and stable from 1995 to 2015, and there was no significant trend change [13,14].

For the phenomenon of stable overflow transport of the Nordic Sea during the last two decades, there have been studies explaining it from different aspects. Based on the model results, Olsen et al. [4] pointed out that the upper interface of the overflow water in the Nordic Sea declined from 1948 to 2005, which would cause a decrease in the pressure difference on both sides of the GSR. However, the rising sea level of the Nordic Sea offsets this effect, resulting in no trend change in the total pressure difference on both sides, making the overflow flux stable. Some other studies showed that the circulation of the Atlantic waters in the Nordic Sea has a greater impact on overflow changes, and the impact of weakened convection has been concealed [15]. Zhang and Thomas [16] believed that the Arctic Ocean, rather than the Greenland Sea, is the northern end of the mean Atlantic Meridional Overturning Circulation (AMOC). They further pointed out that the deep convection of the Labrador Sea and the Greenland Sea contribute the least to the mean AMOC, and AMOC may not be significantly weakened by the closure of the deep convections. However, some other studies still believed that the Greenland Sea is the main source area of the densest overflow water into the North Atlantic after 2005 and is the main ventilation area of the deepest layer in the North Atlantic [7,17]. Other studies suggest that the volume of the dense water above the GSR sill depth in the Nordic Seas is sufficient to supply decades of overflow transport without dense water production [1–3]. The premise in such estimations, however, is that all dense water above the sill depth is freely available for overflow transport. However, basin-scale oceanic circulation is nearly geostrophic and its streamlines are basically the same as the isobaths. The vast majority of the dense water is stored inside the closed geostrophic contours in the deep basin and thus is not freely available for overflow transport [18]. Therefore, an external force or a non-geostrophic mechanism is required to help transport the interior water mass to the boundary current. The numerical simulation results of Yang and Pratt [19] show that 80%–85% of the dense water above the GSR sill depth in the Nordic Seas is not freely available for overflow transport, and the amount of the dense water freely available to overflow accounts for only 15%–20%. Therefore, the Nordic Seas has a relatively small capacity as a dense water reservoir and thus the overflow transport is sensitive to climate changes.

In short, there is still controversy about the reasons for the stable overflow flux in the past two decades. Based on the satellite altimeter data and the reanalysis hydrological data, this paper will analyze the changes in the barotropic pressure and baroclinic pressure on both sides of the GSR and then discuss the reasons for the long-term stable flux of the Nordic Sea overflow by the hydraulic control theory.

The structure of this paper is as follows. Chapter 2 introduces the data and the method for calculating the pressure. Chapter 3 evaluates the credibility of the EN4 data to calculate the pressure by comparing the measured hydrological data and the overflow flux results. Chapter 4 mainly analyzes the spatial distribution of the change trends of the positive pressure, baroclinic pressure, and total pressure on both sides; the change characteristics of the pressure difference on both sides of the GSR; the changes in depth of the overflow water interface in the Nordic Sea; and then analyzes the reasons for stable overflow flux from 1993 to 2015. Chapter 5 mainly analyzes the correlation between the positive pressure and baroclinic pressure on both sides of the GSR and the role of the changes in the properties of the upper seawater. Chapter 6 is the conclusion.

2. Data and Methodology

2.1. Satellite Altimeter Data

The Sea Level Anomaly (SLA) data in this paper is monthly averaged data from 1993 to present of the DUACS 2014 database from the French Space Research Center (CNES), which is merged with multi-satellite altimetry data (Available online: http://www.aviso. oceanobs.com/duacs/ (accessed on 5 May 2016)). The data use Mercator projection with a horizontal resolution of $1/4° \times 1/4°$, and are corrected by atmospheric pressure correction, tide correction, and dry tropospheric correction. The SLA data of DUACS 2014 are based on the Mean Sea Surface (MSS) from 1993 to 2012. Since the mean sea level is the height on a fixed earth reference ellipsoid, the SLA contains sea level change signals caused by the relative crustal movement during this period, mainly as the Glacial Isostatic Adjustment (GIA). Tamisiea and Mitrovica [20] gave the distribution map of the GIA effect on the sea level change measured by the altimeter (their Figure 3b), and their results showed that the GIA has an effect of no more than 0.15 mm/yr on the sea level change trend in the Nordic Sea, so it can be ignored here. Volkov and Pujol [21] verified through field measurement that AVISO's altimeter data can be used to study sea level changes and surface currents in the Nordic Sea.

2.2. Hydrological Data

The hydrological data in this paper are the monthly averaged reanalysis data of the EN4 hydrological data set of the Met Office [22]. The data are obtained from a large amount of observational data, mainly including WOD data (World Ocean Database), GTSPP data (Global Temperature and Salinity Profile Project), Argo data, and ASBO (Arctic Synoptic Basin-wide Observations) data, which have high credibility. The horizontal resolution of the data is $1° \times 1°$, and the coverage area is $1°$ E–$360°$ E and $83°$ S–$89°$ N. The data has 42 vertical layers, and the thickness of the water layer ranges from 10 m in the upper layer to 300 m in the deep ocean.

Since the grid of SLA is inconsistent with the hydrological data grid, the SLA data needs to be interpolated to the same grid point as the hydrological data. The interpolation method is used to obtain the mean value of the 16 SLA grid points in each hydrological grid. If there are more than 8 missing values in a SLA data grid, the mean value at that point is also assigned as missing.

The hydrological observation data of the "Mike" station (Ocean Weather Ship Station Mike, here referred to as OWS-M station) comes from the European Ocean Observatory Network (Euro SITES, Available online: http://www.eurosites.info/stationm.php (accessed on 26 December 2016)). The station is located in the center of the Norwegian Sea ($66°$ N, $02°$ E) and provided long-term ocean and meteorological profile data almost daily from October 1948 to November 2009.

2.3. Method for Calculating Pressure

Based on the hydrostatic assumption, the pressure at a certain depth without considering the atmospheric pressure is:

$$P = P_{trop} + P_{clin} = \rho_0 g \zeta + g \int_z^0 \rho dz \tag{1}$$

where P_{trop} represents the barotropic pressure and P_{clin} is the baroclinic pressure; $\rho_0 = 1028$ kg/m^3 is the surface seawater density; $g = 9.8$ m/s^2 is the gravitational acceleration; ζ is the sea level height, and here is taken as SLA; ρ is the seawater density, which is derived from the temperature and salt data of EN4; z is the calculated pressure depth, and unless otherwise specified, it is taken as 840 m, which is the maximum depth of the GSR sill. Actually, the mean sea surface level is not horizontal, and the spatial difference is huge. However, since this article focuses on the temporal change of pressure rather than the absolute value, taking ζ as the Sea Level Anomaly (SLA) will not affect the final analysis result.

3. Applicability Analysis of EN4 Data

3.1. Comparison with Observations at OWS-M Station

Analysis of the data from the OWS-M station shows that there are enough data for P_{clin} calculations every month above 1000 m depth. The monthly and annual mean results of P_{clin} calculated from the data are shown in Figure 2. Comparing the results of EN4 data with the results of EN4 data, their annual mean change curves have a high degree of overlap. The decline rate of annual mean P_{clin} at the OWS-M station during 1949–2009 was $-0.55 \pm 0.26 \times 10^2$ Pa/dec, while the EN4 data was $-0.3 \pm 0.26 \times 10^2$ Pa/dec, consistently showing a downward trend. Considering the annual mean P_{clin} at the OWS-M station had some missing data, the difference in the decline rate between the two data sets may have been caused by the missing data.

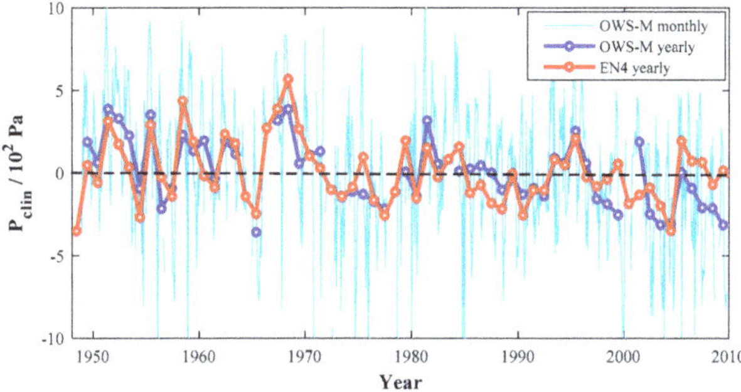

Figure 2. The P_{clin} anomaly from OWS-M data and EN4 data. Note: The OWS-M station data are the daily profile observation results. Firstly, the monthly mean temperature and salinity values at different depths are obtained by averaging, and then the monthly mean P_{clin} (at 840 m depth) is calculated from temperature and salinity; when the cumulative observation level of a month is less than 10 layers or the observation depth is less than 80 data 0 m, the month will be treated as a missing measurement. The annual mean P_{clin} of the OWS-M station is obtained from the monthly data average. A year is regarded as a missing year if there are more than 4 months in the year of missing annual mean data and is not shown.

3.2. Compared with the Observed Overflow Transport

Based on mooring ADCP and temperature and salinity observations, Hansen and Østerhus [3] found that there was no significant trend change in FBC overflow flux from 1995 to 2005, and the trend change did not exceed 0.2 Sv, which is only 10% of the mean flow. Figure 3a–c shows that the SLAs on both sides of the GSR are both increased during this period, while the P_{clin} at the depth of 840 m is decreased at the same time. The final P obtained has no remarkable trend change in the Norwegian Sea or in the Icelandic Sea. Therefore, the trend changes characteristics of the FBC overflow obtained from SLA and EN4 data are more credible.

The measured data show that the trend of DS overflow flux in the 15-year period from 1996 to 2011 is -0.4 Sv. However, the trend is below the 70% confidence level of the *t*-Test, so it is not significant [23]. Here the pressure at the depth of 640 m (approximately the depth of DS) on the north and south sides of the DS increases by the same magnitude, and the pressure difference between the two sides is basically unchanged. The spatial distributions of P_{clin} and P at the depth of 640 m are basically the same as in Figure 2a–c, which is not shown separately here. Thus, the calculation results here are consistent with the observation.

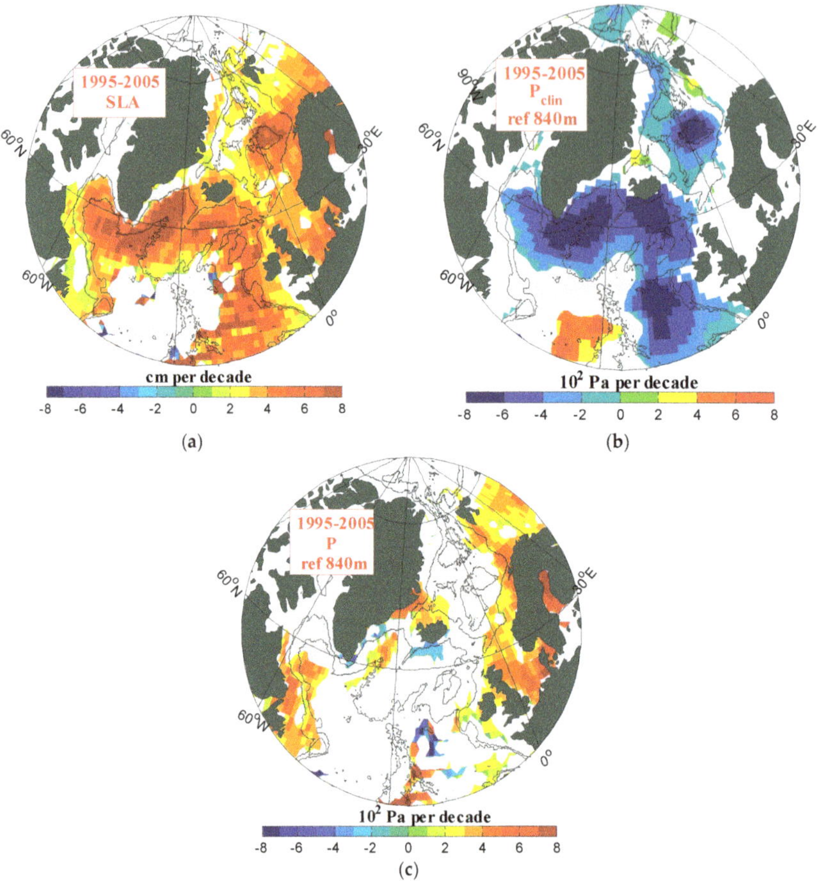

Figure 3. The change trend of SLA (**a**), P_{clin} (**b**), and P (**c**) from 1995 to 2005. The gray area represents the sea area that has not passed the 95% significance test; the white area represents the sea area where the period of missing data is longer than 7 years; the black lines are the 1000 and 3000 m isobaths. The pressure is calculated based on a reference depth of 840 m.

4. Change from 1993 to 2015

In the published literatures, the observation of GSR overflow flux is available until 2015 [14]. Since the Nordic Sea overflow is hydraulically controlled, the pressure difference on both sides of the GSR can be used to analyze the long-term changes of the overflow flux. The depth of the deepest GSR sill is about 840 m on the FBC, which can be used to calculate the pressure difference [4]. As shown in Figure 4, from 1993 to 2015 the SLAs of the Nordic Sea and the North Atlantic subpolar region near the GSR basically increased at the same rate; P_{clin} mostly declined in the south of GSR, increased near DS in the north of GSR, and had no significant change in the south of Norwegian Basin. The pressure difference in the west of Iceland had a clear upward trend; the pressure difference to the east of Iceland was basically unchanged. This means that DS overflow increased, while FBC overflow and IFR overflow did not change much. Therefore, the total overflow in the Nordic Sea slightly increased.

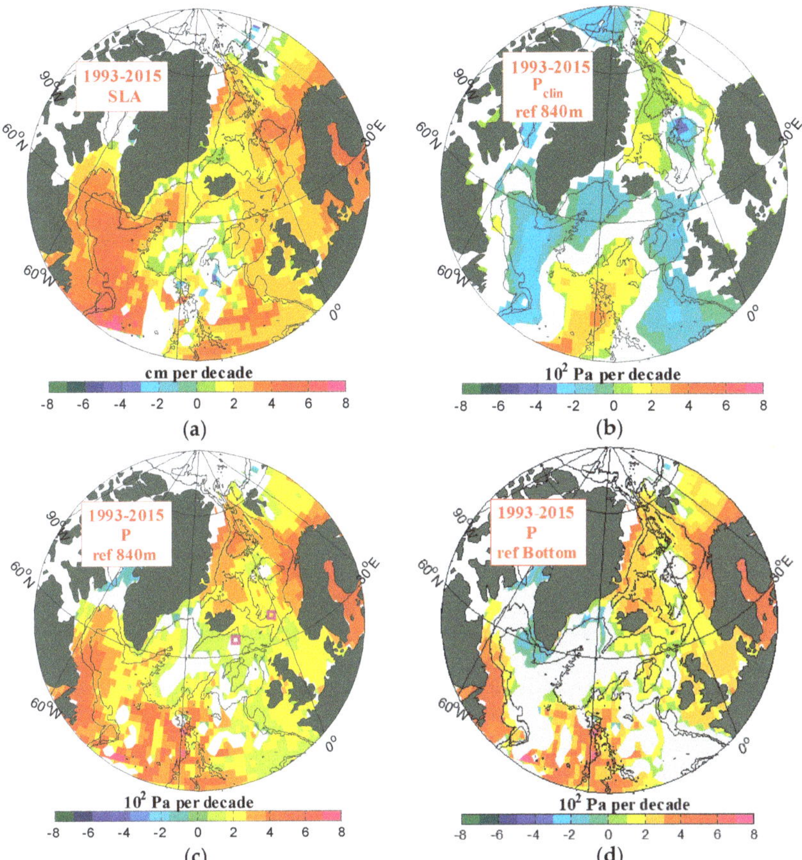

Figure 4. SLA (**a**), Pclin_840 (**b**), P_840 (**c**), P_bottom (**d**) changes trends from 1993 to 2015. The gray area represents the sea area that has not passed the 95% significance test; the white area represents the sea area where the period of missing data is longer than 7 years. The black lines are the 1000 and 3000 m isobaths; the purple square represents the selected point for calculating the pressure difference between the two sides of the GSR in panel (**c**).

In the Nordic Sea, the depth of $\sigma_\Theta = 27.8 \text{ kg/m}^3$ ($D_{27.8}$) (Figure 5a,b, the upper interface of the overflow water) is more consistent with the spatial distribution of the change rate of P_{clin}, indicating that changes in the properties of the upper seawater necessarily indicate the adjustments of the upper interface of the overflow water. Especially in the Nordic Sea, the sinking of $D_{27.8}$ in the Lofoten Basin is about 100 m/dec, which may be directly caused by the reduction of deep convection in the Greenland Sea [5,6] or the weakening of other dense water production. When the total overflow transport flux remains unchanged, the amount of overflow water flowing out of the Lofoten Basin almost remains unchanged. Therefore, the reduction of dense water supply leads to the rapid sinking of the overflow water interface in the Lofoten Basin. There, the rapid decline of P_{clin} and the rapid rise of P_{trop} occur at the same time, while P is basically unchanged, indicating that the above changes are probably caused by the change of physical property of the upper water.

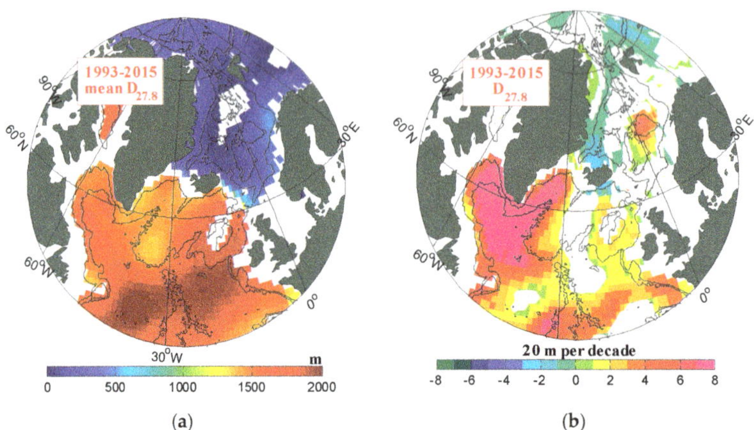

Figure 5. The spatial distribution of the mean value (**a**) and change rate (**b**) of $D_{27.8}$ from 1993 to 2015. The $D_{27.8}$ is the depth of $\sigma_\Theta = 27.8$ kg/m^3.

The Labrador Sea is a fast-sinking area of $D_{27.8}$, with a sinking rate up to 160 m/dec or more, which is consistent with the reported weakening of convection there [24,25]. In the modern climate, the Nordic Sea overflow, entrainment process, and Labrador Sea convection provides about 1/3 of the deep branch of the radial overturning circulation in the Atlantic Ocean [4,26]. Under the condition that the overflow of the Nordic Sea is relatively stable, the convection in the Labrador Sea is significantly weakened, which may be the main reason for the significant weakening of the AMOC near 25° N [24,27]. However, some relatively new observational evidence has indicated that the deep convection of the Labrador Sea has the smallest total contribution to the subpolar overturning circulation [28,29].

The convection of the Labrador Sea is significantly weakened, which causes the upper interface of the dense water to sink quickly. Since the upper interface of the overflow water in the Labrador Sea is deeper than 1500 m, its impact on the P_{clin} at 840 m is small, and the decrease rate of P_{clin} at 840 m depth is only -2×10^2 Pa/dec. The deepening of the overflow water in the Labrador Sea means that warming and freshening of the entire water column causes a large increase in SLA, which is consistent with the calculated results. The greater the bottom depth is, the greater the increase of SLA is. However, there is no significant trend change in the mass of the entire water column from surface to the bottom (Figure 4d).

Although the changes in properties of seawater can ensure the mass conservation of the whole water column, the compression or expansion of the water column caused by the change of properties of seawater will lead to the change of the mass ratio of the upper and lower water column at a certain depth. Therefore, in hydrostatic balance, the pressure change of seawater at a certain depth may be caused by the change of properties of seawater below this depth, and the change of properties of seawater above this depth has no effect on it. The different changing trends of the pressure at 840 m depth and the seabed depth in the Labrador Sea and the Irminge Sea in the south of Greenland (Figure 4b,d) show this effect.

Two points have been selected at the upstream and downstream ends of the FBC to construct the temporal variations of pressure difference. Based on the overflow water sources in different overflow channels and combining the location given by Olsen et al. [4], we selected (64 N, 2 W) and (62 N, 15 W) to estimate the FBC overflow flux (the location is shown in Figure 1). It can be seen from Figure 6 that the inter-annual variation characteristics of ΔP_{trop} and ΔP_{clin} obtained in this paper are very consistent with Figure 2 of Olsen et al. [4]. Both results showed the minimum values of ΔP_{clin} and ΔP in 1995, and the relative maximum values of ΔP_{trop} and ΔP in 2003; from 1993 to 2005, ΔP_{clin} and ΔP were rapidly rising and ΔP_{trop} had no significant change trend. At the same time, the inter-annual variation of ΔP calculated here is about 2×10^2 Pa and about 10% of the mean ΔP, which is basically consistent with the observed inter-annual variation of FBC overflow [3]. In short, the pressure difference between the two points selected in this paper can be used to estimate the FBC overflow flux change. From the spatial distribution map of the SLA trend rate (Figure 4), it can be seen that the trend rate of SLA has a good spatial continuity in the sea areas near both sides of the GSR, and the results would not be significantly changed due to slight difference in the selection of the grid location.

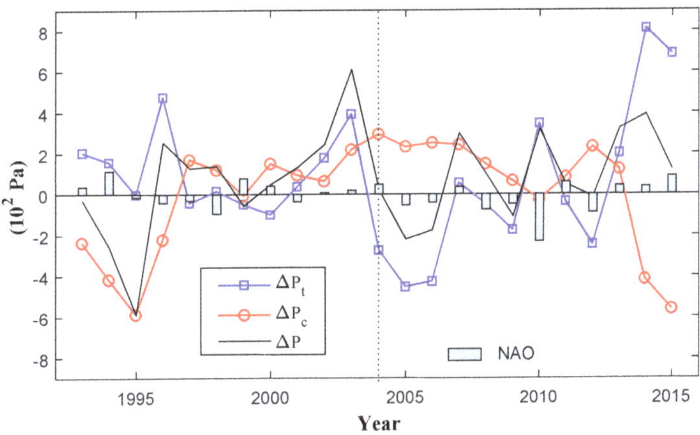

Figure 6. The variations of annual mean of ΔP_{trop}, ΔP_{clin}, and ΔP. The positions of the selected grid points to calculate pressure difference on both sides of the GSR are shown in Figure 1.

Specifically, ΔP_{trop} experienced a slow decline with fluctuation from 1993 to 2005 and reached the minimum value in the past 23 years before 2005. ΔP_{trop} increased with fluctuation from 2005 to 2013 and rose rapidly from 2013 to 2014; after that, it fell back. The year 2014 had the maximum value of ΔP_{trop} in the past 23 years (Figure 6). ΔP_{clin} first decreased slightly in the period of 1993–1997, then increased before 2004, and reached the maximum value in the past 23 years before 2004; then it decreased slowly in the period of 2004–2013, and decreased rapidly in the last two years. ΔP_{clin} in 2015 reached the minimum value in the past 23 years. ΔP was basically at an average level in 1993, followed by a relatively large fluctuation. After experiencing the minimum value in 1995 and the maximum value in 2003, it basically returned to the mean level in 2015. The linear regression of the annual mean ΔP results in a change rate of 1.6×10^2 Pa/dec. Olsen et al. [4] gave a linear coefficient of FBC overflow flux change (Δq) and ΔP of k = 10^{-3} Sv/Pa. Using this linear coefficient, we obtain the FBC overflow enhancement rate of about 0.16 Sv/dec, which is quite small relative to the mean FBC overflow flux (2.9 Sv). At the same time, the linearly increasing trend of ΔP failed the 95% confidence test but passed the 90% confidence test. In fact, ΔP in 2015 was only about 1×10^2 Pa larger than in 1993, which is quite small.

The changes in these three parameters have no significant correspondence with NAO, and most of the wind stress curl changes in the Nordic Sea are related to NAO [30]. This indicates that the interannual sea level changes are not mainly driven by wind stress, but more likely are the result of changes in the properties of the upper seawater.

5. Relationship between Barotropic Pressure and Baroclinic Pressure

Olsen et al. [4] concluded that ΔP_{trop} and ΔP_{clin} on both sides of the FBC have a correlation lag of about three years, and analyzed the mechanism of the correlation as follows: due to wind stress, the sea level difference on both sides of the FBC increases (decreases) and the ΔP on both sides increases (decreases) through the barotropic pressure effect. Then, the overflow transport is enhanced (weakened), causing the iso-density interface in the Norwegian Basin to sink (rise) and the ΔP_{clin} gradually decreases (increases); and then ΔP gradually decreases (increases) until recovers to normal level. This feedback mechanism could help ΔP remain stable, which means the overflow transport is stable. They use a simplified two-layer model to express the mechanism as:

$$P_{trop} = \rho_0 g \Delta h \tag{2}$$

$$\Delta P_{clin} = -g \Delta \rho \Delta D \tag{3}$$

$$\Delta P_{clin} = -g \Delta \rho \Delta D = \frac{-g \Delta \rho k T}{A} \Delta P_{trop} \tag{4}$$

where $\rho_0 = 1.025 \times 10^3$ kg/m^3 is the surface seawater density, g = 9.8 kg/m^3 is the gravitational acceleration, and $\Delta \rho = 0.5$ kg/m^3 is the density difference between overflow water and upper water body. Linear regression coefficient of overflow flux change (Δq) and pressure difference (ΔP) is k = 10^{-3} Sv/Pa. A is the contact area between the overflow layer and the upper layer in the Nordic Sea, or rather the area of the Norwegian Sea deeper than 500 m, which is equal to 5.8×10^{11} m^2 [4]. T is the time for the high-density water interface to sink ΔD after the barotropic pressure disturbance, which is also the time for ΔP to restore to the initial state. The calculated T is approximately equal to three years.

The monthly mean variation of ΔP_{trop}, ΔP_{clin}, and ΔP was constructed based on EN4 and SLA data, and the correlation between ΔP_{trop} and ΔP_{clin} lagging or leading in different months was analyzed (Figure 7). When ΔP_{trop} is about three months ahead of ΔP_{clin}, the negative correlation between them is the largest (-0.59). Olsen et al. [4] defined the horizontal spatial area occupied by overflow water as the seabed deeper than 500 m. However, the dense water in the Norwegian Sea is not freely available for overflow transport, and the dense water in the center of the basin, which occupies most of the area, is circulated by the boundary oceanic circulation. Therefore, the size of the effective overflow area is much smaller than that of the ocean basin. Based on the feedback mechanism of Olsen et al. [4] and the lag time obtained in this article, the horizontal spatial range of available overflow water upstream of the FBC can be estimated to be 0.5×10^{11} m^2, which is about 1/12 of the value given by Olsen et al. [4]. This ratio is close to the percentage of available overflow water in the total overflow water in the Nordic Sea (80%~85%) obtained by other studies [19].

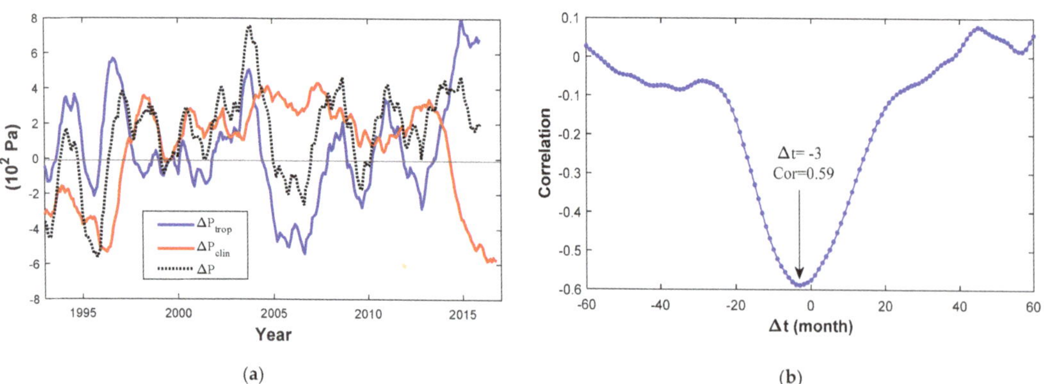

Figure 7. The variation (**a**) and the correlation for different lag time lengths (**b**) of ΔP_{trop} and ΔP_{clin}.

The high P_{clin} correlation with the station in the southern part of the Norwegian Sea (64 N, 2 W) is limited to a small area in the southern part of the Norwegian Sea (the area with a correlation greater than 0.8 in Figure 8). The P_{clin} in this area has a good consistency of change, which can be considered as the available overflow area upstream of the FBC overflow. The area with a correlation greater than 0.8 is about 1.8×10^{11} m^2, and the area with a correlation greater than 0.9 is 0.9×10^{11} m^2. It is more likely that the changes of ΔP_{trop} and ΔP_{clin} are both dominated by changes in seawater properties, so the largest negative correlation between them basically has no lead or lag (Figure 9).

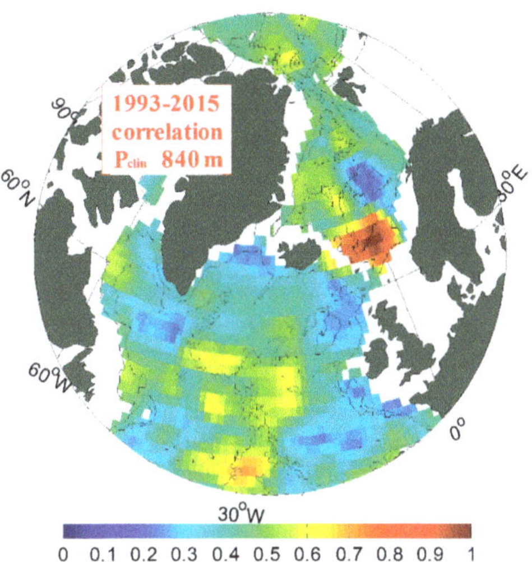

Figure 8. The correlation between P_{clin} and P_{clin} at selected stations in the Norwegian Sea. The time series of the correlation analysis has undergone a 12-month moving average processing. The correlation here is the maximum correlation within 5 years of lead or lag time. The selected stations in the Norwegian Sea are shown in Figure 1 as a purple square. The time series of ΔP_{trop}, ΔP_{clin}, and ΔP are all carried on a 12-month moving mean to remove seasonal fluctuations in this figure. For the sake of comparison, ΔP_{clin} and ΔP are shown as anomalies.

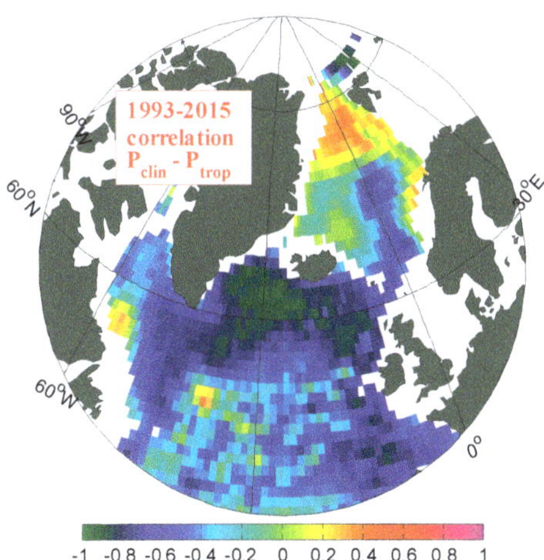

Figure 9. The correlation between P_{clin} and P_{trop}.

Changes in the properties of the upper seawater will cause the reverse change of P_{clin} and P_{trop}, while the total pressure will not change due to the unchanged seawater quality. Therefore, the sea area with a stronger negative correlation between P_{clin} and P_{trop} indicates that changes in the properties of the upper seawater play a greater role in the changes of both there. It can be seen from Figure 9 that there is a strong negative correlation between P_{clin} and P_{trop} in the southern sea area of GSR, the correlation coefficient is close to -1.0, while the total pressure at this place has no trend change characteristics (Figure 4), which shows that the changes of P_{clin} and P_{trop} are mainly caused by the changes in the properties of the upper seawater. In the Nordic Sea north of GSR, this negative correlation is not so strong. Among them, in the Norwegian Sea, P_{clin} and P_{trop} have a certain negative correlation, indicating that the change in the properties of the upper seawater is one of the important factors which cause the changes in the two. There are other processes that lead to the increase in the quality of the upper seawater, which causes a slight increase trend in the total pressure (Figure 4). The negative correlation between P_{clin} and P_{trop} is no longer significant in other areas of the Nordic Sea except the Norwegian Sea. In the Icelandic Sea, the correlation between P_{clin} and P_{trop} is poor and the SLA increases significantly, which leads to a significant increase in the total pressure (Figure 4). There is a weak positive correlation between the two in the Greenland Sea, indicating that the changes of P_{clin} and P_{trop} in this area are mainly affected by other processes.

Under hydrostatic assumption, changes in the density of seawater above 840 m depth will not change the hydrostatic pressure at this depth. To change the pressure at this depth, it needs to change the mass of the water column at this depth. There are two ways. One is to change the absolute mass of the water column, or to change the sea level through wind stress curl, runoff input, sea-air material flux, and other factors. The other is to change the relative mass of the water column by changing the density of the deep layer, causing the column to expand or contract. The mass percentage of the water column above the 840 m depth can change the entire water column.

At present, most ocean numerical models are based on Boussinesq approximation, which cannot reflect sea level changes caused by changes in seawater properties. When the density of the sea layer in Northern Europe decreases, the pressure obtained by simulation decreases, which in turn leads to the weakening of simulated overflow [31]. It can be seen from the results of this paper that the steric effect contributes to most of the sea level trend changes in the sea area surrounding the GSR and has a significant impact on the long-term changes in overflow transport. Therefore, the simulation and prediction of long-term changes in overflow requires the use of non-Boussinesq ocean models, considering the impact of changes in seawater properties on SLA.

6. Conclusions

The Nordic Sea overflow is hydraulically controlled; the changes of the overflow flux depend only on the pressure difference at the depth of the overflow outlet on both sides of the GSR. Based on the satellite altimeter data and the reanalysis hydrological data, we obtained a slight increase in the pressure difference between the two sides of the GSR from 1995 to 2015. However, this trend is not significant and is more consistent with the observed stable overflow flux. Among them, the barotropic pressure and baroclinic pressure in the southern sea area of the GSR have a very good negative correlation (correlation coefficient is close to -1.0). The changes in both are basically caused by the changes in the properties of the upper seawater, and the total pressure there is only a slight increasing trend. The barotropic pressure and baroclinic pressure of the Norwegian Sea in the northern part of the GSR have a certain negative correlation (correlation coefficient is about -0.6), indicating that changes in the properties of the upper seawater are important factors that cause changes in the barotropic and baroclinic pressures in the sea area, and other processes can also lead to a slight increase in the barotropic pressure there. While the correlation between the barotropic pressure and the barotropic pressure in the Icelandic Sea is poor, the barotropic pressure increases significantly which leads to a significant increase in the total pressure there.

By selecting two representative points, the barotropic pressure difference and baroclinic pressure difference on both sides of the FBC are constructed. The changes in the barotropic pressure and baroclinic pressure on both sides of the FBC are more likely caused by the changes in the properties of the local upper seawater. The total pressure difference caused no significant trend changes characteristics between 1993–2015, which is consistent with the observation of stable overflow flux.

In the Nordic Sea, the area with the fastest sinking of the overflow water upper interface is the Lofton Basin, with a sinking speed of more than 100 m/dec, indicating that the storage of overflow water there is rapidly decreasing. The physical processes that produce dense water, such as deep convection in the Greenland Sea, are weakening, and the source of overflow provided is reducing, leading to warming and lightening of the upper layer of the Norwegian Sea and sinking of the upper interface of the overflow water. However, the changes in the properties of the upper seawater in the Norwegian Sea cannot reduce upstream pressure in the depth of the sill to weaken overflow transport. Therefore, it will cause the upper interface of upstream overflow water to further decrease. In the future, when the depth of the overflow water upper interface in the Nordic Sea is less than the depth of the sill on the GSR, the overflow may greatly slow down or even experience a hiatus. This is worthy of close attention and further study.

Author Contributions: Conceptualization, W.S.; methodology, W.S. and N.L.; software, W.S. and X.L.; formal analysis, W.S.; writing—original draft preparation, W.S.; writing—review and editing, N.L.; supervision, X.L.; Funding acquisition, W.S. and X.L. All authors have read and agreed to the published version of the manuscript.

Funding: This research was funded by the National Natural Science Foundation of China, grant number 41806219 and 41806002, the National key research and development program, grant number 2018YFC1407602 and 2019YFC1408405, and the Youth Science and Technology Star Project of Dalian (2020RQ020).

Institutional Review Board Statement: The study did not involve humans or animals.

Informed Consent Statement: We choose to exclude this statement as the study did not involve humans.

Data Availability Statement: The sea level anomaly (SLA) data in this paper are monthly averaged data from 1993 to present of the DUACS 2014 database from the French Space Research Center (CNES), which is merged with multi-satellite altimetry data (http://www.aviso.oceanobs.com/duacs/ accessed on 24 September 2021). The monthly mean reanalysis data of the EN4 hydrological data set are from Met Office Hadley Centre observations data sets (https://www.metoffice.gov.uk/hadobs/en4/ accessed on 24 September 2021).

Conflicts of Interest: The authors declare no conflict of interest.

References

1. Käse, R.H. A Riccati model for Denmark Strait overflow variability. *Geophys. Res. Lett.* **2006**, *33*, 21. [CrossRef]
2. Hansen, B.; Østerhus, S. North Atlantic–nordic seas exchanges. *Prog. Oceanogr.* **2000**, *45*, 109–208. [CrossRef]
3. Hansen, B.; Østerhus, S. Faroe bank channel overflow 1995–2005. *Prog. Oceanogr.* **2007**, *75*, 817–856. [CrossRef]
4. Olsen, S.M.; Hansen, B.; Quadfasel, D.; Østerhus, S. Observed and modelled stability of overflow across the Greenland-Scotland ridge. *Nature* **2008**, *455*, 519–522. [CrossRef]
5. Latarius, K.; Quadfasel, D. Seasonal to inter-annual variability of temperature and salinity in the Greenland Sea Gyre: Heat and freshwater budgets. *Tellus A Dyn. Meteorol. Oceanogr.* **2010**, *62*, 497–515. [CrossRef]
6. Tverberg, V.; Pavlov, V.; Hansen, E.; O'Dwyer, J. Decadal trends in exchange of heat, salt and deep water between the Arctic Ocean and the Nordic Seas. In Proceedings of the ICES Annual Science Conference, 2001, Oslo, Norway, 26–29 September 2001.
7. Brakstad, A.; Våge, K.; Håvik, L.; Moore, G.W.K. Water mass transformation in the Greenland Sea during the period 1986–2016. *J. Phys. Oceanogr.* **2019**, *49*, 121–140. [CrossRef]
8. Lauvset, S.K.; Brakstad, A.; Våge, K.; Olsen, A.; Jeansson, E.; Mork, K.A. Continued warming, salinification and ox-ygenation of the Greenland Sea gyre. *Tellus Dyn. Meteorol. Oceanogr.* **2018**, *70*, 1–9. [CrossRef]
9. Solomon, S. The physical science basis: Contribution of Working Group I to the fourth assessment report of the Intergovernmental Panel on Climate Change. Intergovernmental Panel on Climate Change (IPCC). *Clim. Chang.* **2007**, *2007*, 996.

10. Guemas, V.; Salas-Mélia, D. Simulation of the Atlantic meridional overturning circulation in an atmosphere-ocean global coupled model. Part II: Weakening in a climate change experiment: A feedback mechanism. *Clim. Dyn.* **2008**, *30*, 831–844. [CrossRef]

11. Brodeau, L.; Koenigk, T. Extinction of the northern oceanic deep convection in an ensemble of climate model simulations of the 20th and 21st centuries. *Clim. Dyn.* **2016**, *46*, 2863–2882. [CrossRef]

12. Heuzé, C. North Atlantic deep water formation and AMOC in CMIP5 models. *Ocean Sci.* **2017**, *13*, 609–622. [CrossRef]

13. Chafik, L.; Rossby, T. Volume, heat, and freshwater divergences in the subpolar North Atlantic suggest the Nordic seas as key to the state of the meridional overturning circulation. *Geophys. Res. Lett.* **2019**, *46*, 4799–4808. [CrossRef]

14. Østerhus, S.; Woodgate, R.; Valdimarsson, H. Arctic Mediterranean exchanges: A consistent volume budget and trends in transports from two decades of observations. *Ocean Sci.* **2019**, *15*, 379–399. [CrossRef]

15. Eldevik, T.; Nilsen, J.E.Ø.; Iovino, D.; Olsson, K.A.; Sandø, A.B.; Drange, H. Observed sources and variability of Nordic seas overflow. *Nat. Geosci.* **2009**, *2*, 406–410. [CrossRef]

16. Zhang, R.; Thomas, M. Horizontal circulation across density surfaces contributes substantially to the long-term mean northern Atlantic Meridional Overturning Circulation. *Nat. Commun.* **2021**, *2*, 1–12.

17. Huang, J.; Pickart, R.S.; Huang, R.X.; Lin, P.; Brakstad, A.; Xu, F. Sources and upstream pathways of the densest overflow water in the Nordic Seas. *Nat. Commun.* **2020**, *11*, 1–9. [CrossRef]

18. Nøst, O.A.; Isachsen, P.E. The large-scale time-mean ocean circulation in the Nordic Seas and Arctic Ocean estimated from simple dynamics. *J. Mar. Res.* **2003**, *61*, 175–210. [CrossRef]

19. Yang, J.; Pratt, L.J. On the effective capacity of the dense-water reservoir for the Nordic seas overflow: Some effects of topography and wind stress. *J. Phys. Oceanogr.* **2013**, *43*, 418–431. [CrossRef]

20. Tamisiea, M.E.; Mitrovica, J.X. The moving boundaries of sea level change: Understanding the origins of geographic variability. *Oceanography* **2011**, *24*, 24–39. [CrossRef]

21. Volkov, D.L.; Pujol, M.I. Quality assessment of a satellite altimetry data product in the Nordic, Barents, and Kara sea. *J. Geophys. Res. Oceans* **2012**, *117*, C3. [CrossRef]

22. Good, S.A.; Martin, M.J.; Rayner, N.A. EN4: Quality controlled ocean temperature and salinity profiles and monthly objective analyses with uncertainty estimates. *J. Geophys. Res. Oceans.* **2013**, *118*, 6704–6716. [CrossRef]

23. Jónsson, S.; Valdimarsson, H. Water mass transport variability to the North Icelandic shelf, 1994–2010. *ICES J. Mar. Sci.* **2012**, *69*, 809–815. [CrossRef]

24. Robson, J.; Hodson, D.; Hawkins, E.D.; Sutton, R. Atlantic overturning in decline? *Nat. Geosci.* **2014**, *7*, 2–3. [CrossRef]

25. Yang, Q.; Dixon, T.H.; Myers, P.G.; Bonin, J.; Chambers, D.; Van Den Broeke, M.R.; Mortensen, J. Recent increases in Arctic freshwater flux affects Labrador Sea convection and Atlantic overturning circulation. *Nat. Commun.* **2016**, *7*, 1–8. [CrossRef]

26. Bailey, D.A.; Rhines, P.B.; Häkkinen, S. Formation and pathways of North Atlantic Deep Water in a coupled ice-ocean model of the Arctic-North Atlantic Oceans. *Clim. Dyn.* **2005**, *25*, 497–516. [CrossRef]

27. Srokosz, M.A.; Bryden, H.L. Observing the Atlantic Meridional Overturning Circulation yields a decade of inevitable surprises. *Science* **2015**, *348*, 6241. [CrossRef]

28. Lozier, M.S.; Li, F.; Bacon, S.; Bahr, F.; Bower, A.S.; Cunningham, S.A.; de Jong, M.F.; de Steur, L.; de Young, B.; Fischer, J.; et al. A sea change in our view of overturning in the subpolar North Atlantic. *Science* **2019**, *363*, 516–521. [CrossRef]

29. Petit, T.; Lozier, M.S.; Josey, S.A.; Cunningham, S. Atlantic deep water formation occurs primarily in the iceland basin and irminger sea by local buoyancy forcing. *Geophys. Res. Lett.* **2020**, *47*, e2020GL091028. [CrossRef]

30. Serra, N.; Kaese, R.H.; Kähl, A.; Stammer, D.; Quadfasel, D. On the low-frequency phase relation between the Denmark Strait and the Faroe-Bank Channel overflows. *Tellus Dyn. Meteorol. Oceanogr.* **2010**, *62*, 530–550. [CrossRef]

31. Chen, X.Y.; Qiao, F.L.; Wang, X.H.; Zhao, W.; Yuan, Y.L. On the study on boussinesq approximation and hydrostatic approximation in ocean circulation. *Adv. Mar. Sci.* **2004**, *22*, 91–96.

Article

A Probabilistic Model for Maximum Rainfall Frequency Analysis

Maurycy Ciupak [1], Bogdan Ozga-Zieliński [1], Tamara Tokarczyk [1,*] and Jan Adamowski [2]

[1] Hydrology and Water Resources Engineering Department, Institute of Meteorology and Water Management—National Research Institute, Ul. Podleśna 61, 01-673 Warszawa, Poland; Maurycy.Ciupak@imgw.pl (M.C.); Bogdan.Ozga-Zielinski@imgw.pl (B.O.-Z.)

[2] Department of Bioresource Engineering, McGill University, 21 111 Lakeshore, Sainte-Anne-de-Bellevue, QC H9X3V9, Canada; Jan.Adamowski@mcgill.ca

* Correspondence: Tamara.Tokarczyk@imgw.pl

Abstract: As determining the probability of the exceedance of maximum precipitation over a specified duration is critical to hydrotechnical design, particularly in the context of climate change, a model was developed to perform a frequency analysis of maximum precipitation of a specified duration. The PMAXTP model (Precipitation MAXimum Time (duration) Probability) harbors a pair of computational modules fulfilling different roles: (i) statistical analysis of precipitation series, and (ii) estimation of maximum precipitation for a specified duration and its probability of exceedance. The input data consist of homogeneous 30-element series of precipitation values for 16 different durations: 5, 10, 15, 30, 45, 60, 90, 120, 180, 360, 720, 1080, 1440, 2160, 2880, and 4320 min, obtained through Annual Maximum Precipitation (AMP) and Peaks-Over-Threshold (POT) approaches. The statistical analysis of the precipitation series includes: (i) detecting outliers using the Grubbs-Beck test; (ii) checking for the random variable's independence using the Wald-Wolfowitz test and the Anderson serial correlation coefficient test; (iii) checking the random variable's stationarity using nonparametric tests, e.g., the Kruskal-Wallis test and Spearman rank correlation coefficient test for trends of mean and variance; (iv) identifying the trend of the random variables using correlation and regression analysis, including an evaluation of the form of the trend function; and (v) checking for the internal correlation of the random variables using the Anderson autocorrelation coefficient test. To estimate maximum precipitations of a specified duration and with a specified probability of exceedance, three-parameter theoretical probability distributions were used: a shifted gamma distribution (Pearson type III), a log-normal distribution, a Weibull distribution (Fisher-Tippett type III), a log-gamma distribution, as well as a two-parameter Gumbel distribution. The best distribution was selected by: (i) maximum likelihood estimation of parameters; (ii) tests of the hypothesis of goodness of fit of the theoretical probability distribution function with the empirical distribution using Pearson's χ^2 test; (iii) selection of the best-fitting function within each type according to the criterion of minimum Kolmogorov distance; (iv) selection of the most credible probability distribution function from the set of various types of best-fitting functions according to the Akaike information criterion; and (v) verification of the most credible function using single-dimensional tests of goodness of fit: the Kolmogorov-Smirnov test, the Anderson-Darling test, the Liao-Shimokawa test, and Kuiper's test. The PMAXTP model was tested on data from two meteorological stations in northern Poland (Chojnice and Bialystok) drawn from a digital database of high-resolution precipitation records for the period of 1986 to 2015, available for 100 stations in Poland (i.e., the Polish Atlas of Rainfall Intensities (PANDa)). Values of maximum precipitation with a specified probability of exceedance obtained from the PMAXTP model were compared with values obtained from the probabilistic Bogdanowicz-Stachý model. The comparative analysis was based on the standard error of fit, graphs of the density function for the probability of exceedance, and estimated quantile errors. The errors of fit were lower for the PMAXTP compared to the Bogdanowicz-Stachý model. For both stations, the smallest errors were obtained for the quantiles determined on the basis of maximum precipitation POT using PMAXTP.

Citation: Ciupak, M.; Ozga-Zieliński, B.; Tokarczyk, T.; Adamowski, J. A Probabilistic Model for Maximum Rainfall Frequency Analysis. *Water* **2021**, *13*, 2688. https://doi.org/10.3390/w13192688

Academic Editor: Renato Morbidelli

Received: 1 September 2021
Accepted: 22 September 2021
Published: 28 September 2021

Publisher's Note: MDPI stays neutral with regard to jurisdictional claims in published maps and institutional affiliations.

Keywords: annual maximum precipitation; peaks-over-threshold methods; statistical analysis; maximum precipitation frequency analysis; gamma; Weibull; log-gamma; log-normal; Gumbel distributions; nonparametric tests

1. Introduction

A frequency analysis of values of maximum precipitation of a specified duration and probability of exceedance is an essential part of engineering [1]. Given the significant impact of maximum precipitation on various spheres of human activity (e.g., the economy, agriculture, industry, and the environment), such an analysis is widely applied, particularly in the context of observed climate change [2,3].

A widely used tool in the statistical description of rare meteorological (climatic) events is the extreme value theorem (EVT). Two probability distributions are used when employing the EVT: the generalized extreme value distribution (GEV) and the generalized Pareto distribution (GPD) [4,5]. Encompassing three families of distributions (Gumbel (G), Fréchet (F), and Weibull (WE)), the GEV distribution offers the advantage of high accuracy of fit to observed precipitation data [6]. Commonly used methods for the estimation of the unknown parameters of theoretical probability distributions include: maximum likelihood, L-moments, and the Bayesian method [7–9]. Ragulina and Reitan [10] proposed a Bayesian hierarchical model approach to the selection of a GEV distribution, where Bayesian inference was applied both to parameter estimation and model selection. For most locations in Japan investigated by Yuan et al. [11], a log-Pearson type 3 distribution (LGA) proved to be the best-fitting theoretical probability distribution for annual maximum hourly precipitation data. Młyński et al. [12] found that among the G, GA, WE, log-normal (LN), and GEV distributions, the latter best described annual maximum daily precipitation in Poland's upper Vistula basin.

An assumption of the EVT is that the random variables subjected to analysis show stationarity, i.e., the statistical properties of the mechanism generating these variables remain unchanged over time. Such conditions are rarely encountered in nature, and extreme events are increasingly of a nonstationary nature. In the case of maximum precipitation, its natural variation is overlaid by changes in climate and human intervention in land use (e.g., reduction in soil drainage). In this situation, time series of maximum precipitation values exhibit non-stationarity in the form of long-term trends and/or periodic fluctuations. In recent years, it has become increasingly common to analyze the frequency of nonstationary phenomena using the theory of nonstationary extreme value (NSEV). Katz et al. [13] extended the traditional approach to a frequency analysis to deal with nonstationary cases, where it is assumed that there is a constant probability of the occurrence of an extreme event with values that vary with time. Likewise, Adlouni et al. [14] developed a method for estimating a GEV distribution under nonstationary conditions. Parameters of the distribution were estimated by the maximum likelihood method (MLM), and the covariance of the observed variables was included in the parameters of the probability distribution.

Another approach, used in engineering practice for estimating values of maximum precipitation with a specified duration and probability of exceedance, is regionalization. In Poland, Bogdanowicz and Stachý [15,16] used a clustering procedure for a series of annual maximum precipitation values to distinguish three precipitation regions. In these regions, annual maximum values were described using a WE extreme value distribution. Satisfying the assumptions of independence, stationarity, and identity of probability distribution, Shahzadi et al. [17] used a regional analysis of flooding frequency and a Monte Carlo method to divide the territory of Pakistan into three homogeneous subregions. The estimation of parameters followed the L-moments method, while quantile estimation was carried out using GA, GEV, GPA, generalized normal (GNO), and generalized logistic (GLO) distributions.

Quantiles of an extreme value distribution are usually estimated directly from a random sample of annual maximum precipitation (AMP) values. In view of the shortness of the time series, alternative solutions were used, thereby enabling statistical inference to be carried out based on a broader set of information than the annual maxima. Examples include analyses of seasonal maxima and models of annual maxima with different seasonal variances. In these models, the probabilistic description is usually based on mixed distributions. Earlier research on mixed distributions assumed the same probability density function for the distinguished seasons (homogeneous mixed distributions). An example of this approach is the two-population general extreme value distribution (TPGEV), based on the assumption of GEV-GEV distributions [18], gamma-gamma distributions (GA-GA), and log-normal-log-normal distributions (LN-LN) [19,20]. However, hydrometeorological variables are composed of different types of probability density functions.

Numerous studies on non-homogeneous mixed distributions have led to an improvement of the characteristics of the analyzed variables through the use of two-component models, such as the mixed gamma-Gumbel distribution (GA-G) [21] or the two-component generalized extreme value distribution (TCGEV) composed of a GEV and a Gumbel (G) distribution [22]. A GA-GP mixed distribution, incorporating a gamma distribution [23] and generalized Pareto distribution (GP), is commonly used. It serves mainly to model meteorological situations featuring both dry and wet periods. Another approach to the frequency analysis of maximum precipitation is the determination of the relationship between the intensity of precipitation and duration, and between duration and frequency of occurrence. For the modeling of two-dimensional dependences, the use of copula functions is recommended as a method of estimation of a two-dimensional distribution function [24,25]. In recent years, analyses have been made of a multidimensional dependence structure of extreme precipitation event variables using vine copula functions. The method involves the step-by-step mixing of two-dimensional copulas, which leads to a simplification of the estimation of multidimensional distribution functions [26].

Although there have been many attempts at using models for nonstationary series of extreme events [27–34], engineering practice shows that the assumption of the stationarity of time series is still widely adopted.

The purpose of this paper is to present the PMAXTP model for a frequency analysis of maximum precipitation with a specified duration and probability of exceedance, together with the results of testing the model against data from two meteorological stations located in northern Poland: Chojnice and Białystok. Values of maximum precipitation with a specified duration and probability of exceedance were estimated for two time series: (i) a 30-year series of annual maximum precipitation (AMP) values from the period 1986–2015 and (ii) a 30-element series of maximum precipitation values from the period 1986–2015 obtained by means of peaks-over-threshold (POT) analysis. The 30 highest values from the obtained set were used for further analyses. Computations were performed for 16 different durations: 5, 10, 15, 30, 45, 60, 90, 120, 180, 360, 720, 1080, 1440, 2160, 2880, and 4320 min. The results given by the PMAXTP model were compared with those obtained with the probabilistic Bogdanowicz-Stachý model of maximum precipitation [15,16], which is in common use in Polish engineering practice.

2. Problem Formulation and Methodology

The PMAXTP model for a frequency analysis of maximum precipitation with a specified duration and probability of exceedance was developed with the use of the method of alternative events (MAE), which serves to compute annual maximum flows with a specified probability of exceedance [35]. The overall scheme of the PMAXTP model is shown in Figure 1. The model contains two computational modules, one that performs a statistical analysis of series of precipitation data, and another that estimates maximum precipitation with a given duration and probability of exceedance. The latter includes an estimation of parameters of the distributions by the maximum likelihood method, verification of goodness of fit by Pearson's χ^2 test, selection of the best-fitting probability distribution function within each distribution type according to the criterion of minimum Kolmogorov distance, selection of the most credible function according to the Akaike information criterion (AIC), and determination of the quantile confidence interval with regard to the randomness of the series of observations. The results returned by the PMAXTP model are values of maximum precipitation with a specified duration τ (min) $\in$ {5, 10, 15, 30, 45, 60, 90, 120, 180, 360, 720, 1080, 1440, 2160, 2880, 4320} and a given probability of exceedance p (%) $\in$ {99.9, 99.5, 99, 98.5, 98, 95, 90, 80, 70, 60, 50, 40, 30, 20, 10, 5, 3, 2, 1, 0.5, 0.3, 0.2, 0.1, 0.05, 0.03, 0.02, 0.01}.

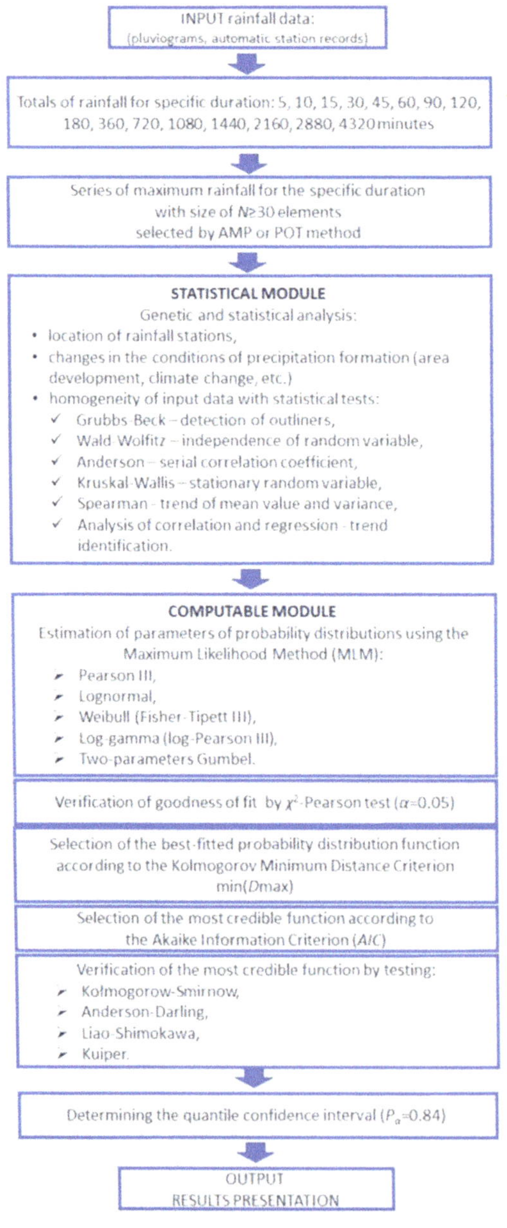

Figure 1. Overall scheme of the PMAXTP model.

An analysis of the homogeneity of the random variables of series of maximum precipitation with different durations was performed by genetic (physical) methods and by statistical methods [35,36]. The identification of the trend of the analyzed random variables and evaluation of the form of the trend function were carried out by correlation and regression analysis, where the dependent variable is the maximum precipitation selected by the AMP or POT method, and the independent variable is the time (τ). The correlation was analyzed using the nonparametric Spearman rank correlation test [37] and the parametric Pearson linear correlation coefficient test [38]. In regression analysis, the global Fisher-Snedecor F-test [39] tests three equivalent null hypotheses: the significance of the slope, the significance of the coefficient of determination, and the significance of the linear relationship between the analyzed variables. Verification is performed for the null hypothesis that the independent variable (time τ) has no effect on the analyzed dependent variable, which here is the maximum precipitation (P_τ^{AMP} and P_τ^{POT}). An evaluation of the form of the trend function is performed using scatter plots of the analyzed random variables with respect to time (τ). These provide a visual assessment and an evaluation of the form of the trend function: linear, power, exponential, etc.

The internal correlation of the analyzed random variable was checked using the Anderson autocorrelation coefficient test [40]. This analysis identifies the occurrence of periodic fluctuations and their effect on the variation of the analyzed variables. The results are presented numerically and graphically for a specified lag, with an indication of the autocorrelation coefficients and an evaluation of white noise (standard error) for the confidence level assumed (α).

The computation of the maximum precipitation with a specified probability of exceedance is performed using probabilistic models of the properties of the random variables P_τ^{AMP} and P_τ^{POT}. An analysis of the properties of random maximum precipitations served as the basis for the acceptance of potential probability distribution models: e.g., G, GA, LN, log-gamma (LGA), and WE. The first four models are three-parameter distributions with the following parameters: α ($\alpha > 0$), λ ($\lambda > 0$) or μ ($\mu > 0$), and ε ($\varepsilon \leq x \leq +\infty$), representing, respectively, the parameters of scale, shape, and position, i.e., the lower (left-hand) limit of the probability distribution (see details in Appendix A).

The PMAXTP model assumes that each type of distribution is represented by a family of functions $f_i(x)$, shifted with respect to each other, each of which has a certain fixed lower limit (ε_i) satisfying $0 \leq \varepsilon_i < \min\limits_{1 \leq j \leq n} (x_j)$, where n is the size of the random sample. The value of ε_i may take values ranging from 0 up to the minimum value of the variable (X) in the random sample ($x_1, x_2, \cdots, x_n$). Hence, the lower limit (ε_i) of the ith specific function in the family of a selected type of distributions is the discriminant of that function within the family, and is not subject to estimation. In the G distribution, described by Equations (A9) and (A10) in Appendix A, only two parameters appear: the scale α and the shape μ.

The parameters of probability density functions were estimated by the MLM using dedicated software [41]. The procedure was as follows:

(i)   Estimation of parameters of four types of functions belonging to the probability distribution families GA, WE, LGA, and LN for a fixed value and range of variation of the distribution lower limit ε_i for the ith function belonging to the family of the selected probability distribution. In the case of the G distribution, the parameters are estimated for a single function; there is no distribution lower limit (ε).

(ii)   Obtainment of i sets of estimated values of parameters for each selected probability distribution function by the solution of systems of equations according to explicit formulas, or the determination of a set of parameter values using Brent's or Newton's numerical methods [42].

(iii)   Check of the goodness of fit of the selected theoretical distribution with the empirical distribution using Pearson's χ^2 test [43] at a significance level $\alpha = 0.05$.

(iv)   Formation of a set of noncontradictory probability distribution functions from all probability distribution functions for which the hypothesis of goodness of fit was not rejected. Sets of noncontradictory functions are formed separately for each selected probability distribution function type: GA, WE, LGA, and LN.

(v)   Selection of the best-fitting function within each distribution type. For each theoretical distribution type used, there may exist many noncontradictory functions with different lower limit values ε_i. A single function is selected for each distribution type (GA, WE, LGA, LN) according to the criterion of minimum Kolmogorov distance, $\min(D_{max})$ [35,44]. The probability distribution function for which, within a given distribution type, the Kolmogorov distance $D_{\max}$ attains its minimum value is called the best-fitting function in the sense of the Kolmogorov distance criterion. These single functions, identified for each of the distribution types used, form the set of best-fitting functions.

(vi)   Selection of the most credible probability distribution function from the set of best-fitting functions of particular types (GA, WE, LGA, LN, G), performed by computing the value of the Akaike information criterion (AIC) [45] for each of those functions. The most credible function is taken to be the function with the smallest AIC value.

(vii)   Verification of the most credible distribution of maximum precipitation values, P_τ^{AMP} and P_τ^{POT}, was based on nonparametric tests used to analyze the goodness-of-fit of a theoretical mathematical model to an empirical model. The verification of the distributions was concentrated on their tail part. The tails of the distributions are significant in terms of the occurrence of extreme values of the random variable, that is, values with a very low probability of exceedance. Thus, to evaluate the goodness-of-fit of the distributions, the following single-dimensional statistical tests were used: the Kolmogorov-Smirnov test ($D_{\mathrm{K\text{-}S}}$) [46,47], the Anderson-Darling test ($D_{\mathrm{A\text{-}D}}$) [48], the Liao-Shimokawa test ($D_{\mathrm{L\text{-}S}}$) [49], and Kuiper's test (D_{K}) [50]. (For details, see Appendix B.) The $D_{\mathrm{K\text{-}S}}$ test may be used for the verification of large deviations of a theoretical cumulative probability distribution from the empirical distribution. The $D_{\mathrm{A\text{-}D}}$ test is sensitive to deviations in the tail part, while the $D_{\mathrm{L\text{-}S}}$ test represents a weighted mean distance between the theoretical and empirical probability distributions in the whole range of the analyzed random variable, and is regarded as the most suitable for verification of the Gumbel and Weibull distributions [49]. The D_{K} test was used to verify the goodness-of-fit of the distribution in its central part, as well as in the lower and upper parts of the tail of the distribution.

(viii) Selection of a probabilistic model, performed by comparing the estimated quantile errors resulting from the randomness of the sample of maximum precipitations with a specified duration τ selected by the AMP and POT methods (P_τ^{AMP} and P_τ^{POT}).

3. Study Area and Data

The PMAXTP model was tested on data from two meteorological stations located in Poland: Chojnice and Bialystok (Figure 2, black hexagons). The choice of stations was based on the availability of long series of historical data and current meteorological observations.

Data were drawn from the Rain-Brain database, created under the Development and Implementation of a Polish Atlas of Rainfall Intensities (PANDa) project [51] carried out in 2016 and 2017 by Poland's Institute of Meteorology and Water Management—National Research Institute (IMGW—PIB). Under the PANDa project, a series of depths of precipitation having specific durations were subjected to qualitative assessment, including a comparison of digital records with analog data (from Hellmann rain gauges), and information was drawn from a system of ground-based radars operating in the measurement and observation network of the IMGW—PIB. The observations were verified with respect to the occurrence of meteorological configurations which might cause rainfall of a given quantity in specified pressure conditions, characteristic of the analyzed region.

The study was based on the 30 highest precipitation depth values for 16 specified durations, $\tau = \{5, 10, 15, 30, 45, 60, 90, 120, 180, 360, 720, 1080, 1440, 2160, 2880, 4320\}$ (minutes) for the two precipitation stations mentioned above.

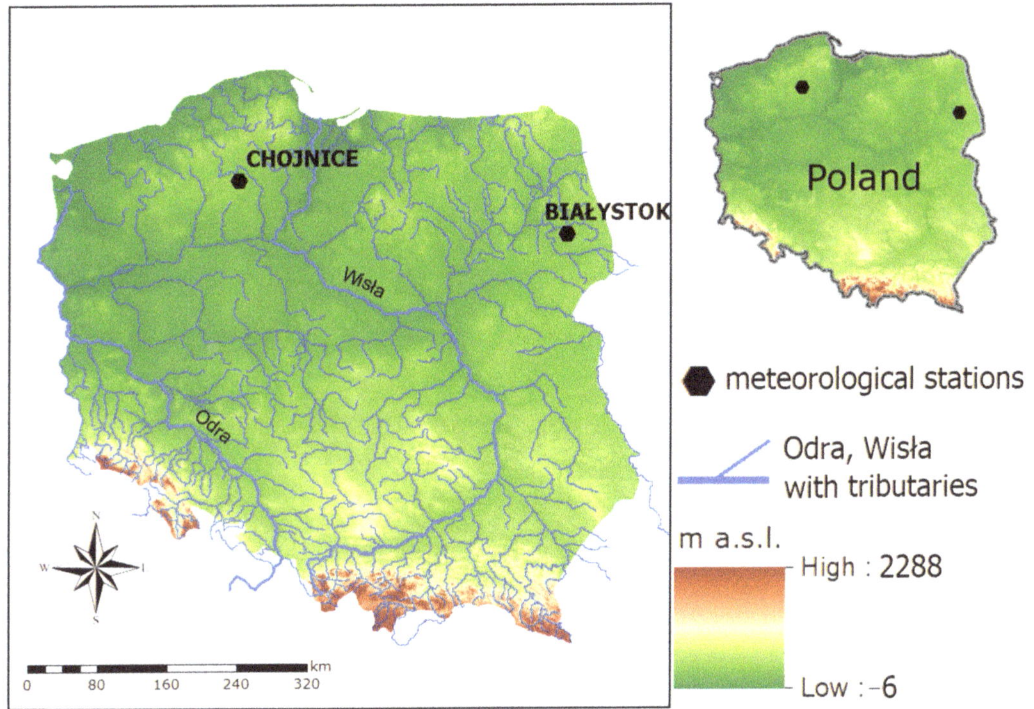

Figure 2. Location of the Chojnice and Bialystok meteorological stations in Poland.

Two methods were used to select maximum precipitation values: AMP [1,2,52] and POT [53]. Under the AMP method, a single maximum precipitation value was selected for the year, independent of its duration. A defect of the AMP method is that it fails to take into account all the high precipitation depth values occurring in a given year. In the POT method, it is possible to take into account all high precipitation depth values in a given year, i.e., the method selects these values that exceed a threshold determined a priori. The analyses were based on events with values not less than $P_{\min,\tau}^{POT} = 3.5\tau^{0.275}$ [51]. Thus, threshold values P_{τ}^{MAX} (mm) were set for precipitation with specified durations (τ), as given in Table 1 [51]. The subsequent analyses used 30-element series of maximum precipitation data, selected by both methods.

Table 1. Minimum quantity of precipitation $P_{\min,\tau}^{POT}$ (mm) taken as a threshold in the POT method.

τ (min)	5	10	15	30	45	60	90	120	180	360	720	1080	1440	2160	2880	4320
$P_{\min,\tau}^{POT}$ (mm)	5.4	6.6	7.4	8.9	10.0	10.8	12.1	13.1	14.6	17.7	21.4	23.9	25.9	28.9	31.3	35.0

4. Results and Discussion

4.1. Results of Analysis of Homogeneity for the PMAXTP Model

An analysis was made of the genetic, time, and measurement homogeneity of the precipitation series from the stations in Chojnice and Bialystok. Based on a visual assessment of the measurement series and information contained in IMGW—PIB reports (Meteorological Yearbooks and Precipitation Yearbooks Report [51]), no significant factors were found that might have an impact on the genetic homogeneity of the series of maximum precipitation values observed in the years 1986–2015.

An analysis was made of the statistical properties of the series of precipitation measurements from Chojnice and Bialystok using nonparametric significance tests [35,36]. The results are presented in Tables 2–6. Tables 2 and 3 contain the results of outlier detection using the Grubbs-Beck test [54,55], checking for the independence of the analyzed random variable using the Wald-Wolfowitz test (Test of Series) and Anderson serial autocorrelation coefficient test [40,55,56], and checking the stationarity of the analyzed random variable using the Kruskal-Wallis test and Spearman rank correlation coefficient test for the trends of mean and variance [57,58]. The final column of Tables 2 and 3 indicates genetically and statistically homogeneous series of maximum precipitation data selected by the AMP and POT methods.

In the case of P_τ^{AMP}, the Grubbs-Beck test detected outliers for precipitation with the duration $\tau = 360$ and $\tau = 720$ min, at both the Chojnice station (Table 2) and the Bialystok station (Table 3). In Tables 2 and 3, for a positive test result (+), the number of the outlier in the chronological sequence and the quantity of precipitation are also given. For the P_τ^{POT} series at Chojnice (Table 2), outliers were detected for $\tau \in \{15, 30\}$ and $\tau \in \{120, \ldots, 4320\}$ min, while at Bialystok (Table 3), outliers were detected for $\tau \in \{5, \ldots, 15\}$, $\tau \in \{60, \ldots, 360\}$ and $\tau \in \{2160, \ldots, 4320\}$ min. Based on the theorem developed by Neyman and Scott [59] stating that the families of LN, G, and WE distributions—these being the distributions assumed as potential models describing the maximum precipitation values—are entirely susceptible to the occurrence of outliers in a random sample, it was concluded that the occurrence of the detected outliers should be considered entirely natural, and such elements were not removed from the measurement series.

Table 2. Results of nonhomogeneity analysis of AMP and POT precipitation series from Chojnice meteorological station; $(-)/(+)$ denotes, respectively, negative and positive test results; $\sqrt{}$—denotes homogenous series.

τ (min)	Grubbs-Beck Test ±Outliers (mm)		Test of Series		Kruskal-Wallis Test		Spearman Rank Correlation Test for Trend of Mean		for Trend of Variance		Homogeneity of Precipitation P_τ^{MAX}	
	AMP	POT	AMP	POT	AMP	POT	AMP	POT	AMP	POT	AMP	POT
5	$(-)$	$(-)$	$(-)$	$(-)$	$(-)$	$(-)$	$(-)$	$(-)$	$(-)$	$(-)$	$\sqrt{}$	$\sqrt{}$
10	$(-)$	$(-)$	$(-)$	$(-)$	$(-)$	$(-)$	$(-)$	$(-)$	$(-)$	$(-)$		
15	$(-)$	$(+) [5] = 24.5$	$(-)$	$(-)$	$(+)$	$(-)$	$(+)$	$(-)$	$(+)$	$(-)$		$\sqrt{}$
30	$(-)$	$(+) [4] = 33.7$	$(-)$	$(-)$	$(+)$	$(-)$	$(+)$	$(-)$	$(+)$	$(-)$		$\sqrt{}$
45	$(-)$	$(-)$	$(-)$	$(-)$	$(+)$	$(-)$	$(+)$	$(-)$	$(+)$	$(+)$		
60	$(-)$	$(-)$	$(-)$	$(-)$	$(+)$	$(-)$	$(+)$	$(-)$	$(-)$	$(+)$		
90	$(-)$	$(-)$	$(-)$	$(-)$	$(+)$	$(-)$	$(+)$	$(-)$	$(-)$	$(-)$		$\sqrt{}$
120	$(-)$	$(+) [19] = 42.9$	$(-)$	$(-)$	$(+)$	$(-)$	$(+)$	$(-)$	$(-)$	$(-)$		$\sqrt{}$
180	$(-)$	$(+) [19] = 48.4$	$(-)$	$(-)$	$(+)$	$(-)$	$(+)$	$(-)$	$(+)$	$(-)$		$\sqrt{}$
360	$(+) [25] = 60.3$	$(+) [24] = 60.3$	$(-)$	$(-)$	$(+)$	$(-)$	$(+)$	$(-)$	$(+)$	$(-)$		$\sqrt{}$
720	$(+) [4] = 11.8$ $[25] = 67.7$	$(+) [24] = 67.6$	$(-)$	$(-)$	$(-)$	$(-)$	$(-)$	$(-)$	$(-)$	$(-)$	$\sqrt{}$	$\sqrt{}$
1080	$(+) [4] = 11.8$	$(+) [24] = 71.9$	$(-)$	$(-)$	$(-)$	$(-)$	$(-)$	$(-)$	$(-)$	$(-)$	$\sqrt{}$	$\sqrt{}$
1440	$(-)$	$(+) [25] = 71.9$	$(-)$	$(-)$	$(-)$	$(-)$	$(-)$	$(-)$	$(-)$	$(-)$	$\sqrt{}$	$\sqrt{}$
2160	$(-)$	$(+) [20] = 80.5$	$(-)$	$(-)$	$(-)$	$(-)$	$(-)$	$(-)$	$(-)$	$(-)$	$\sqrt{}$	$\sqrt{}$
2880	$(-)$	$(+) [22] = 87.2$	$(-)$	$(-)$	$(-)$	$(-)$	$(-)$	$(-)$	$(-)$	$(-)$	$\sqrt{}$	$\sqrt{}$
4320	$(-)$	$(+) [21] = 87.9$	$(-)$	$(-)$	$(-)$	$(-)$	$(-)$	$(-)$	$(-)$	$(-)$	$\sqrt{}$	$\sqrt{}$

Table 3. Results of nonhomogeneity analysis of AMP and POT precipitation series from Bialystok meteorological station; $(-)/(+)$ denotes, respectively, negative and positive test results; $\sqrt{}$—denotes homogenous series.

τ (min)	Grubbs-Beck Test $\pm$Outliers (mm)		Test of Series		Kruskal-Wallis Test		Spearman Rank Correlation Test				Homogeneity of Precipitation P_τ^{MAX}	
							for Trend of Mean		for Trend of Variance			
	AMP	POT	AMP	POT	AMP	POT	AMP	POT	AMP	POT	AMP	POT
5	$(-)$	$(+)$ [15] = 15.5	$(-)$	$(-)$	$(+)$	$(-)$	$(+)$	$(-)$	$(-)$	$(-)$		$\sqrt{}$
10	$(-)$	$(+)$ [15] = 22.3	$(-)$	$(-)$	$(+)$	$(-)$	$(+)$	$(-)$	$(-)$	$(-)$		$\sqrt{}$
15	$(-)$	$(+)$ [17] = 24.6	$(-)$	$(-)$	$(+)$	$(+)$	$(+)$	$(-)$	$(-)$	$(+)$		
30	$(-)$	$(-)$	$(-)$	$(-)$	$(+)$	$(+)$	$(+)$	$(-)$	$(-)$	$(+)$		
45	$(-)$	$(-)$	$(-)$	$(-)$	$(+)$	$(-)$	$(+)$	$(-)$	$(-)$	$(-)$		$\sqrt{}$
60	$(-)$	$(-)$	$(-)$	$(-)$	$(+)$	$(-)$	$(+)$	$(-)$	$(-)$	$(-)$		$\sqrt{}$
90	$(-)$	$(+)$ [22] = 42.0	$(-)$	$(-)$	$(+)$	$(-)$	$(+)$	$(-)$	$(-)$	$(-)$		$\sqrt{}$
120	$(-)$	$(+)$ [23] = 47.7	$(-)$	$(-)$	$(+)$	$(-)$	$(+)$	$(-)$	$(-)$	$(+)$		
180	$(-)$	$(+)$ [23] = 52.2	$(-)$	$(-)$	$(+)$	$(-)$	$(+)$	$(-)$	$(-)$	$(-)$		$\sqrt{}$
360	$(+)$ [4] = 10.89 [25] = 67.70	$(+)$ [23] = 67.7	$(-)$	$(-)$	$(+)$	$(+)$	$(+)$	$(+)$	$(-)$	$(-)$		
720	$(+)$ [25] = 73.90	$(+)$ [21] = 73.9	$(-)$	$(-)$	$(+)$	$(-)$	$(+)$	$(-)$	$(-)$	$(-)$		$\sqrt{}$
1080	$(-)$	$(+)$ [20] = 79.6	$(-)$	$(-)$	$(+)$	$(-)$	$(+)$	$(-)$	$(-)$	$(-)$		$\sqrt{}$
1440	$(-)$	$(+)$ [23] = 84.50	$(-)$	$(-)$	$(+)$	$(+)$	$(+)$	$(+)$	$(-)$	$(-)$		
2160	$(-)$	$(+)$ [21] = 101.30	$(-)$	$(-)$	$(+)$	$(-)$	$(+)$	$(-)$	$(-)$	$(-)$		$\sqrt{}$
2880	$(-)$	$(+)$ [20] = 106.20	$(-)$	$(-)$	$(+)$	$(-)$	$(+)$	$(-)$	$(-)$	$(-)$		$\sqrt{}$
4320	$(-)$	$(-)$	$(-)$	$(-)$	$(+)$	$(-)$	$(+)$	$(-)$	$(-)$	$(-)$		$\sqrt{}$

For all observed values of maximum precipitation P_τ^{AMP} and P_τ^{POT} (Tables 2 and 3), the Wald-Wolfowitz test (Test of Series) and the Anderson serial correlation coefficient test showed that the analyzed measurement series were random and formed a simple sample, i.e., the random variables were independent variables. The significance level $\alpha = 0.05$ used in the test took account of the size of the random sample, $n = 30$. For series of length greater than 30, a lower value may be taken as the test significance level (e.g., $\alpha = 0.01$). For the detection of outliers with the Grubbs-Beck test, the higher value $\alpha = 0.10$ was used, on the assumption that series of measurements of meteorological phenomena may be characterized by greater anthropogenic impact.

The stationarity of the measurement series was checked using the Kruskal-Wallis test and Spearman rank correlation test for the trends of the mean and variance. According to the Kruskal-Wallis test, in the P_τ^{AMP} series from both Chojnice and Bialystok, jumps in the mean were detected, with the exception of the observations for $\tau = 5$ and $\tau \in \{720, \ldots, 4320\}$ min at Chojnice. In the case of the P_τ^{POT} precipitation values, most of the observations were stationary, with the exception of $\tau = 5$ at Chojnice and $\tau \in \{15, 30\}$ and $\tau = 1440$ min at Bialystok.

The Spearman's rank correlation test for the trends of mean and variance revealed nonstationarity mainly for the P_τ^{AMP} precipitation values. In the case of P_τ^{POT}, nonstationary observations were the exception. For example, in the observations from Chojnice for $\tau = 10$ min and $\tau \in \{45, 60\}$ min, a trend was detected in the mean and variance, respectively, while for the Bialystok data, such trends were detected, respectively, for $\tau \in \{360, 1440\}$ and $\tau = 120$ min.

The results of correlation testing and the identification of the trend of maximum precipitation for the AMP and POT series are given in Tables 4–6. The identification of the trend of the analyzed random variables was performed using the nonparametric Spearman rank correlation test [37] and the parametric Pearson linear correlation coefficient test [38]. An analysis was made of the correlation between the studied random variables (P_τ^{AMP} and P_τ^{POT}) and the time variable τ (Table 4). Positive and negative values indicate upward and downward trends, respectively. Spearman's coefficient also indicates the strength of the trend. The closer the values are to 1.0, the stronger is the relationship between the analyzed random variable and the time variable τ. Pearson's coefficient indicates proportionality, that is, linear dependence between variables, while Spearman's coefficient indicates any monotonic relationship, even if nonlinear. Figures shown in bold type in Table 4 indicate significant correlations, with the probability $p \leq 0.05$. Strong dependences between the observed maximum precipitation values and the independent variable τ were recorded in the case of P_τ^{AMP} at both Chojnice and Bialystok.

Table 4. Correlations between the maximum precipitation variables and time τ for the Chojnice and Bialystok stations. Bold values of Spearman's rank correlation and Pearson's linear correlation coefficients are significant at $p < 0.05$ for $n = 30$, where n is the size of the sample.

τ (min)	5	10	15	30	45	60	90	120	180	360	720	1080	1440	2160	2880	4320
Nonparametric Spearman rank correlation coefficient test for CHOJNICE station																
P_τ^{AMP}	0.277	0.309	**0.396**	**0.481**	**0.452**	**0.472**	**0.439**	**0.495**	**0.516**	**0.458**	0.198	0.100	0.136	0.112	0.206	0.220
P_τ^{POT}	0.175	**−0.449**	−0.181	−0.270	−0.019	−0.169	−0.129	−0.046	−0.319	0.036	−0.201	−0.203	−0.226	−0.326	−0.285	−0.340
Parametric Pearson linear correlation coefficient test for CHOJNICE station																
P_τ^{AMP}	0.267	0.297	0.290	0.292	0.303	0.335	**0.388**	**0.434**	**0.425**	**0.387**	0.299	0.186	0.159	0.101	0.189	0.210
P_τ^{POT}	0.209	−0.299	−0.255	−0.331	−0.124	−0.235	−0.142	−0.068	−0.222	0.090	0.046	−0.045	−0.114	−0.205	−0.145	−0.193
Nonparametric Spearman rank correlation coefficient test for BIAŁYSTOK station																
P_τ^{AMP}	**0.584**	**0.552**	**0.553**	**0.524**	**0.471**	**0.477**	**0.482**	**0.458**	**0.482**	**0.454**	**0.470**	**0.415**	**0.458**	**0.433**	**0.417**	**0.366**
P_τ^{POT}	0.181	0.007	0.222	0.227	0.042	0.056	0.137	0.271	0.194	**0.434**	0.315	0.236	**0.407**	0.251	0.067	0.353
Parametric Pearson linear correlation coefficient test for BIAŁYSTOK station																
P_τ^{AMP}	**0.448**	**0.490**	**0.489**	**0.427**	**0.399**	**0.396**	**0.428**	**0.411**	**0.466**	**0.468**	**0.486**	**0.457**	**0.456**	**0.451**	**0.434**	**0.423**
P_τ^{POT}	0.131	0.054	0.174	0.115	0.018	0.006	0.102	0.168	0.195	**0.375**	**0.368**	0.304	**0.371**	0.228	0.129	**0.364**

The form of the trend function was assessed using regression analysis (Tables 5 and 6), where the dependent variable is the maximum precipitation and the independent variable is the time τ. Tables 5 and 6 give the results of the regression analysis, including the following indicators: Pearson's correlation coefficient r, the coefficient of determination r^2, the Fisher-Snedecor global F-test [60], the test probability p resulting from the latter test, the size of the random sample n, and the standard error of estimation $S(E)$. Statistically significant regression coefficients for the analyzed variables are identified according to the criterion for statistical significance adopted in the model, with $\alpha = 0.05$. This means that the regression coefficients are significant for a test probability $p \leq 0.05$.

The global F-test tests three equivalent null hypotheses: H_0: $\beta_1 = 0$ (significance of the slope); H_0: $r^2 = 0$ (significance of the coefficient of determination); and H_0: $y = \beta_1 x + \beta_0$ (significance of the linear relationship between the analyzed variables), where β_1 is the slope; β_0 is a free term; and x and y denote the independent and dependent variables, respectively. Verification is made of the null hypothesis that the independent variable x (in Tables 5 and 6, the independent variable is time, τ) does not influence the analyzed dependent variable y (in Tables 5 and 6, the dependent variables are $P_5^{AMP}, \ldots, P_{4320}^{AMP}$ and $P_5^{POT}, \ldots, P_{4320}^{POT}$). If, in the course of verification, the null hypothesis is rejected, the regression coefficient is assessed as significant, meaning that τ has a significant influence on the analyzed dependent variable. Examples of random variables with no trend and showing a trend are given in Tables 5 and 6, respectively, for observations from Chojnice and Bialystok.

Table 5. Results of simple regression analysis for the Chojnice station, where the dependent variables are P_τ^{AMP} and P_τ^{POT}, and the independent variable is time (τ), for $n = 30$, where r is Pearson's correlation coefficient; r^2 is the coefficient of determination; $F(1,n)$ is the Fisher-Snedecor test; $S(E)$ is the standard error of estimation; and p (p-value) is the value of the test probability. Bold type indicates significance of regression parameters, namely the existence (for $p \leq 0.05$) of a significant linear trend coefficient.

τ	P_τ^{AMP}—CHOJNICE					P_τ^{POT}—CHOJNICE				
	r	r^2	$F(1,n = 28)$	$S(E)$	p	r	r^2	$F(1,n = 28)$	$S(E)$	p
5	0.266	0.071	2.142	2.278	0.154	0.208	0.043	1.274	1.374	0.268
10	0.297	0.088	2.718	3.780	0.110	0.299	0.089	2.751	2.641	0.108
15	0.296	0.087	2.691	4.872	0.112	0.255	0.065	1.949	3.754	0.173
30	0.292	0.085	2.609	6.916	0.117	0.331	0.109	3.447	5.396	0.073
45	0.302	0.092	2.824	7.244	0.103	0.124	0.015	0.440	5.757	0.512
60	0.335	0.112	3.545	7.244	0.070	0.234	0.055	1.634	5.596	0.211
90	0.388	0.151	4.967	7.399	**0.034**	0.142	0.020	0.577	5.762	0.453
120	0.434	0.188	6.513	7.623	**0.016**	0.067	0.004	0.128	5.983	0.722
180	0.425	0.181	6.181	8.089	**0.019**	0.221	0.049	1.447	6.489	0.238
360	0.386	0.149	4.926	9.047	**0.034**	0.090	0.008	0.229	7.484	0.635
720	0.298	0.089	2.744	9.857	0.108	0.046	0.002	0.059	7.524	0.808
1080	0.185	0.034	0.998	12.241	0.326	0.045	0.002	0.056	9.465	0.813
1440	0.158	0.025	0.726	13.263	0.401	0.113	0.012	0.366	10.337	0.550
2160	0.101	0.010	0.289	15.139	0.594	0.204	0.042	1.226	11.338	0.277
2880	0.188	0.035	1.035	15.589	0.317	0.145	0.021	0.604	11.986	0.443
4320	0.209	0.043	1.287	16.387	0.266	0.193	0.037	1.086	12.190	0.306

Table 6. Results of simple regression analysis for the Bialystok station, where the dependent variables are P_τ^{AMP} and P_τ^{POT}, and the independent variable is time (τ), for $n = 30$, where r is Pearson's correlation coefficient; r^2 is the coefficient of determination; $F(1,n)$ is the Fisher-Snedecor test; $S(E)$ is the standard error of estimation; and p (p-value) is the value of the test probability. Bold type indicates significance of regression parameters, namely the existence (for $p \leq 0.05$) of a significant influence of the variable τ on the analyzed dependent variable.

τ	P_τ^{AMP}—BIALYSTOK					P_τ^{POT}—BIALYSTOK				
	r	r^2	$F(1,n = 28)$	$S(E)$	p	r	r^2	$F(1,n = 28)$	$S(E)$	p
5	0.489	0.240	8.846	2.407	**0.006**	0.253	0.064	1.915	1.840	0.177
10	0.4901	0.240	8.879	3.352	**0.006**	0.083	0.007	0.197	2.979	0.660
15	0.489	0.239	8.826	4.147	**0.006**	0.197	0.039	1.137	3.454	0.295
30	0.425	0.181	6.232	5.768	**0.019**	0.061	0.004	0.104	4.854	0.749
45	0.399	0.159	5.309	6.816	**0.028**	0.106	0.011	0.323	5.407	0.574
60	0.397	0.157	5.248	6.825	**0.029**	−0.017	0.0003	0.008	5.124	0.928
90	0.427	0.183	6.269	7.513	**0.018**	0.113	0.012	0.363	5.866	0.551
120	0.409	0.167	5.628	8.285	**0.024**	0.142	0.020	0.576	6.456	0.454
180	0.465	0.216	7.729	8.195	**0.009**	0.145	0.021	0.605	6.837	0.443
360	0.466	0.217	7.773	9.875	**0.009**	0.301	0.091	2.801	8.684	0.105
720	0.487	0.237	8.708	10.828	**0.006**	0.368	0.135	4.390	10.032	**0.045**
1080	0.459	0.212	7.513	12.186	**0.010**	0.376	0.142	4.624	11.615	**0.040**
1440	0.458	0.210	7.465	13.717	**0.011**	0.367	0.134	4.350	12.633	**0.046**
2160	0.454	0.206	7.267	16.796	**0.012**	0.287	0.083	2.531	15.312	0.123
2880	0.436	0.191	6.600	18.190	**0.016**	0.193	0.037	1.088	16.566	0.306
4320	0.426	0.182	6.217	21.715	**0.018**	0.359	0.129	4.152	19.319	0.515

Values shown in bold type in Tables 5 and 6 indicate the presence of a significant influence of time τ on the analyzed random variable. In these cases, the estimated regression slope coefficients β_1 are significantly different from zero. At Chojnice, the observations of maximum precipitation showed a trend only in the case of P_τ^{AMP} for the durations $\tau \in \{90, \ldots, 360\}$ min. At Bialystok, however, in all of the analyzed observations of maximum precipitation P_τ^{AMP} and in three cases of P_τ^{POT} ($\tau \in \{720, \ldots, 1440\}$ min), an upward trend was detected. The test probability p determined for the computed regression coefficients was below the assumed significance level $\alpha = 0.05$.

An assessment of the form of the trend function (linear, power, exponential, etc.) was made using scatter plots of the analyzed random variables with respect to time τ (Figure 3). The scatter plots of P_{10}^{AMP}, P_{30}^{AMP}, and P_{60}^{AMO} showed a clear linear upward trend, while those for the variables P_{10}^{POT}, P_{30}^{POT}, and P_{60}^{POT} showed, respectively, small upward and downward trends. In this case, the slope β_1 was close to 0, and the test probabilities (P_{10}^{POT}: $p = 0.660$; P_{30}^{POT}: $p = 0.749$; P_{60}^{POT}: $p = 0.928$) were substantially higher than the significance level $\alpha = 0.05$ used in the analysis. In the annual data, seasonal (monthly or daily) fluctuations were not analyzed. If the analyzed series of values of P_{τ}^{AMP} or P_{τ}^{POT} contain a trend or periodic fluctuations, they cannot be used as an input in the computational procedures of the PMAXTP method.

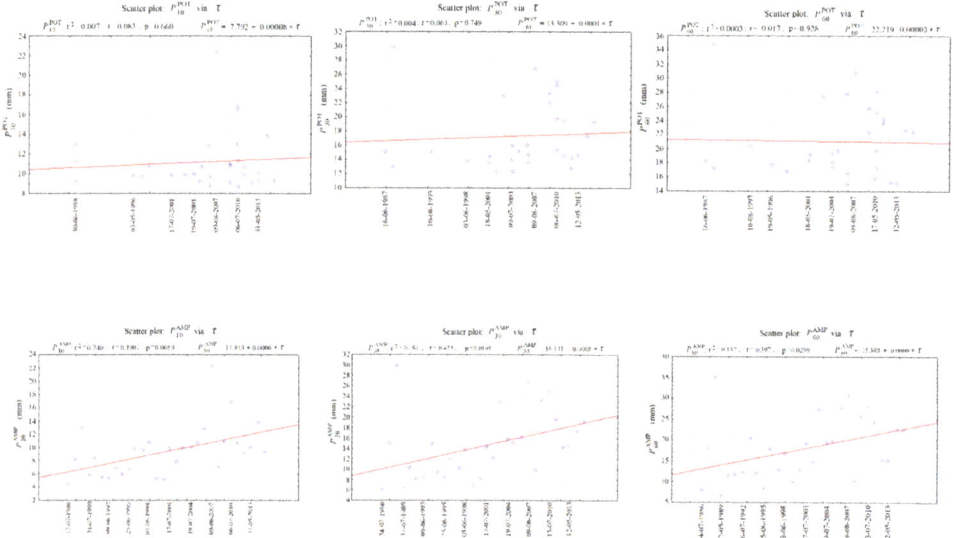

Figure 3. Scatter plots of dependent random variables observed at the Bialystok station: P_{10}^{POT}, P_{ao}^{POT}, P_{60}^{POT} and P_{10}^{AMP}, P_{30}^{AMP}, P_{60}^{AMP} with respect to the independent variable time (τ), with indication of the simple regression equation, coefficient of determination (r^2), linear correlation coefficient (r), and test probability (p) compared with the assumed significance level $\alpha < 0.05$.

An analysis was made of the internal correlation of the series of random variables P_{τ}^{AMP} and P_{τ}^{POT} using Anderson's test [40]. An autocorrelation analysis was performed for lags up to 25 (Figure 4). The greatest autocorrelation coefficients were detected for P_{1080}^{AMP} with $lag = 1$ ($\rho = 0.358$) and for P_{90}^{POT} with $lag = 4$ ($\rho = 0.417$). Other autocorrelation values were not large and lay within the confidence interval for the assumed significance level $\alpha = 0.05$. This is a sufficient condition to conclude a lack of correlation; that is, that the analyzed random variables are independent. An analysis of the autocorrelation plots (Figure 4) also showed an absence of periodic fluctuations.

Nonhomogeneity analysis, performed using genetic and statistical methods, showed that most of the observations of maximum precipitation selected by the POT method satisfied the homogeneity requirements, except for the observations for duration $\tau = \{10, 45, 60\}$ min at Chojnice and $\tau = \{15, 30, 120, 1440\}$ min at Bialystok (Tables 2 and 3). Most of the maximum precipitation observations selected by the AMP method are nonhomogeneous; exceptions are the P_{τ}^{AMP} observations from Chojnice with duration $\tau = 5$ and $\tau = \{720, \ldots, 4320\}$ min.

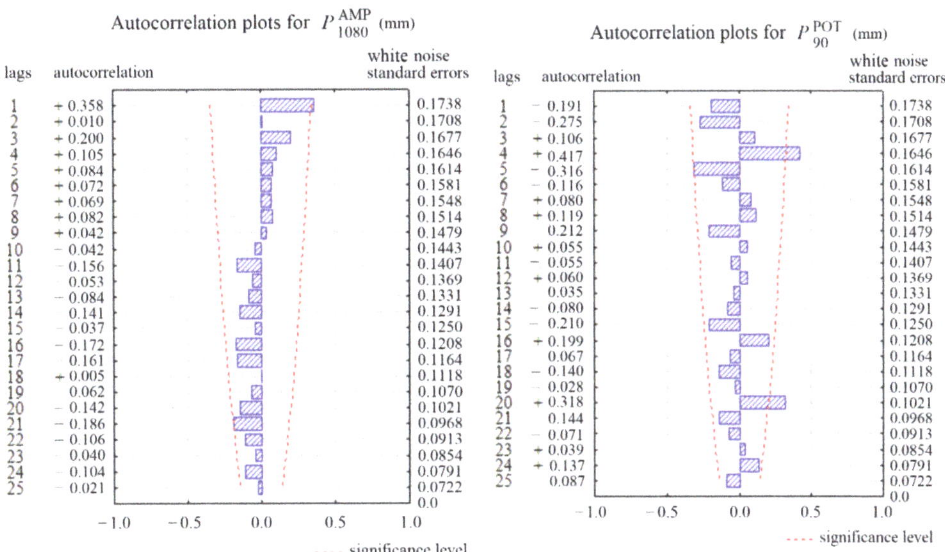

Figure 4. Autocorrelation function of random variables observed at Bialystok: P^{POT}_{10}, P^{POT}_{30}, and P^{POT}_{60} for lags of up to 25 elements in a series, with indication of autocorrelation coefficients, calculated white noise (standard error), and confidence level α.

4.2. Computation of Maximum Precipitation with Specified Probability of Exceedance Using the PMAXTP Method

Parameters of the probability distributions of the analyzed random variables were estimated for the two adopted methods of selection of maximum precipitations, P^{AMP}_{τ} and P^{POT}_{τ} (for details, see Section 2). The most credible distribution was selected for the analyzed random variable by minimizing the value of the Akaike information criterion (AIC) [45]. Calculations were performed for three-parameter (α, λ or μ, ε; Equations (A1), (A3), (A5) and (A7) in Appendix A) probability distributions GA, WE, LGA, and LN, and for the two-parameter (α, μ; Equation (A9) in Appendix A) G distribution. Sample results obtained at each stage of the procedure are given in Table 7. The most credible theoretical probability distribution for precipitation P^{AMP}_5 at the Chojnice station was found to be GA, while for P^{POT}_5, it was found to be WE. At the Bialystok station, the most credible theoretical distribution for P^{POT}_5 was determined to be LGA.

Verification of the distributions of maximum precipitation identified as most credible at the meteorological stations in Chojnice and Bialystok was performed by means of nonparametric tests of goodness of fit: D_{K-S}, D_{A-D}, D_{L-S}, and D_K (defined by Equations (A11)–(A14) in Appendix B). For purposes of inference, a significance level of $\alpha = 0.05$ was arbitrarily selected. This is a consequence of the fact that the value of the significance level of a test is closely related to the size (length) of the random sample on whose basis the parameters of the theoretical distributions are estimated. In the present analysis, the series contained $n = 30$ elements, which means that the significance level can be taken to be at most $\alpha = 0.05$. Verification was performed for the most credible theoretical probability distributions, which are shown in Table 8 for maximum precipitation with specified duration τ, together with the results obtained in single-dimensional statistical tests and the critical values, respectively for P^{AMP}_{τ} and P^{POT}_{τ} at the Chojnice station and P^{POT}_{τ} at Bialystok. All of the tests failed to reject the null hypothesis on the goodness of fit of the theoretical distribution with the empirical distribution, for the analyzed variables P^{AMP}_{τ} and P^{POT}_{τ}, with the exception of the D_{A-D} test in relation to the maximum precipitation P^{POT}_{90} at Chojnice (value shown in bold type in Table 8). The least of the maximum distances between values of the theoretical and empirical cumulative probability distributions, particularly in the tail part, was situated decidedly below the critical value of the D_{A-D} test defined at a significance level of $\alpha = 0.05$, which signifies rejection of the hypothesis of the goodness of fit of the theoretical and empirical distributions.

Table 7. Sample results of the procedure to select probability distributions for maximum precipitation values P_τ^{AMP} and P_τ^{POT} for $\tau = 5$ min. GA—gamma distribution; WE—Weibull; LN—log-normal; LGA—log-gamma; G—Gumbel; χ^2—Pearson's χ^2 goodness-of-fit test; min(D_{max})—Kolmogorov's minimum distance criterion. Bold values represent the most credible distributions according to the Akaike information criterion (AIC).

Precipitation		Probability Distribution							
		Type	Parameters				χ^2 $\chi^2_{(\alpha_{kr}=0.05)} = 7.815$	min(D_{max})	AIC
			α	λ	μ	ε			
CHOJNICE	P_5^{AMP}	GA	1.321	3.642	-	0.1	0.831	0.500	**138.738**
		WE	4.534	1.678	-	2.4	0.569	0.496	139.551
		LN	0.379	-	1.806	0.1	1.090	0.514	139.268
		LGA	0.036	113.463	-	0.1	1.182	0.552	140.164
		G	2.009	-	5.394	-	0.977	0.499	139.215
	P_5^{POT}	GA	0.732	3.356	-	5.2	5.292	0.579	100.323
		WE	2.172	1.481	-	5.7	5.751	0.549	**98.064**
		LN	0.360	-	1.231	4.0	5.268	0.601	101.625
		LGA	0.048	12.131	-	4.2	5.194	0.603	101.709
		G	1.022	-	7.031	-	6.013	0.667	102.049
BIAŁYSTOK	P_5^{POT}	GA	1.224	1.683	-	5.8	5.293	0.500	103.137
		WE	2.087	1.188	-	5.9	5.751	0.553	102.860
		LN	0.805	-	0.566	0.1	5.269	0.479	103.267
		LGA	0.107	2.967	-	5.6	5.194	0.473	**102.737**
		G	1.079	-	7.153	-	6.014	0.643	107.837

Table 8. Results of tests of fit of the theoretical probability distributions for P_τ^{AMP} and P_τ^{POT}, where $\tau = \{5, 10, 15, 30, 45, 60, 90, 120, 180, 360, 720, 1080, 1440, 2160, 2880, 4320\}$ (min). D_{K-S}—Kolmogorov-Smirnov test, D_{A-D}—Anderson-Darling test, D_{L-S}—Liao-Shimokawa test, D_K—Kuiper's test, significance level $\alpha = 0.05$. The value in bold type indicates rejection of the hypothesis of goodness of fit to the empirical distribution according to the statistic D_{A-D} at $\alpha = 0.05$.

| τ | CHOJNICE | | | | | | | | | | BIALYSTOK | | | | |
	AMP	D_{K-S}	D_{A-D}	D_{L-S}	D_K	POT	D_{K-S}	D_{A-D}	D_{L-S}	D_K	POT	D_{K-S}	D_{A-D}	D_{L-S}	D_K
5	GA	0.091	0.332	0.714	0.160	WE	0.100	0.472	0.819	0.195	LGA	0.086	0.323	0.764	0.168
10	-					-					LN	0.084	0.245	0.691	0.158
15	-					WE	0.117	0.268	0.668	0.206	-				
30	-					WE	0.108	0.285	0.678	0.185	-				
45	-					-					WE	0.078	0.259	0.647	0.155
60	-					-					WE	0.080	0.194	0.621	0.157
90	-					GA	0.150	**1.098**	1.199	0.278	WE	0.085	0.186	0.607	0.163
120	-					GA	0.062	0.119	0.516	0.123	-				
180	-					LGA	0.087	0.155	0.537	0.155	GA	0.103	0.253	0.666	0.195
360	-					LGA	0.071	0.184	0.613	0.133	-				
720	GA	0.183	0.749	1.016	0.308	LGA	0.088	0.324	0.771	0.173	WE	0.093	0.265	0.655	0.186
1080	G	0.124	0.485	0.836	0.245	WE	0.105	0.423	0.819	0.189	WE	0.091	0.263	0.677	0.161
1440	G	0.125	0.443	0.825	0.229	WE	0.109	0.286	0.676	0.172	-				
2160	G	0.080	0.118	0.514	0.124	GA	0.078	0.257	0.662	0.153	WE	0.071	0.164	0.591	0.136
2880	WE	0.079	0.209	0.650	0.157	GA	0.103	0.470	0.831	0.204	WE	0.087	0.217	0.616	0.166
4320	WE	0.072	0.166	0.575	0.145	GA	0.107	0.401	0.762	0.211	WE	0.107	0.276	0.680	0.192

$\alpha_{cr.} = 0.05$ for: $D_{K-Scr.} = 0.242$; $D_{A-Dcr.} = 0.795$; $D_{L-Scr.} = 1.505$; $D_{Kcr.} = 0.317$.

The results obtained from the PMAXTP model for the values of maximum precipitation with a specified probability of exceedance were compared with the results from the Bogdanowicz-Stachý model [1,2]. In the latter model, the procedure for computing the values of maximum precipitation with a specified probability of exceedance p consisted of:

(i)   regionalization of maximum precipitation;

(ii)   estimation of parameters of the probability distribution function depending on the identified region and selected duration.

The procedure of the Bogdanowicz-Stachý model conforms to the recommendations of the World Meteorological Organization [61]. The input data originated from 20 meteorological stations situated in latitudinal strips running along the coast, lake districts, lowland parts, and southern upland parts of Poland. Mountain areas were omitted, due to the absence of stations monitoring precipitation at all altitudes. The maximum quantity of precipitation with a specified duration and specified probability of exceedance was determined using the formula (A16) in Appendix C, taking account of the regionalization of the meteorological stations in Chojnice and Bialystok.

Quantile values determined using the PMAXTP and Bogdanowicz-Stachý models were compared using statistical and graphical measures. According to the regionalization carried out by Bogdanowicz and Stachý, the Chojnice meteorological station belongs to the north-west region for precipitation with durations in the range <5, >60 min, to the central region for durations in the range <60, >720 min, and to the southern/coastal region for durations in the range <720, >4320 min. The Bialystok station, located in the northeast of Poland, belongs to the central region irrespective of the duration of precipitation being considered.

For a comparison of the results given by the two models, i.e., PMAXTP and Bogdanowicz-Stachý, various statistical measures can be used [62]. In our study, we used the standard error of fit S(E), which is shown in Table 9. The error is given by the following formula [63]:

$$S(E) \; = \; \sqrt{\dfrac{\sum\limits_{i=1}^{i=m}\left(P_{\tau,i}^{MAX} - \hat{P}_{\tau,i}^{MAX} \right)}{m - l}} \tag{1}$$

where $P_{\tau,i}^{MAX}$ is the observed maximum precipitation selected by the AMP or POT method for a specified duration (τ); $\hat{P}_{\tau,i}^{MAX}$ is the estimated maximum precipitation from the PMAXTP or Bogdanowicz-Stachý model; m = 30 is the size of the random sample formed from empirical quantiles for m = 30 selected probabilities $p \in \{96.8, 93.6, 90.3, 87.1, 83.9, 80.7, 77.4, 74.2, 70.9, 67.7, 64.5, 61.3, 58.1, 54.8, 51.6, 48.4, 45.2, 41.9, 38.7, 35.5, 32.3, 29.0, 25.8, 22.6, 19.4, 16.1, 12.9, 9.7, 6.5, 3.2\}$ % and the corresponding theoretical distributions computed using the PMAXTP and Bogdanowicz-Stachý methods. Finally, l is the number of parameters of the theoretical probability distribution according to the density function (Equations (A1), (A3), (A5), (A7) and (A9) in Appendix A).

Computations of the error S(E) were performed separately for specified durations τ of maximum precipitation. The value of the standard error of fit increased with increasing values of τ for both models. The smallest errors were obtained for the quantiles determined from the maximum precipitation values selected using the POT method and the PMAXTP model. An exception was the quantiles determined for the AMP values at the Chojnice station for duration τ equal to 720 and 4320 min. The errors of fit of the theoretical to the empirical distributions in the Bogdanowicz-Stachý model for precipitation values selected by the AMP method were on average 210% greater than those obtained with the PMAXTP model, and for the POT precipitation values, the errors were 300% greater. The most frequently selected most credible theoretical probability distribution for random samples of both AMP and POT maximum precipitation values, and for both the Bialystok and the Chojnice stations, was the WE distribution.

Figures 5–10 show a comparison of the functions for the probability of exceedance of maximum precipitations P_{τ}^{AMP} or P_{τ}^{POT} determined using the models, for Chojnice (Figures 5–7) and Bialystok (Figures 8–10). The plots contain density functions of probability distributions computed only for homogeneous observations of precipitation selected by the AMP and POT methods, in accordance with the results shown in Tables 2, 3 and 8. The diagrams show comparisons of: (i) the most credible probability functions for maximum precipitation determined by the PMAXTP model for the AMP observation series (orange solid line) and for maximum precipitation selected by the POT method (blue solid line); (ii) upper limits of confidence intervals (orange and blue dotted lines); (iii) observations of AMP and POT maximum precipitation (orange and blue squares); and (iv) the probability function determined using the probabilistic Bogdanowicz-Stachý model (red solid line).

Table 9. Comparison of the PMAXTP and Bogdanowicz-Stachý methods for P_τ^{AMP} and P_τ^{POT}, where $\tau = \{5, 10, 15, 30, 45, 60, 90, 120, 180, 360, 720, 1080, 1440, 2160, 2880, 4320\}$ (min), using the standard error of fit S(E). The comparison refers to the maximum precipitation values computed for the meteorological station in Chojnice (P_τ^{AMP} and P_τ^{POT}) and in Bialystok (P_τ^{POT}). Values in bold type are the smallest errors S(E) obtained separately for the Chojnice and Bialystok stations.

	CHOJNICE								BIAŁYSTOK			
	PMAXTP				B&S				PMAXTP		B&S	
τ	AMP		POT		AMP		POT		POT		POT	
	Distrib.	S(E)	Distrib.	S(E)	Distrib.	S(E)	Distrib.	S(E)	Distrib.	S(E)	Distrib.	S(E)
5	GA	0.484	WE	**0.263**	WE	0.801	WE	1.380	LGA	**0.780**	WE	2.139
10									LN	**0.639**	WE	2.613
15			WE	**0.742**			WE	2.135				
30			WE	**1.418**			WE	2.654				
45									WE	**0.778**	WE	3.037
60									WE	**0.737**	WE	3.902
90			GA	**1.992**			WE	4.761	WE	**1.133**	WE	4.424
120			GA	**1.325**			WE	5.469				
180			LGA	**1.558**			WE	5.557	GA	**1.999**	WE	5.209
360			LGA	**2.902**			WE	6.095				
720	GA	**4.113**	LGA	4.236	WE	7.246	WE	7.713	WE	**2.869**	WE	5.854
1080	G	3.341	WE	**3.287**	WE	8.441	WE	7.814	WE	**2.610**	WE	6.280
1440	G	3.359	WE	**2.847**	WE	10.430	WE	8.588				
2160	G	2.924	GA	**2.383**	WE	11.139	WE	8.866	WE	**3.543**	WE	7.704
2880	WE	3.083	GA	**2.761**	WE	12.845	WE	10.442	WE	**3.646**	WE	8.593
4320	WE	**2.720**	GA	2.738	WE	14.218	WE	11.486	WE	**4.648**	WE	11.634

At the Chojnice station, for practically all of the analyzed durations of maximum precipitation, the quantile values from the Bogdanowicz-Stachý model are markedly higher than the observed precipitations and values of corresponding quantiles from the PMAXTP model, in relation to the maximum precipitations selected both by the AMP method (orange squares and solid line) and by the POT method (blue squares and solid line). The differences between the quantiles are particularly visible in the central region and in the region of the upper tails of the probability distributions. Similar maximum quantile values were obtained for precipitation with duration $\tau = \{15, 30, 180, 1080\}$ min. At Chojnice, the AMP values were described by the models GA and WE, while for description of the POT maximum precipitation values, the WE distribution was selected for short durations τ, and GA and LGA for medium and long durations.

At the Bialystok station, in the case of maximum precipitations with duration $\tau = \{5, 45, 60, 90, 180\}$ min (Figures 8 and 9), the quantile values determined using the Bogdanowicz-Stachý model (red solid line) are markedly higher than the corresponding quantiles obtained using the PMAXTP model for the maximum precipitations determined by the POT method (blue squares and solid line). Differences between quantiles are particularly visible in the central region and in the region of the upper tails of the probability distributions. The closest results for quantiles of POT maximum precipitations calculated using the PMAXTP method and from the Bogdanowicz-Stachý model were obtained for precipitation with duration $\tau = \{720, 1080\}$ min (Figure 9). For maximum precipitation with such durations, the most credible theoretical distribution was WE, while for short durations, $\tau = \{5, 10\}$ min, the respective distributions were LGA and GA. For maximum precipitation selected by the POT method with duration $\tau = \{2160, \dots, 4320\}$ min, the Bogdanowicz-Stachý model returned markedly lower quantile values than the PMAXTP method.

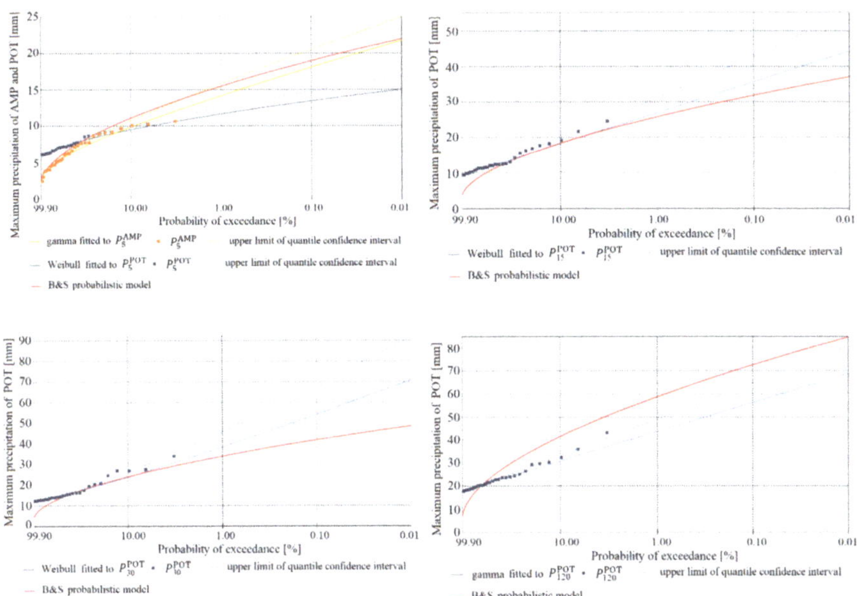

Figure 5. Plots of functions of probability of exceedance for the random variables P_τ^{AMP} and/or P_τ^{POT}, where $\tau = \{5, 15, 30, 120\}$ min, for the most credible probability distributions, with indicated upper limits of quantile confidence intervals according to the PMAXTP method, compared with the model of Bogdanowicz and Stachý, for the Chojnice meteorological station.

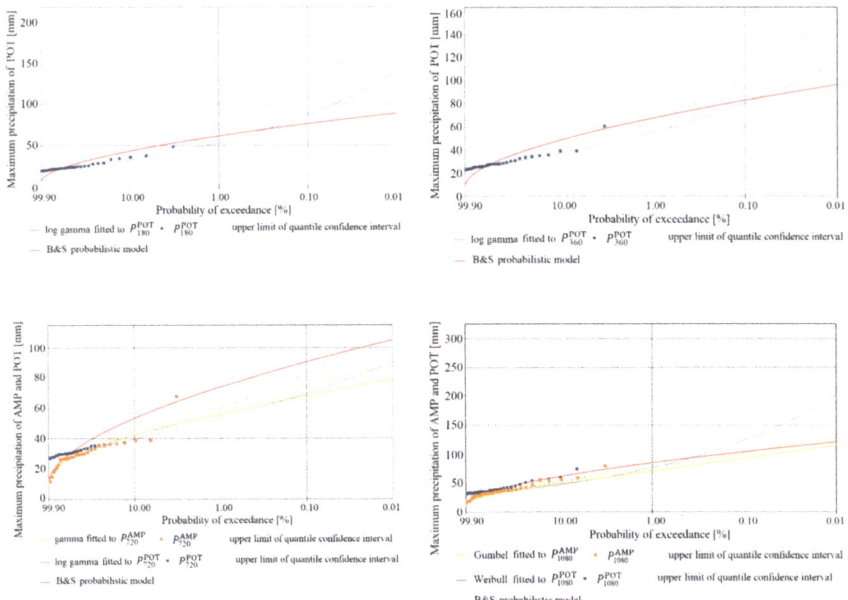

Figure 6. Plots of functions of probability of exceedance for the random variables P_τ^{AMP} and/or P_τ^{POT}, where $\tau = \{180, 360, 720, 1080\}$ min, for the most credible probability distributions, with indication of upper limits of quantile confidence intervals according to the PMAXTP method, compared with the model of Bogdanowicz and Stachý, for the Chojnice meteorological station.

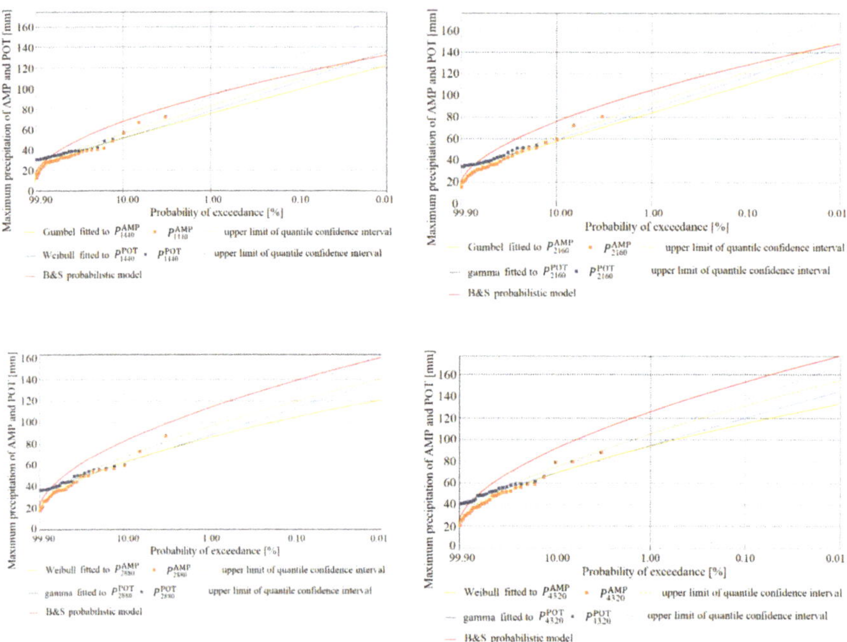

Figure 7. Plots of functions of probability of exceedance for the random variables P_τ^{AMP} and P_τ^{POT}, where $\tau = \{1440, 2160, 2880, 4320\}$ min, for the most credible probability distributions, with indication of upper limits of quantile confidence intervals according to the PMAXTP method, compared with the model of Bogdanowicz and Stachý, for the Chojnice meteorological station.

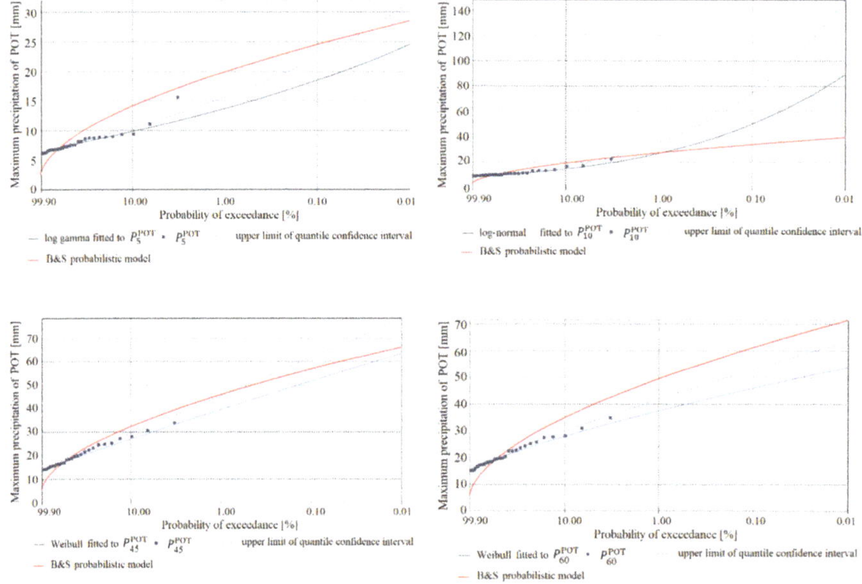

Figure 8. Plots of functions of probability of exceedance for the random variables P_τ^{POT} where $\tau = \{5, 10, 45, 60\}$ min, for the most credible probability distributions, with indication of upper limits of quantile confidence intervals according to the PMAXTP method, compared with the model of Bogdanowicz and Stachý, for the Bialystok meteorological station.

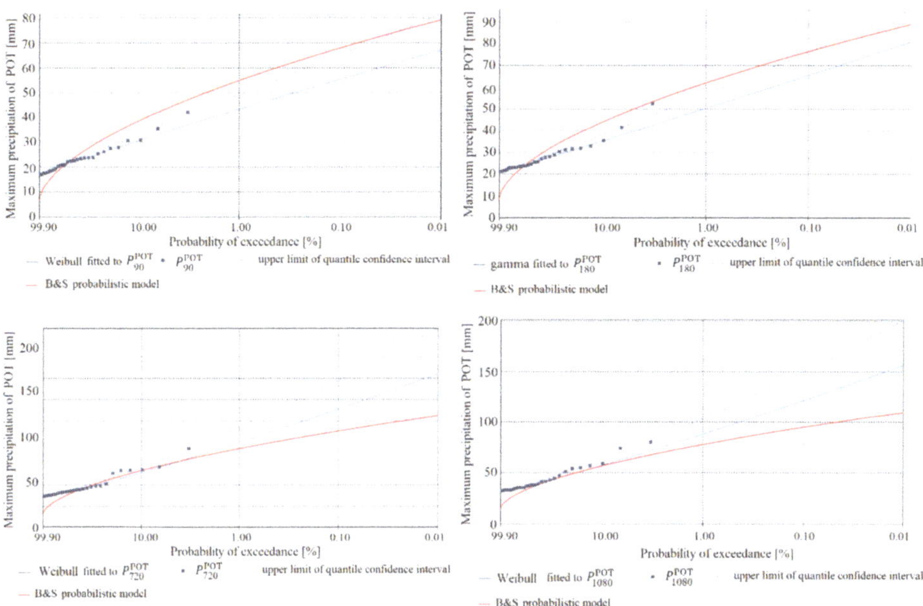

Figure 9. Plots of functions of probability of exceedance for the random variables P_τ^{POT} where $\tau = \{90, 180, 720, 1080\}$ min, for the most credible probability distributions, with indication of upper limits of quantile confidence intervals according to the PMAXTP method, compared with the model of Bogdanowicz and Stachý, for the Bialystok meteorological station.

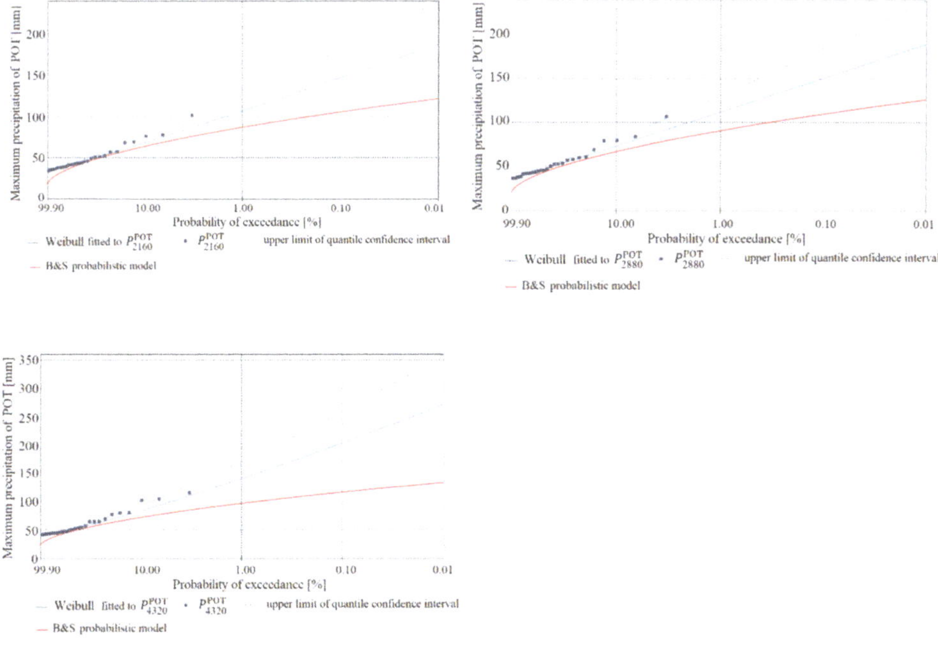

Figure 10. Plots of functions of probability of exceedance for the random variables P_τ^{POT} where $\tau = \{2160, 2880, 4320\}$ min, for the most credible probability distributions, with indication of upper limits of quantile confidence intervals according to the PMAXTP method, compared with the model of Bogdanowicz and Stachý, for the Bialystok meteorological station.

The final element of the verification of maximum precipitation values was a comparison of the estimated quantile error resulting from the randomness of the sample of maximum precipitations computed using the PMAXTP model for the random variables P_τ^{AMP} and P_τ^{POT} at the meteorological station in Chojnice (Figures 11–13) and for P_τ^{POT} at the meteorological station in Bialystok (Figures 14–16).

The largest errors for values of maximum precipitation with high probabilities, such as 99.0 and 99.9, at the Chojnice station were observed for maximum precipitations selected using the AMP method (Figure 11 for $\tau = 5$, Figure 12 for $\tau = \{720, 1080\}$, Figure 13 for $\tau = \{1440, \ldots, 2160\}$ min)—markedly higher errors for the AMP series than for the POT series at Chojnice. The largest errors for values of maximum precipitation with low probabilities, such as 0.01 and 0.001, were recorded for the Chojnice station (Figure 12 for $\tau = \{720, 1080\}$ and Figure 13 for $\tau = \{1440, \ldots, 2160\}$ min) for POT precipitations (markedly higher errors for the POT series than for the AMP series at Chojnice). The smallest differences in the quantile error in the entire range of theoretical occurrence of maximum precipitation were observed at Chojnice (Figure 13 for $\tau = \{2880, 4320\}$ min).

Calculations were made for 100 total rainfall measuring sites in Poland (Figure 17). Calculated characteristics of maximum rainfall totals, i.e., quantile values for $p(\%) \in \{99.9, 99.5, 99, 98.5, 98, 95, 90, 80, 70, 60, 50, 40, 30, 20, 10, 5, 3, 2, 1, 0.5, 0.3, 0.2, 0.1, 0.05, 0.03, 0.02, 0.01\}$ of a specified duration, $\tau(\text{min}) \in \{5, 10, 15, 30, 45, 60, 90, 120, 180, 360, 720, 1080, 1440, 2160, 2880, 4320\}$, upper limits of the confidence interval and quantile errors were interpolated by the Thiessen Polygons (TP) method, which allowed for the assignment of certain areas for which measuring sites are representative as well as for the proportional division and distribution of sites within Poland. Higher resolution calculations can be achieved using Gaussian geostatistical simulation models [64] that accept any simple kriging model [65] or residual kriging model [66].

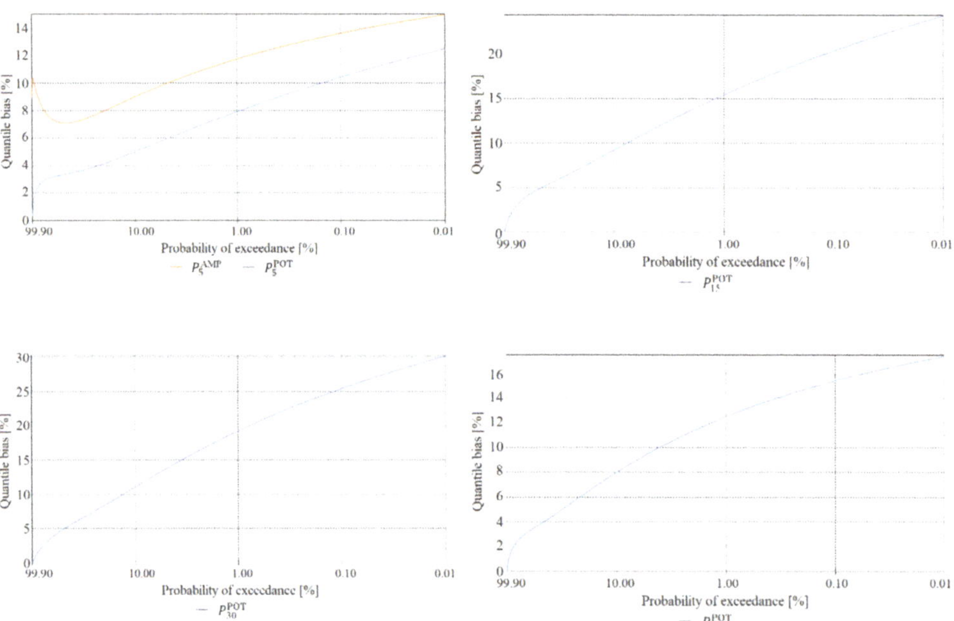

Figure 11. Comparison of estimated values of quantile error resulting from the randomness of the sample of maximum precipitations computed using the PMAXTP model for the random variables P_τ^{AMP} and P_τ^{POT} with durations $\tau = \{5, 15, 30, 120\}$ min, for the Chojnice meteorological station.

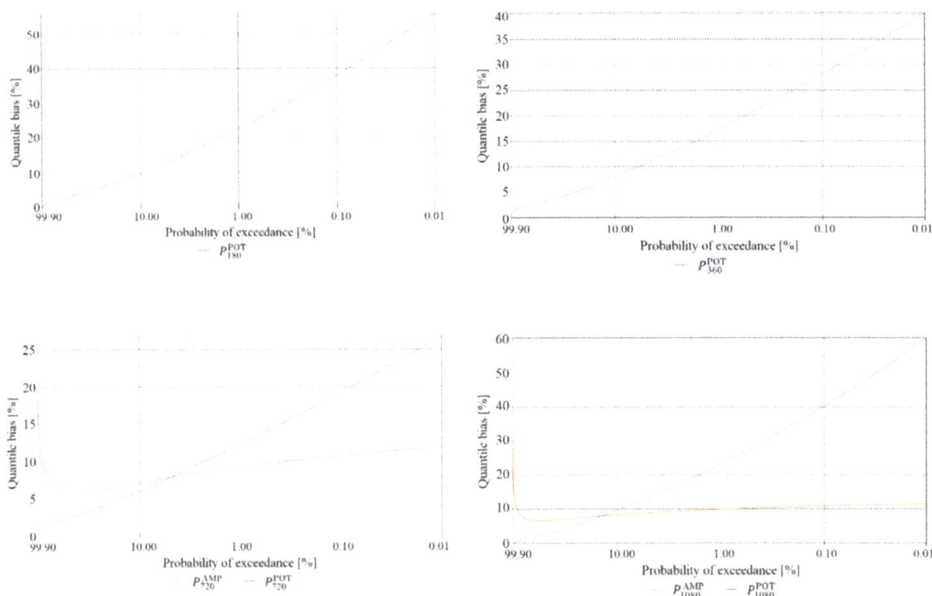

Figure 12. Comparison of estimated values of quantile error resulting from the randomness of the sample of maximum precipitations computed using the PMAXTP method for the random variables P_τ^{AMP} and P_τ^{POT} with durations $\tau = \{180, 360, 720, 1080\}$ min, for the Chojnice meteorological station.

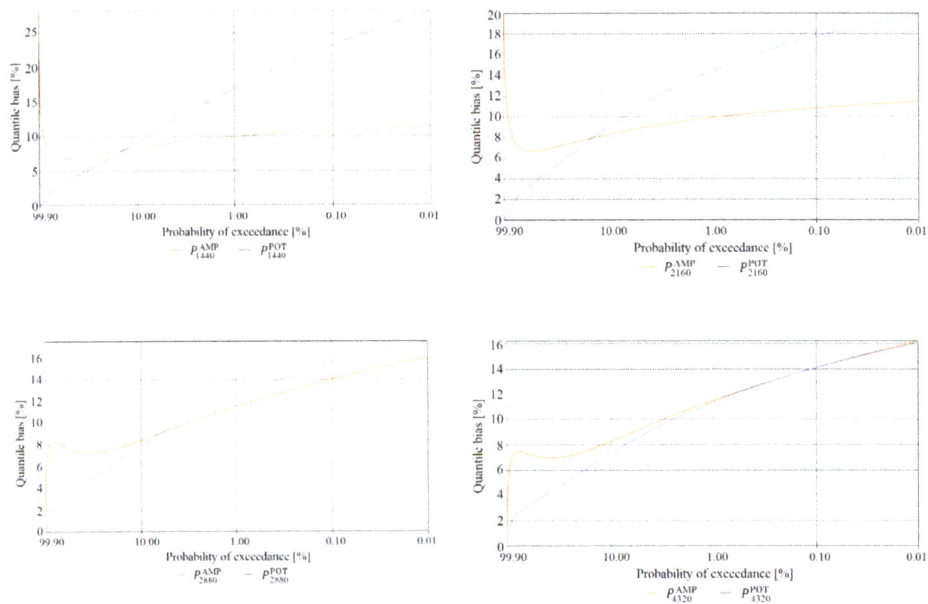

Figure 13. Comparison of estimated values of quantile error resulting from the randomness of the sample of maximum precipitations computed using the PMAXTP method for the random variables P_τ^{AMP} and P_τ^{POT} with durations $\tau = \{1440, 2160, 2880, 4320\}$ min, for the Chojnice meteorological station.

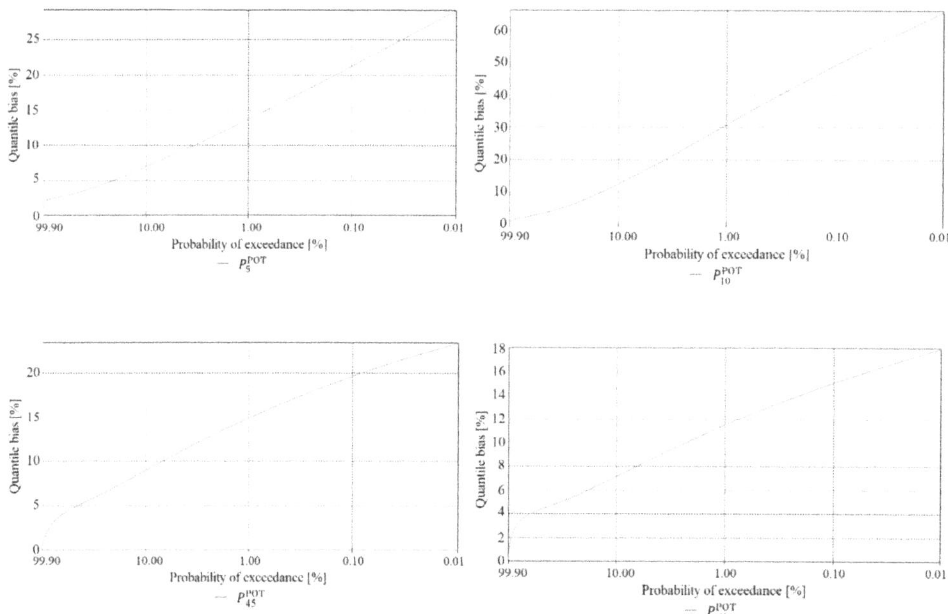

Figure 14. Comparison of estimated values of quantile error resulting from the randomness of the sample of maximum precipitations computed using the PMAXTP model for the random variable P_τ^{POT} with durations $\tau = \{5, 10, 45, 60\}$ min, for the Bialystok meteorological station.

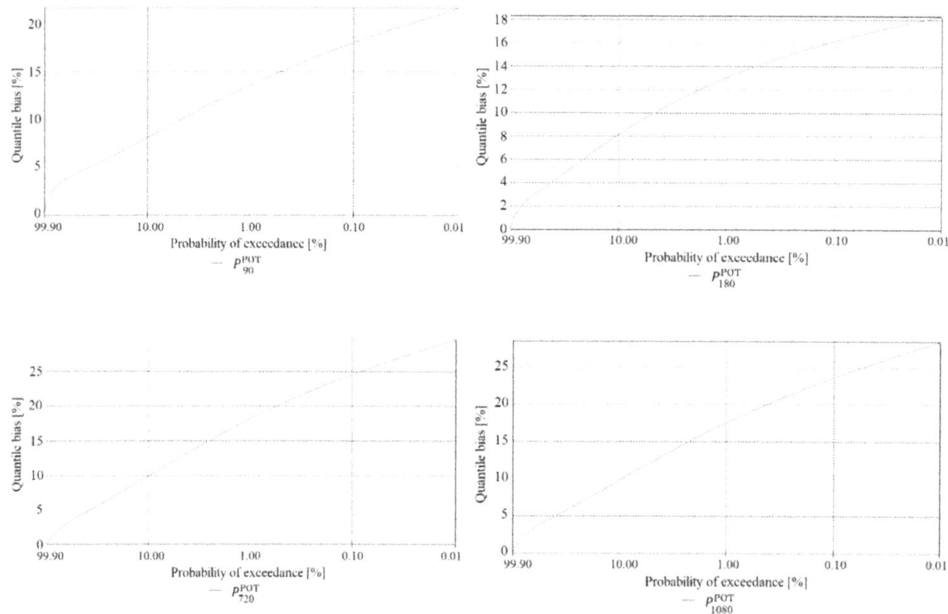

Figure 15. Comparison of estimated values of quantile error resulting from the randomness of the sample of maximum precipitations computed using the PMAXTP model for the random variable P_τ^{POT} with durations $\tau = \{90, 180, 720, 1080\}$ min, for the Bialystok meteorological station.

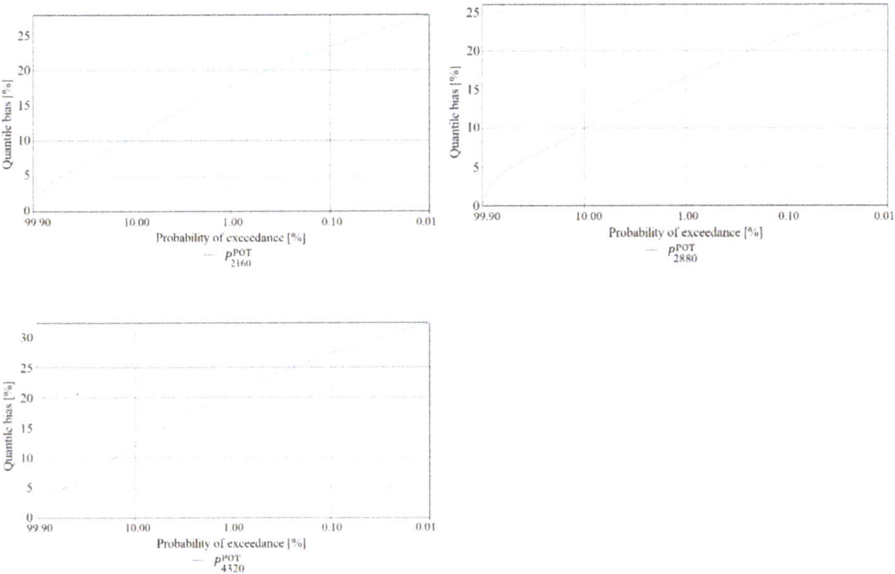

Figure 16. Comparison of estimated values of quantile error resulting from the randomness of the sample of maximum precipitations computed using the PMAXTP model for the random variable P_τ^{POT} with durations $\tau = \{2160, 2880, 4320\}$ min, for the Bialystok meteorological station.

Interpolation also can be performed using the Inverse Distance Weighted (IDW) method, which uses a linearly weighted set of sampling points to determine mesh node values by using reverse weighted distance values. The weight is a function of the inverse distance, and the interpolated surface should be a variable surface depending on the position of the point [67]. An example of interpolating the maximum precipitation value P_τ^{AMP} with a duration of $\tau = 30$ min with a probability $p = 1\%$ calculated using the IDW method is shown in Figure 18 (left part).

The IDW is a deterministic interpolation method because it is directly based on surrounding measured values. Another example is the set of geostatistical methods, such as the Kriging methods (right part of Figure 18), which include autocorrelation, which represents the statistical relationship between the measured points, thus providing a certain measure of reliability or accuracy of the forecast. The Kriging method is most suitable when one knows that there is spatial distance correlation or directional deviation in the data being analyzed.

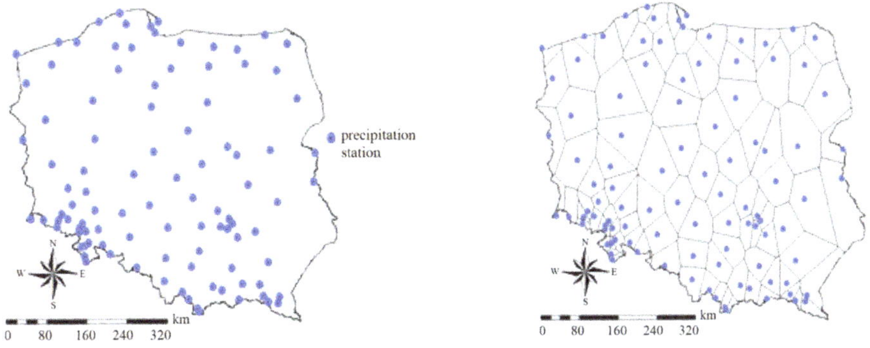

Figure 17. Thiessen Polygons based on precipitation measurement sites in Poland.

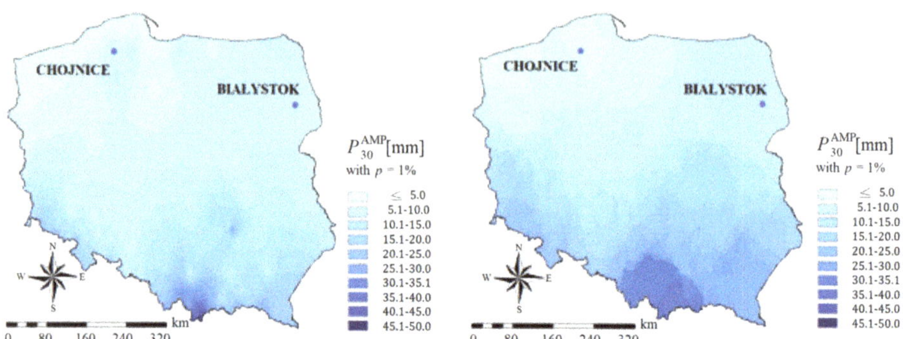

Figure 18. Interpolation of maximum precipitations computed using the PMAXTP model for the random variable P_τ^{AMP} with durations $\tau = 30$ min with probability of exceedance $p = 1\%$ using IDW method (**left** part) and kriging method (**right** part) for the Bialystok and Chojnice meteorological stations.

5. Conclusions

This paper described the PMAXTP model for a frequency analysis of maximum precipitation with a specified duration. It consists of two modules: statistical and computational. The first step selects values of maximum precipitation of a specified duration, which is conducted using two different methods: Annual Maximum Precipitation (AMP) and Peaks-Over-Threshold (POT). The advantage of the POT method is that it selects a larger number of observations of precipitation with the highest values in a given year, which leads to a better estimation of the characteristics of maximum precipitation with a specified duration and probability of exceedance. This is a significant issue in the design of drainage structures, particularly when they are at high risk of damage. The statistical module enables an analysis of the homogeneity of the series of measurements of maximum precipitation that serve as the input to the computational module, in which the mathematical models used for parameter estimation require a simple random sample, that is, one that satisfies the assumptions of independence and stationarity.

The computational module enables the selection of the best (the most credible) theoretical probability distribution by means of: (i) estimation of the parameters of four types of distributions belonging to the families gamma (GA), Weibull (WE), log-gamma (LGA), log-normal (LN), and Gumbel function (G); (ii) test of the hypothesis of goodness of fit of the theoretical probability distribution function with the empirical distribution using Pearson's χ^2 test; (iii) selection of the best-fitting function in each distribution type according to the criterion of minimum Kolmogorov distance; (iv) selection of the most credible distribution function from the set of best-fitting functions of various types; and (v) verification of the most credible distributions of precipitations P_τ^{AMP} and P_τ^{POT} using the single-dimensional tests D_{K-S}, D_{A-D}, D_{L-S}, and D_K.

The PMAXTP model was tested on data from two meteorological stations in Poland (Chojnice and Bialystok) representing two regions characterized by different spatial variability of maximum precipitation. The results were compared with those given by the Bogdanowicz-Stachý model—which to date has frequently been used in engineering practice in Poland—based on estimated values of the quantile error resulting from the randomness of the sample of maximum precipitation values computed for the tested stations.

In general, the errors of fit for the theoretical to the empirical distribution for the PMAXTP model were lower than the errors for the Bogdanowicz-Stachý model. The smallest errors were obtained for the quantiles determined on the basis of maximum precipitation POT using the PMAXTP model for both analyzed stations.

The following detailed conclusions may be drawn from the results:

- Most of the observations of maximum precipitation selected by the POT method satisfied the requirement of homogeneity, with the exception of the observations with durations $\tau = \{10, 45, 60\}$ min at Chojnice and $\tau = \{15, 30, 120, 1440\}$ min at Bialystok.
- Most of the observations selected by the AMP method did not satisfy the requirement of homogeneity, with the exception of the observations with durations $\tau = 5$ min and $\tau = \{720, \ldots, 4320\}$ min at Chojnice.
- Errors of fit of the theoretical to the empirical distributions for the Bogdanowicz-Stachý model were on average 210% higher than the errors for the PMAXTP model in the case of the precipitation P_τ^{AMP}, and 300% higher in the case of P_τ^{POT}.
- The smallest errors were obtained for the quantiles determined on the basis of observations of maximum precipitation P_τ^{POT} obtained using the PMAXTP model.
- For the meteorological station in Chojnice, practically all of the quantile values determined by the Bogdanowicz-Stachý model were markedly higher than those obtained by the PMAXTP model and the quantiles of the empirical precipitations P_τ^{AMP} and P_τ^{POT}, while for the station in Bialystok, the Bogdanowicz-Stachý model gave higher quantile values for $\tau = \{5, \ldots, 180\}$ min and markedly lower values for $\tau = \{2160, \ldots, 4320\}$ min.
- The greatest errors for the low quantiles, i.e., the values of maximum precipitation that are exceeded with high probability, were observed for the precipitation values for P_τ^{AMP}, and the greatest errors for high quantiles, i.e., the values of maximum precipitation that are exceeded with low probability, were observed for the precipitation values for P_τ^{POT}.

Author Contributions: Conceptualization, B.O.-Z. and M.C.; methodology, B.O.-Z. and M.C.; software, M.C. and B.O.-Z.; validation, B.O.-Z., T.T. and J.A.; formal analysis, B.O.-Z. and M.C.; investigation, M.C. and B.O.-Z.; resources, Institute of Meteorology and Water Management–National Research Institute; data curation, Institute of Meteorology and Water Management–National Research Institute; writing—original draft preparation, M.C., B.O.-Z., T.T. and J.A.; writing—review and editing, M.C., B.O.-Z., T.T. and J.A.; visualization, M.C.; supervision, B.O.-Z. and T.T.; project administration, B.O.-Z.; funding acquisition, B.O.-Z. All authors have read and agreed to the published version of the manuscript.

Funding: This research received no external funding.

Institutional Review Board Statement: Not applicable.

Informed Consent Statement: Not applicable.

Acknowledgments: The authors acknowledge the financial and data support provided by the Polish Hydrological and Meteorological Service at the Institute of Meteorology and Water Management—National Research Institute.

Conflicts of Interest: The authors declare no conflict of interest.

Appendix A. The Density Function $f(x)$ and the Quantile Function x_p

The density function $f(x)$ and the quantile function x_p of the three-parameter GA distribution are written as [68]:

$$f(x) = \frac{(x - \varepsilon)^{\lambda - 1}}{\alpha^\lambda \Gamma(\lambda)} \exp\left(-\frac{x - \varepsilon}{\alpha}\right) \tag{A1}$$

$$x_p = \varepsilon + \alpha t_p(\lambda) \tag{A2}$$

where $\Gamma(\lambda) = \int_0^\infty t^{\lambda - 1} \exp(-t)dt$ is Euler's gamma function; x is an observation of the random variable X; x_p is a quantile of the theoretical GA distribution; and $t_p(\lambda)$ is a quantile of the standardized gamma distribution, with probability of exceedance p.

The WE distribution is defined as [68]:

$$f(x) = \frac{\lambda}{\alpha}\left(\frac{x-\varepsilon}{\alpha}\right)^{\lambda-1} \exp\left[-\left(\frac{x-\varepsilon}{\alpha}\right)^{\lambda}\right] \tag{A3}$$

$$x_p = \alpha[-\ln(1-(1-p))]^{\frac{1}{\lambda}} + \varepsilon \tag{A4}$$

The LGA distribution [69] is represented by the equations:

$$f(x) = \frac{(\ln x - \ln \varepsilon)^{\lambda-1}}{\alpha^{\lambda}\Gamma(\lambda)x} \exp\left(-\frac{\ln x - \ln \varepsilon}{\alpha}\right) \tag{A5}$$

$$x_p = \varepsilon \exp[\alpha t_p(\lambda)] \tag{A6}$$

The log-normal distribution (LN) [70] is represented as:

$$f(x) = \frac{1}{(x-\varepsilon)\alpha\sqrt{2\pi}} \exp\left[-\frac{1}{2}\left(\frac{\ln(x-\varepsilon)-\mu}{\alpha}\right)^2\right] \tag{A7}$$

$$x_p = \exp\left[\mu + \frac{\alpha\sqrt{2}}{\mathrm{erf}(2(1-p)-1)}\right] + \varepsilon \tag{A8}$$

where: $\mathrm{erf}(\dots)$ is the Gauss error function, and other symbols have the same meanings as above, except that x_p denotes a quantile of the theoretical WE, LGA, and LN distributions, respectively.

The Gumbel distribution [71] is written as:

$$f(x) = \frac{1}{\alpha}\exp\left[-\frac{x-\mu}{\alpha} - \exp\left(-\frac{x-\mu}{\alpha}\right)\right] \tag{A9}$$

$$x_p = -\alpha \ln[-\ln(1-p)] + \mu \tag{A10}$$

where x_p is a quantile of the theoretical G distribution.

Appendix B. The Goodness-of-Fit Tests

The following are nonparametric goodness-of-fit tests used to test the goodness of fit of a mathematical model (theoretical distribution) with observations (empirical distribution).

The Kolmogorov-Smirnov statistic D_{K-S} [46]:

$$D_{K-S} = \max_{1<i\leq n}(\hat{\delta}_i), \text{gdzie}: \hat{\delta}_i = \max\left[\frac{i}{n} - F_0(x_i;\hat{\theta}), F_0(x_i;\hat{\theta}) - \frac{i-1}{n}\right] \tag{A11}$$

where n is the size of the random sample, and $F_0(x_i;\hat{\theta})$ is the distribution function of the theoretical probability distribution for the estimated parameter vector $\hat{\theta}$.

The Anderson-Darling statistic D_{A-D} [48]:

$$D_{A-D} = -n - \frac{1}{n}\sum_{i=1}^{n}\left\{(2i-1)lnF_0(x_i;\hat{\theta}) + (2n+1-2i)ln(1-F_0(x_{n+1-i};\hat{\theta}))\right\} \tag{A12}$$

The Liao-Shimokawa statistic D_{L-S} [49]:

$$D_{L-S} = \frac{1}{\sqrt{n}}\sum_{i=1}^{n}\frac{\max\left[\frac{i}{n} - F_0(x_i;\hat{\theta}), F_0(x_i;\hat{\theta}) - \frac{i-1}{n}\right]}{\sqrt{F_0(x_i;\hat{\theta})\left[1 - F_0(x_i;\hat{\theta})\right]}} \tag{A13}$$

The Kuiper statistic D_K [50]:

$$D_K = \max_{1 < i \le n} \left(\hat{\delta}_i^+ \right) + \max_{1 < i \le n} \left(\hat{\delta}_i^- \right) \tag{A14}$$

where $\hat{\delta}_i^+ = \max\left[\frac{i}{n} - F_0\left(x_i; \hat{\theta}\right) \right]; \hat{\delta}_i^- = \max\left[F_0\left(x_i; \hat{\theta}\right) - \frac{i-1}{n} \right].$

Appendix C. Formulas Used in the Probabilistic Model of Maximum Precipitation of Bogdanowicz and Stachý Model

The Weibull probability distribution (extreme value type 3, EV3), $f(x)$, and quantile of maximum precipitation x_p are given as follows [1,2]:

$$f(x) = \frac{\lambda}{\theta - \varepsilon} \left[\frac{x - \varepsilon}{\theta - \varepsilon} \right]^{\lambda - 1} \exp\left\{ - \left[\frac{x - \varepsilon}{\theta - \varepsilon} \right]^{\lambda} \right\} \tag{A15}$$

$$x_p = \varepsilon + \alpha(-\ln p)^{\frac{1}{\lambda}} \tag{A16}$$

where ε is the lowest bound; $\varepsilon(\tau) = 1.42\tau^{0.33}$; θ is the quantile with probability of exceedance $1/e = 0.367\ldots$; λ is a shape parameter; and $\alpha = \theta - \varepsilon$ is a scale parameter.

References

1. Salas, J.D.; Obeysekera, J. Revisiting the concepts of return period and risk for nonstationary hydrologic extreme events. *J. Hydrol. Eng.* **2014**, *19*, 554–568. [CrossRef]
2. Derbile, E.K.; Kasei, R.A. Vulnerability of crop production to heavy precipitation in north-eastern Ghana. *Int. J. Clim. Chang. Strateg. Manag.* **2012**, *4*, 36–53. [CrossRef]
3. Tian, Q.; Li, Z.; Sun, X. Frequency analysis of precipitation extremes under a changing climate: A case study in Heihe River basin, China. *J. Water Clim. Chang.* **2020**, *12*, 772–786. [CrossRef]
4. Lazoglou, G.; Anagnostopoulou, C. An Overview of Statistical Methods for Studying the Extreme Rainfalls in Mediterranean. *Proceedings* **2017**, *1*, 4132. [CrossRef]
5. Acero, F.J.; Parey, S.; Garcia, J.A.; Dacunha-Castelle, D. Return Level Estimation of Extreme Rainfall over the Iberian Peninsula: Comparison of Methods. *Water* **2018**, *10*, 179. [CrossRef]
6. Jenkinson, A.F. The frequency distribution of the annual maximum (or minimum) of meteorological elements. *Q. J. R. Meteorol. Soc.* **1955**, *81*, 158–171. [CrossRef]
7. Lee, S.H.; Maeng, S.J. Frequency analysis of extreme rainfall using L-moment. *Irrig. Drain.* **2003**, *52*, 219–230. [CrossRef]
8. Ahmed, A.O.M.; Al-Kutubi, H.S.; Ibrahim, N.A. Comparison of the Bayesian and maximum likelihood estimation for Weibull Distribution. *J. Math. Stat.* **2010**, *6*, 100–104. [CrossRef]
9. Teimouri, M.; Nadarajah, S. Comparison of Estimation methods for the Weibull distribution. *Stat. J. Theor. Appl. Stat.* **2011**, *47*, 93–109. [CrossRef]
10. Ragulina, G.; Reitan, T. Generalized extreme value shape parameter and its nature for extreme precipitation using long time series and the Bayesian approach. *Hydrol. Sci. J.* **2017**, *62*, 863–879. [CrossRef]
11. Yuan, J.; Emura, K.; Farnham, C.; Alam, M.A. Frequency analysis of annual maximum hourly precipitation and determination of best fit probability distribution for regions in Japan. *Urban Clim.* **2018**, *24*, 276–286. [CrossRef]
12. Młyński, D.; Wałęga, A.; Petroselli, A.; Tauro, F.; Cebulska, M. Estimating Maximum Daily Precipitation in the Upper Vistula Basin, Poland. *Atmosphere* **2019**, *10*, 43. [CrossRef]
13. Katz, R.W.; Parlang, M.B. Naveau, Statistics of extremes in hydrology. *Adv. Water Resour.* **2002**, *25*, 1287–1304. [CrossRef]
14. Adlouni, S.E.; Ouarda, T.B.M.J.; Zhang, X.; Roy, R.; Bobée, B. Generalized maximum likelihood estimators for the nonstationary generalized extreme value model. *Water Resour. Res.* **2005**, *43*, 6–13. [CrossRef]
15. Bogdanowicz, E.; Stachý, J. *Maksymalne Opady Deszczu w Polsce—Charakterystyki Projektowe (Maximum Rainfalls in Poland—Design Values)*; Instytut Meteorologii i Gospodarki Wodnej: Warszawa, Poland, 1998; Volume 23, pp. 55–75.
16. Bogdanowicz, E.; Stachý, J. Maximum Rainfall in Poland—A Design Approach. In Proceedings of the Symposium on the Extremes of the Extremes: Extraordinary Floods, Reykjavik, Iceland, July 2000; pp. 15–18. Available online: https://www.researchgate.net/publication/295420771_Maximum_rainfall_in_Poland_-_A_design_approach (accessed on 22 September 2021).
17. Shahzadi, A.; Akhter, A.S.; Saf, B. Regional Frequency Analysis of Annual Maximum Rainfall in Monsoon Region of Pakistan using L-moment. *Pak. J. Stat. Oper. Res.* **2015**, *9*, 111–136. [CrossRef]
18. Raynal-Villasenor, J.A. Maximum likelihood parameter estimators for the two populations GEV distributions. *IJRRAS* **2012**, *11*, 350–357.
19. Alila, Y.; Mtiraoui, A. Implications of heterogeneous flood-frequency distributions on traditional stream-discharge prediction techniques. *Hydrol. Process.* **2002**, *16*, 1065–1084. [CrossRef]

20. Evin, G.; Merleau, J.; Perreault, L. Two-component mixture of normal, gamma and Gumbel distributions for Hydrological applications. *Water Resour. Res.* **2011**, *47*, W08525. [CrossRef]
21. Shin, J.Y.; Lee, T.; Ouarda, T.B. Heterogeneous Mixture Distributions for Modeling Multisource Extreme Rainfalls. *J. Hydrometeorol.* **2015**, *16*, 2639–2656. [CrossRef]
22. Rulfová, Z.; Buishand, A.; Roth, M.; Kyselý, J. A two-component generalized extreme value distribution for precipitation frequency analysis. *J. Hydrol.* **2016**, *534*, 659–668. [CrossRef]
23. Kleiber, W.; Katz, R.W.; Rajagopalan, B. Daily spatiotemporal precipitation simulation using latent and transformed Gaussian Processes. *Water Resour. Res.* **2012**, *48*, W01523. [CrossRef]
24. Evin, G.; Favre, A.-C. A new rainfall model based on the Neyman-Scott process using cubic copulas. *Water Resour. Res.* **2008**, *44*. [CrossRef]
25. Gyasi-Agyei, Y. Copula-based daily rainfall disaggregation model. *Water Resour. Res.* **2011**, *47*, W07535. [CrossRef]
26. Vernieuwe, H.; Vandenberghe, S.; De Baets, B.; Verhoest, N.E.C. A continuous rainfall model based on vine copulas. *Hydrol. Earth Syst. Sci.* **2015**, *19*, 2685–2699. [CrossRef]
27. Strupczewski, W.G.; Singh, V.P.; Feluch, W. Non-stationary approach to at-site flood frequency modelling. Part I. Maximum likelihood estimation. *J. Hydrol.* **2001**, *248*, 123–142. [CrossRef]
28. Strupczewski, W.G.; Singh, V.P.; Mitosek, H.T. Non-stationary approach to at-site flood frequency modelling. Part III. Flood analysis of Polish rivers. *J. Hydrol.* **2001**, *248*, 152–167. [CrossRef]
29. Villarini, G.; Smith, J.A.; Napolitano, F. Nonstationary modelling of a long record of rainfall and temperature over Rome. *Adv. Water Resour.* **2010**, *33*, 1256–1267. [CrossRef]
30. Lopez, J.; Frances, F. Non-stationary flood frequency analysis in continental Spanish rivers, using climate and reservoir indices as external covariates. *Hydrol. Earth Syst. Sci.* **2013**, *17*, 3189–3203. [CrossRef]
31. Cheng, L.; AghaKouchak, A.; Gilleland, E.; Katz, R.W. Nonstationary extreme value analysis in a changing climate. *Clim. Chang.* **2014**, *127*, 353–369. [CrossRef]
32. Zhang, D.; Yan, D.; Wang, Y.C.; Lu, F.; Liu, S. GAMLSS—based nonstationary modeling of extreme precipitation in Beijing-Tianjin-Hebei region of China. *Nat. Hazards* **2015**, *77*, 1037–1053. [CrossRef]
33. Debele, S.E.; Strupczewski, W.G.; Bogdanowicz, E. A comparison of three approaches to non-stationary flood frequency analysis. *Acta Geophys.* **2017**, *65*, 863–883. [CrossRef]
34. Serinaldi, F.; Kilsby, C.G. Stationarity is undead: Uncertainty dominates the distribution of extremes. *Adv. Water Resour.* **2015**, *77*, 17–36. [CrossRef]
35. Ozga-Zielinska, M.; Ozga-Zielinski, B.; Brzezinski, J. *Guidelines for Flood Frequency Analysis—Long Measurement Series of River Discharge*; Institute of Meteorology and Water Management (IMGW-PIB): Warszawa, Poland, 2005; Volume 44. Available online: https://serc.carleton.edu/hydromodules/steps/168500.html (accessed on 22 September 2021).
36. Ozga-Zieliński, B. Metody analizy niejednorodności ciągów pomiarowych zjawisk hydrologicznych (Methods of nonhomogeneity analysis of measurement series of hydrological phenomena). *Wiadomości Instututu Meteorologii Gospodarki Wodnej* **1999**, *22*, 13–32.
37. Spearman, C. The proof and measurement of association between two thigs. *Am. J. Psychol.* **1904**, *1*, 45–58. [CrossRef]
38. Pearson, F.R.S.K. On the inheritance of mental disease. *Ann. Hum. Genet.* **1931**, *4*, 362–380. [CrossRef]
39. Box, G.E.P. Non-normality and tests on variances. *Biometrika* **1953**, *40*, 318–335. [CrossRef]
40. Anderson, R.L. Distribution of the serial correlation coefficient. *Ann. Math. Statist.* **1941**, *8*, 1–13. [CrossRef]
41. Ciupak, M.; Ozga-Zielinski, B.; Brzezinski, J. *FFA and PFA Software, Version 1.1. Eng*, Available Free of Charge on Demand from authors by email maurycy.ciupak@imgw.pl or bogdan.ozga-zielinski@imgw.pl; Institute of Meteorology and Water Management: Warsaw, Poland, 2019.
42. Chapra, S.; Canale, R. *Numerical Methods for Engineers*, 5th ed.; McGraw-Hill Science/Engineering/Math: New York, NY, USA, 2006; p. 960.
43. Neyman, J. *Contributions to the Theory of the χ2 Test. Berkeley Symposium on Mathematical Statistics and Probability*; University of California Press: Berkeley, CA, USA, 1949; pp. 239–273.
44. Pratt, I.W.; Gibbons, I.D. *Concepts of Nonparametric Theory*; Springer: New York, NY, USA, 1981. [CrossRef]
45. Akaike, H. A new look at the statistical model identification. *IEEE Trans. Autom. Control* **1974**, *19*, 716–722. [CrossRef]
46. Genest, C.; Rémillard, B.; Beaudoin, D. Godness-of-fit tests for copulas: A review and power study. *Insur. Math. Econ.* **2009**, *44*, 199–213. [CrossRef]
47. Homaei, F.; Najafzadeh, M. A reliability—Based probabilistic evaluation of the wave-induced scour depth around marine structure piles. *Ocean Eng.* **2020**, *196*, 106818. [CrossRef]
48. Anderson, T.W.; Darling, D.A. A Test of Goodness of Fit. *J. Am. Stat. Assoc.* **1954**, *49*, 765–769. [CrossRef]
49. Liao, M.; Shimokawa, T. A new goodness-of-fit test for Type—I extreme value and 2-parameter Weibull distributions with estimated parameters. *J. Stat. Comput. Simul.* **1999**, *64*, 23–48. [CrossRef]
50. Koziol, J.A. A weighted Kuiper statistic for goodness of fit. *Stat. Neerl.* **2008**, *50*, 394–403. [CrossRef]
51. Institute of Meteorology and Water Management (IMGW-PIB). *Internal Report from Completion Parts I and II of PANDa Project*; Institute of Meteorology and Water Management (IMGW-PIB): Gdynia, Poland, 2017; Volume I. (In Polish)
52. Nguyen, T.-H.; Outayek, S.E.; Lim, S.H.; Nguyen, V.-T.-V. A systematic approach to selecting the best probability models for annual maximum rainfalls—A case study using data in Ontario (Canada). *J. Hydrol.* **2017**, *553*, 49–58. [CrossRef]

53. Burszta-Adamiak, E.; Licznar, P.; Zaleski, J. Criteria for identifying maximum rainfalls determined by the peaks-over-threshold (POT) method under the Polish Atlas of Rainfalls Intensities (PANDa) project. *Meteorol. Hydrol. Water Manag. Res. Oper. Appl.* **2019**, *7*, 3–13. [CrossRef]
54. Grubbs, F.E.; Beck, G. Extension of Sample Size and Percentage Points for Significance Tests of Outlying Observations. *Technometrics* **1972**, *14*, 847–854. [CrossRef]
55. US Department of the Interior, Geological Survey. *Guidelines for Determining Flood Flow Frequency—Bulletin 17B of the Hydrology Subcommittee*; Revised September 1981; Editorial Corrections March 1982; Interagency Advisory Committee on Water Data, US Department of the Interior, Geological Survey: Reston, VA, USA, 1982.
56. Pilon, P.J.; Condie, R.; Harvey, K.D. *Consolidated Frequency Analysis Package CFA User Manual for Version 1—DEC PRO Series*; Water Resources Branch, Inland Waters Directorate Environment Canada: Ottawa, ON, Canada, 1985.
57. Dahmen, E.R.; Hall, M.J. *Screening of Hydrological Data: Tests for Stationarity and Relative Consistency*; No. 49; International Livestock Research Institute (ILRI) Publication: Wageningen, The Netherlands, 1990.
58. Sneyers, R. *On the Statistical Analysis of Series of Observations*; TN No.143; WMO No. 415; World Meteorological Organization (WMO): Geneva, Switzerland, 1990.
59. Neyman, J.; Scott, E. *Outlier Proneness of Phenomena and of Related Distributions, Optimizing Methods in Statistics*; Academic Press: New York, NY, USA; London, UK, 1971.
60. Phillips, P.C.B. True Characteristic Function of F Distribution. *Biometrika* **1982**, *69*, 261–264. [CrossRef]
61. World Meteorological Organization (WMO). *Guide to Hydrological Practices*, 6th ed.; WMO No. 168; WMO: Geneva, Switzerland, 2009; Volume II.
62. Barzkar, A.; Najafzadeh, M.; Homaei, F. Evaluation of Drought Events in Various Climatic Conditions using Data-Driven Models and A Reliability-Based Probabilistic Model. *Nat. Hazards* **2021**, 1–22. [CrossRef]
63. Kite, G.W. *Frequency and Risk Analysis in Hydrology*; Water Resources Publications: Littleton, CO, USA, 2019; p. 187.
64. Dietrich, C.R.; Newsam, G.N. A Fast and Exact Method for Multidimensional Gaussian Stochastic Simulations. *Water Resour. Res.* **1993**, *29*, 2861–2869. [CrossRef]
65. Oliver, M.A.; Webster, R. Kriging: A Method of Interpolation for Geographical Information Systems. *Int. J. Geogr. Inf. Syst.* **1990**, *4*, 313–332. [CrossRef]
66. Ustrnul, Z.; Czekierda, D. Construction of the air temperature maps for Poland using GIS. *Archiwum Fotogrametrii Kartografii Teledetekcji* **2003**, *13*, 243–254.
67. Watson, D.F.; Philip, G.M. A Refinement of Inverse Distance Weighted Interpolation. *Geoprocessing* **1985**, *2*, 315–327.
68. Gupta, R.D.; Kundu, D. Exponentiated exponential family; an alternative to gamma and Weibull. *Biom. J.* **2001**, *43*, 117–130. [CrossRef]
69. Bailey, D.H.; Borwein, J.M.; Calkin, N.J.; Girgensohn, R.; Luke, D.R.; Moll, V.H. *Experimental Mathematics in Action*; A K Peters: Wellesley, MA, USA, 2007.
70. Hogg, R.V.; Craig, A.T. *Introduction to Mathematical Statistics*; Macmillan Publishing Co.: New York, NY, USA, 1978.
71. Willemse, W.J.; Kaas, R. Rational reconstruction of frailty-based mortality models by a generalisation of Gompertz'law of mortality. *Insur. Math. Econ.* **2007**, *40*, 468. [CrossRef]

Article

Spatiotemporal Characteristics and Trends of Meteorological Droughts in the Wadi Mina Basin, Northwest Algeria

Mohammed Achite [1,2], Andrzej Wałęga [3,*], Abderrezak Kamel Toubal [1], Hamidi Mansour [2] and Nir Krakauer [4]

1   Laboratory of Water & Environment, Faculty of Nature and Life Sciences, University Hassiba Benbouali of Chlef, Chlef 02180, Algeria; m.achite@univ-chlef.dz (M.A.); toubalabderrezak@gmail.com (A.K.T.)
2   Georessources, Environment and Natural Risks Laboratory, University of Oran, 2 Mohamed Ben Ahmed, P.O. Box 1015, El M'naouer, Oran 31000, Algeria; l_mansou_l@yahoo.fr
3   Department of Sanitary Engineering and Water Management, University of Agriculture in Krakow, Mickiewicza 24/28 Street, 30-059 Krakow, Poland
4   Department of Civil Engineering, The City College of New York, New York, NY 10031, USA; nkrakauer@ccny.cuny.edu
*   Correspondence: andrzej.walega@urk.edu.pl

Abstract: Drought has become a recurrent phenomenon in Algeria in the last few decades. Significant drought conditions were observed during the late 1980s and late 1990s. The agricultural sector and water resources have been under severe constraints from the recurrent droughts. In this study, spatial and temporal dimensions of meteorological droughts in the Wadi Mina basin (4900 km^2) were investigated to assess vulnerability. The Standardized Precipitation Index (SPI) method and GIS were used to detail temporal and geographical variations in drought based on monthly records for the period 1970–2010 at 16 rainfall stations located in the Wadi Mina basin. Trends in annual SPI for stations in the basin were analyzed using the Mann–Kendall test and Sen's slope estimator. Results showed that the SPI was able to detect historical droughts in 1982/83, 1983/84, 1989/90, 1992/93, 1993/94, 1996/97, 1998/99, 1999/00, 2004/05 and 2006/07. Wet years were observed in 1971/72, 1972/73, 1995/96, 2008/09 and 2009/10. Six out of 16 stations had significant decreasing precipitation trends (at 95% confidence), whereas no stations had significant increasing precipitation trends. Based on these findings, measures to ameliorate and mitigate the effects of droughts, especially the dominant intensity types, on the people, community and environment are suggested.

Keywords: drought; trends; SPI; mina basin; Algeria

Citation: Achite, M.; Wałęga, A.; Toubal, A.K.; Mansour, H.; Krakauer, N. Spatiotemporal Characteristics and Trends of Meteorological Droughts in the Wadi Mina Basin, Northwest Algeria. *Water* **2021**, *13*, 3103. https://doi.org/10.3390/w13213103

Academic Editor: Athanasios Loukas

Received: 14 September 2021
Accepted: 31 October 2021
Published: 4 November 2021

Publisher's Note: MDPI stays neutral with regard to jurisdictional claims in published maps and institutional affiliations.

1. Introduction

Drought is a recurring phenomenon that affects a wide variety of sectors, making it difficult to develop a single definition of it. According to a water-resource-oriented definition, which takes into account the water requirements related to biological, economic and social characteristics of a region, drought refers to a condition of severe reduction of water supply availability (compared to a normal value), extending along a significant period of time over a large region [1]. Drought is a complex phenomenon that involves different human and natural factors that determine the risk and vulnerability to it [2].

The particularly strong influence of drought on many sectors is visible in arid and semiarid regions, where water is scarce [3]. Water scarcity can strongly impact the agricultural sector in such regions [4]. In the case of the Mediterranean Basin, much of which is arid or semiarid, the extremely variable precipitation across temporal and spatial scales is influenced by geographical position of the region between two contrasting masses of water: the Atlantic Ocean and the Mediterranean Sea [5–7]. An additional feature determining high variability of precipitation in this region is the presence of various mountain ranges distributed along the coastal areas from east to west [7]. To avoid water scarcity, increased knowledge about variability of meteorological conditions could be used to mitigate the

153

effect of drought, as well as guide various irrigation scheduling and water productivity strategies in arid sandy soils. According to Rossi [8] the drought mitigation measures can be divided into three main categories: (1) water-supply oriented, such as using additional sources of low quality water and improvement of existing water system efficiency, (2) water-demand reduction: restriction of municipal uses and irrigation, pricing, dual distribution system, water recycling and (3) minimalization of drought impact by temporal relocation of water resources, tax relief, and development of warning systems. Knowledge about drought phenomena can also help with sustaining reforestation programs under an eventual increase in aridity [9] and with water resources planning and management via reservoirs to overcome scarcity [10].

Meteorological drought can be assessed using many indicators. For example, Weighted Anomaly Standardized Precipitation Index (WASP) was developed by Lyon [11] to monitor precipitation in the tropical regions. Crop Moisture Index (CMI) is commonly calculated weekly along with the Palmer Drought Severity Index (PDSI) output as a short-term drought indicator of impact on agriculture [12]. Drought Reconnaissance Index (DRI) [13] is based on a simplified water balance equation considering precipitation and potential evapotranspiration. Effective Drought Index (EDI) as a good index for operational monitoring of both meteorological and agricultural drought [14]. Hydro-thermal Coefficient of Selyaninov (HTC) developed by Selyaninov, Bokwa et al. [15] uses temperature and precipitation values, and is sensitive to dry conditions specific to the climate regime being monitored. RPI (Relative Precipitation Index) is the ratio of precipitation sum for the given period and the long-term average for the same period expressed in percent [15]. NOAA Drought Index (NDI) is a precipitation-based index in which the actual precipitation measured is compared with normal values during the growing season [16]. Palmer Drought Severity Index (PDSI) [17] uses monthly temperature and precipitation data along with information on the water-holding capacity of soils. SPEI (Standardized Precipitation Evapotranspiration Index) is a standardized monthly climatic balance computed as the difference between the cumulative precipitation and the potential evapotranspiration [18]. The Standardized Precipitation Index (SPI), developed by McKee et al. [19] in the 1990s, is robust and effective for evaluating meteorological drought and remains a very popular choice among researchers to reveal drought and to estimate duration and intensity of drought events [19]. The SPI has several advantages, as discussed by [20] and [21], over many other drought indices, such as some of those mentioned above. Firstly, it is based only on rainfall, so that in the absence of other hydro-meteorological measurements, drought assessment is still possible. Secondly, SPI can be used to quantify precipitation deficit for multiple timescales, which enables it to assess drought conditions in meteorological, hydrological and agriculture applications. Finally, standardization of the SPI index ensures that the frequency of extreme drought events at any location and any timescale is approximately constant.

Due to its robustness and convenience, SPI has already been widely used to characterize dry and wet conditions in many countries in the Mediterranean region, such as Turkey [22,23], Spain [24,25], Italy [26–30], Iran [31–33], Greece [34–36], Iraq [37–39], and Palestine [40].

In particular, many researchers in North Africa have studied meteorological drought using SPI indices, including in Algeria [41–45], in Morocco [46], and in Tunisia [47–49]. So far, there has not been a study on spatial and temporal variations of meteorological drought, expressed by SPI, in the region of the Wadi Mina basin of northwest Algeria, which is characterized by high intensities of agriculture and presence of forest cover. According to [50], renewable water resources in Algeria are quite low and can be approximated as 19 billion cubic meters per year. In the other words, the water resources are equal 450 cubic meters (m^3) per capita per year and are slightly below the 500 m^3 per capita per year that is recommended as the scarcity threshold indicating a water crisis. Moreover, the water resources have high variability and projections are that rainfall could decrease by more than 20% by 2050, which would result in greatly worsening water shortages in different basins of Algeria [51]. Knowledge on extreme dry conditions is also very important because these can influence not only on water scarcity for agriculture but also on natural ecosystems, mainly forest in the case of the studied basin. For example, Mensah et al. [52] showed that elevated temperatures will further exacerbate the drought impacts on forest ecosystems at sites with precipitation levels equal or smaller than the atmospheric evaporative demand and strong influence of vapor pressure deficits on carbon uptake, and can worsen the decline in soil moisture.

In this work, the objective was to better characterize annual-scale drought patterns over the Wadi Mina basin in order to aid water resource planning. The main objectives are to (1) map characteristics of drought patterns over the basin during 1970–2010, (2) identify any trends in precipitation or in drought characteristics, (3) identify drought years over the observation period, and (4) estimate the return periods for severe drought across the basin.

2. Study Area and Data

2.1. Study Area

The Wadi Mina basin, with an area of 4900 km^2, is located in the northwest of Algeria (Figure 1). The Wadi Mina involves four major tributaries: Wadi Mina, Wadi Haddad, Wadi Abd and Wadi Taht. The climate is continental, with cold winters and hot summers. Mean annual precipitation ranges from about 220 to 400 mm, and most precipitation occurs between November and March. Mean annual temperatures are about 16 °C to 19.5 °C. Almost half the basin is covered by a varying density of vegetation, with in particular 32% of scrub, 35.8% of forests and 20% cereal crops [53].

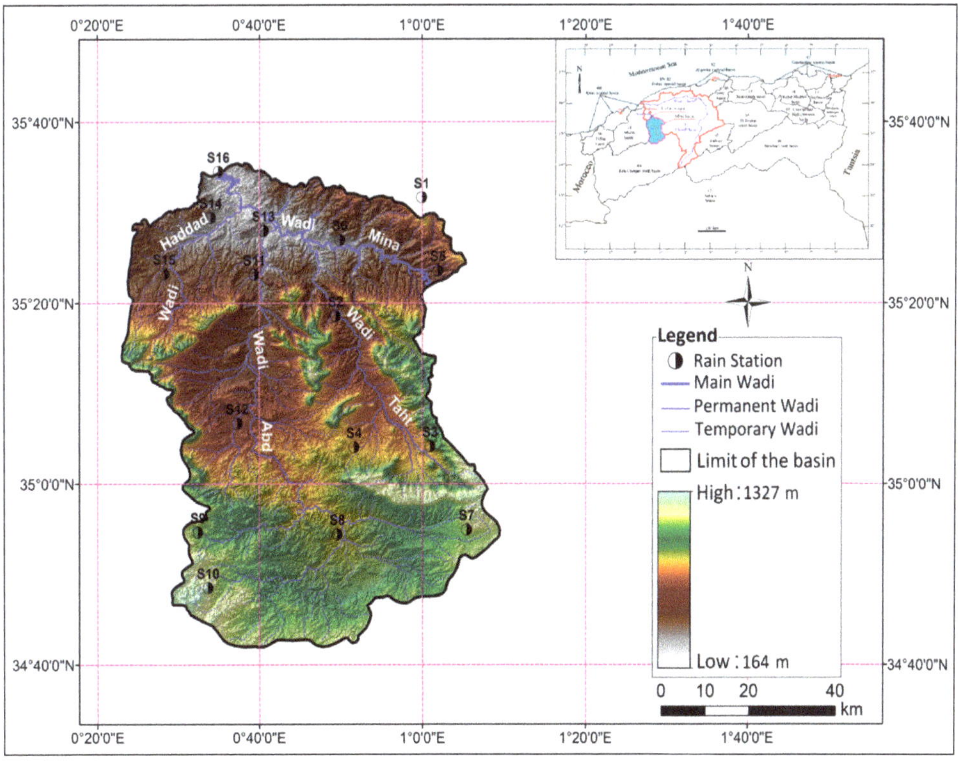

Figure 1. Topography and station distribution for the Wadi Mina basin in northern Algeria.

2.2. Data Used

Monthly precipitation records for a 40-year observation period (September 1970 to August 2010, using water years that go from September to August) are compiled for 16 stations from the Algeria National Agency of Water Resources (Figure 1 and Table 1). These stations constitute a relatively well-distributed network with acceptable spatial density over the basin. To assure quality, data was checked for inhomogeneities using the double mass curve, linear regression and Mann-Whitney test methods. The procedure detected a few inhomogeneities, for which the irregular data were adjusted using data of nearby reliable stations. Rainfall data of these 16 stations were analyzed statistically to evaluate rainfall variability in the study area (Table 2). These preliminary statistical analyses included measure of central tendency (mean, and median), dispersion (standard deviation SD, coefficient of variation CV) and distribution (skewness Cs and kurtosis Ck) (Table 3).

Table 1. Characteristics of rain gauge stations used in the analysis.

Rain Station	ID	Name	Geographical Coordinates		Elevation	Period of Observation
			Longitude (E)	Latitude (N)		
			(°)	(°)	(m)	
S1	12702	Rahuia	1°00′	35°31′	650	
S2	13001	Kef Mahboula	0°49′	35°18′	475	
S3	13002	Frenda	1°01′	35°04′	990	
S4	13004	Ain El Haddid	0°51′	35°04′	829	
S5	13101	Mechra Safa	1°02′	35°23′	655	
S6	13102	Djilali Benamar	0°49′	35°27′	300	
S7	13201	Ain Kermes	1°05′	34°55′	1162	
S8	13202	Rosfa	0°49′	34°54′	960	Septmber 1970–August 2010
S9	13203	Tiricine	0°32′	34°54′	1070	
S10	13204	Sidi Youcef	0°33′	34°48′	1100	
S11	13302	Ain Hamara	0°39′	35°23′	288	
S12	13304	Takmaret	0°37′	35°06′	655	
S13	13306	Oues El-Abtal	0°40′	35°28′	354	
S14	13401	Sidi A.E.K Djilali	0°34′	35°29′	225	
S15	13407	El Hachem	0°28′	35°23′	417	
S16	13410	SMBA	0°35′	35°34′	145	

Table 2. Descriptive statistics of annual rainfall series in the Wadi Mina basin (1970/71–2009/10 water years).

N°	Min (mm)	Max (mm)	Mean (mm)	Median (mm)	SD (mm)	Cv (%)	Cs	Ck
S1	210.00	524.70	352.53	333.10	89.27	25.32	−0.87	0.19
S2	143.00	672.20	343.63	326.85	106.90	31.11	1.06	0.88
S3	221.00	672.90	396.42	388.00	11203	28.26	0.09	0.61
S4	194.80	610.00	312.83	302.65	102.92	32.90	1.60	1.23
S5	197.70	734.40	378.03	366.40	119.22	31.54	1.02	0.88
S6	158.60	645.10	345.38	314.35	120.84	34.99	0.15	0.75
S7	155.70	580.20	323.70	320.80	107.93	33.34	0.25	0.83
S8	77.70	557.00	218.40	187.80	113.76	52.09	2.18	1.55
S9	115.20	561.50	306.84	306.75	104.40	34.02	0.11	0.54
S10	159.20	631.00	294.89	270.40	99.59	33.77	1.76	1.15
S11	164.80	506.40	265.10	260.55	74.97	28.28	3.13	1.51
S12	120.50	413.10	254.25	241.65	73.14	28.77	−0.34	0.57
S13	129.60	558.00	278.65	266.10	84.84	30.45	2.12	1.18
S14	135.60	474.20	254.13	239.55	72.12	28.38	1.33	1.08
S15	152.60	517.00	291.01	276.25	78.85	27.10	0.21	0.57
S16	141.00	436.60	237.97	226.95	63.09	26.51	1.86	1.15

To test for stationarity, the Kwiatkowski–Phillips–Schmidt–Shin test (KPSS) was used [54]. The results of the stationarity test for monthly, seasonal and yearly series are shown in Table 3. All the monthly, seasonal and annual series of the rainfall stations are indicated as showing stationarity (*p*-value is more than 0.05).

Table 3. Results of stationarity tests for the monthly, seasonal and yearly series.

Station	Monthly Series *p*-Value	Seasonal Series *p*-Value	Yearly Series *p*-Value
S1	0.576	0.427	0.412
S2	0.459	0.529	0.425
S3	0.756	0.871	0.345
S4	0.842	0.777	0.310
S5	0.912	0.867	0.610
S6	0.956	0.569	0.524
S7	0.875	0.784	0.459
S8	0.758	0.657	0.351
S9	0.910	0.741	0.301
S10	0.986	0.891	0.295
S11	0.886	0.741	0.287
S12	0.782	0.625	0.254
S13	0.975	0.412	0.210
S14	0.754	0.541	0.354
S15	0.621	0.459	0.311
S16	0.524	0.567	0.421

3. Methodology

3.1. SPI

The standardized precipitation index (SPI) is commonly used to detect meteorological drought. Each drought is characterized by drought intensity (Di), a drought magnitude (Dm) and drought duration (Dd). Run intensity can be either the value of the SPI at any moment (Dint) or the minimum SPI value during a drought event (Dmi) The drought magnitude (Dm) is equal to the accumulated values of below-threshold SPI during each drought event (Figure 2).

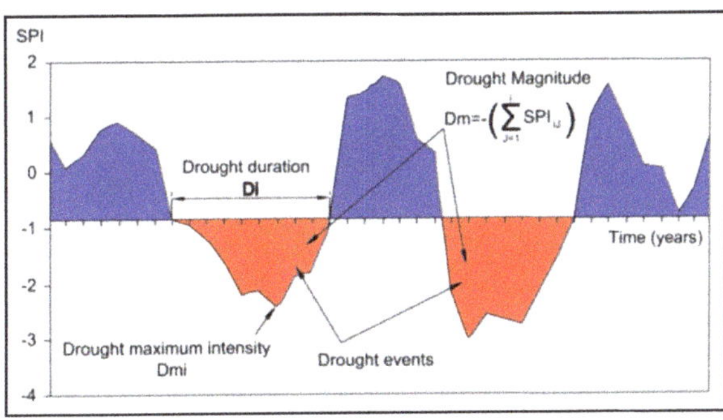

Figure 2. Definition of drought properties based on the SPI index [55].

SPI is mathematically based on the cumulative probability of monthly precipitation amount recorded at the observation post [56,57]. No evaporation estimate is considered, unlike other drought indices such as SPEI. SPI = 0 denotes average (climatological) precipitation, SPI = 1 denotes 1 standard deviation wetter than average, and SPI = −1 denotes 1 standard deviation drier than average. In the case of the presented analysis, the monthly precipitations were aggregated over water years, and finally a yearly SPI (12-month timescale) for each water year was calculated. SPI periods (years) with SPI below the defined threshold are considered drought years, and consecutive drought years are grouped into droughts. The whole period of observation at a meteorological station is used to determine the parameters of a precipitation probability density function, taken to be in the form of a gamma distribution:

$$g(x) = \frac{1}{\beta^{\alpha}\Gamma(\alpha)} x^{\alpha-1} e^{-x/\beta} \tag{1}$$

where α and β are the shape and scale parameters respectively. x is consecutively precipitation and $\Gamma(\alpha)$ is the gamma function. The gamma function defined by the following:

$$\Gamma(a) = \int_0^\infty y^{a-1} e^{-y} dy \tag{2}$$

The alpha and beta parameters of the gamma distribution are estimated from the precipitation time series as

$$\alpha = \frac{1}{4A}\left(1 + \sqrt{1 + \frac{4A}{3}}\right), A = ln(x) - \frac{\sum ln(x_i)}{n}, \beta = \frac{x}{\alpha} \tag{3}$$

where x is the mean value of precipitation quantity; n is the precipitation measurement number; x_i is the quantity of precipitation in a sequence of data.

The cumulative probability can be presented as:

$$G(x) = \int_0^x g(x)dx = \frac{1}{\hat{\beta}^{\hat{a}}\Gamma\left(\hat{\alpha}\right)} \int_0^x x^{\alpha_{pro}-1} e^{-x/\beta_{pro}} dx \tag{4}$$

To allow for the possibility that the precipitation may be zero, a mixture probability distribution is used, for which the cumulative probability becomes

$$H(x) = q + (1-q)G(x) \tag{5}$$

where q is the probability that the quantity of precipitation equals zero.

The calculation of the SPI is presented on the basis of the following equation [20,58]:

$$SPI = \begin{cases} -\left(t - \frac{c_0 + c_1 t + c_2 t^2}{1 + d_1 t + d_2 t^2 + d_3 t^3}\right). & 0 < H(x) \le 0.5 \\ +\left(t - \frac{c_0 + c_1 t + c_2 t^2}{1 + d_1 t + d_2 t^2 + d_3 t^3}\right). & 0.5 < H(x) \le 1.0 \end{cases} \tag{6}$$

where t is determined as

$$t = \begin{cases} \sqrt{ln\left(\frac{1}{(H(x))^2}\right)}. & 0 < H(x) \le 0.5 \\ \sqrt{ln\left(\frac{1}{1-(H(x))^2}\right)}. & 0.5 < H(x) \le 1.0 \end{cases} \tag{7}$$

and c_0. c_1. c_2. d_1. d_2 and d_3 are coefficients whose values are:

$$c_0 = 2.515517. \ c_1 = 0.802853. \ c_2 = 0.010328$$

$$d_1 = 1.432788. \ d_2 = 0.189269. \ d_3 = 0.001308$$

According to McKee et al. [18] different categories and approximate probabilities of wet and dry spells can be considered based on SPI for the timescale of interest, as shown in Table 4. SPI is expected to follow a near-normal (bell curve) distribution, with SPI values near 0 being the most common and high positive or negative SPI (corresponding to very wet or very dry periods, respectively) being rare.

Table 4. Drought classification based on SPI value and corresponding event probabilities based on the approximation that SPI values follow a standard normal distribution.

SPI Values	Drought Category	Probability (%)
2.00 or more	Extremely wet	2.3
1.50 to 1.99	Very wet	4.4
1.00 to 1.49	Moderately wet	9.2
−0.99 to 0.99	Near normal	68.2
−1.00 to −1.49	Moderate drought	9.2
−1.50 to −1.99	Severe drought	4.4
−2.00 or less	Extreme drought	2.3

These probabilities shown in Table 4 are estimates, assuming that SPI is normally distributed. Achieving an approximately standard normal probability distribution is the main motivation behind the transformation of precipitation to SPI.

3.2. Trend Analysis

Trend analysis determines whether the measured values of a variable show a consistent increase or decrease during a time period. Many statistical methods can be used for trend detection in a time series of meteorological and hydrological records. In this study we used simple and accepted methods for evaluating trends, the Mann–Kendall test and Sen's estimator of slope.

The Mann–Kendall method is a widely used non-parametric test for detecting trends in climatological and hydrological time series. It has been suggested by many authors to assess trends in environmental data time series because, unlike least-squares linear regression, it is robust to outlying and extreme values.

The Mann–Kendall test statistic S is given by [59]:

$$S = \sum_{k=1}^{n-1} \sum_{j=k+1}^{n} sgn\left(x_j - x_k\right) \tag{8}$$

where n is the number of data. x are the data values at times j and k ($j > k$) and the sign function is

$$sgn\left(x_j - x_k\right) = sgn\left(R_j - R_i\right) = \begin{cases} +1, & if\left(x_j - x_k\right) > 0 \\ 0, & if\left(x_j - x_k\right) = 0 \\ -1, & if\left(x_j - x_k\right) < 0 \end{cases} \tag{9}$$

The variance of S is computed by

$$Var(S) = \frac{[n(n-1)(2n+5)] - \sum_{i=1}^{m} t_i(t_i - 1)(2t_i + 5)}{18} \tag{10}$$

where t_i is the number of ties of extent i and m is the number of tied rank groups. For n larger than 10, a Z test statistic that, under the null hypothesis of no correlation, approximates a standard normal distribution is computed as the Mann–Kendall test statistic as follows:

$$Z = \begin{cases} \frac{S-1}{\sqrt{Var(S)}}, & if\ S > 0 \\ 0, & if\ S = 0 \\ \frac{S-1}{\sqrt{Var(S)}}, & if\ S < 0 \end{cases} \tag{11}$$

If a linear trend is present in a time series, then the true slope (change per unit time) can be estimated by using a simple non-parametric procedure developed by Sen [60]. The slope estimates of the $n(n-1)/2$ unique pairs of data are first computed by:

$$Q(i,j) = \frac{X_j - X_i}{j - i} \text{ for } i,\, j = 1,\, 2.\ldots.n \tag{12}$$

where x_j and x_i are data values at time j and i ($j > i$). respectively. The median of these N values of Q is Sen's estimator of slope. After sorting the Q values, if N is even, then Sen's estimator is calculated by:

$$Q_{med} = \frac{1}{2}\left(Q_{\frac{N}{2}} + Q_{\frac{N+2}{2}}\right) \tag{13}$$

If N is odd, then Sen's estimator is computed by:

$$Q_{med} = \left(Q_{\frac{N+1}{2}}\right) \tag{14}$$

Sen's estimator Q_{med} provides the rate of change and enables determination of the total change in any variable during the analysis period. Sen's slopes are expressed here as rate of change per 40 years (1970–2010) in mm.

3.3. Drought Charcateristics

3.3.1. Frequency Analysis

Drought frequency (F_i) is the chance of a station being in drought in a given year. This was estimated empirically based on the following formula:

$$F_i = \frac{n}{N}100\% \tag{15}$$

where n—number of years of drought (SPI equal 0 or less), N—number of analyzed years.

3.3.2. Drought Intensity (DI)

Drought intensity (DI) is used to represent the severity of the drought. The drought intensity of a site within a certain period is usually reflected by the SPI value. The more negative the SPI value, the more serious the drought is. Its formula is as follows:

$$D_i = \left(\frac{1}{m}\sum_{i=1}^{m}|SPI_i|\right)j \tag{16}$$

3.3.3. Drought Magnitude (DM)

DM corresponds to the cumulative water deficit over a drought period. DM is the sum of the absolute values of all SPI values (0 or less) during a drought event (Equation (16)):

$$DM = -\sum_{j=1}^{i} SPI_{i,j} \tag{17}$$

3.3.4. Drought Duration (DD)

DD equals the number of time periods between the drought start and its end. In our case, we consider all SPI values below 0 as drought years.

3.4. Return Period of Drought

In addition to computing drought frequencies as empirical probabilities in the 40-year observation record, return periods of severe drought were also computed in this study using the annual maximum series (AMS) approach. The AMS here is based on the time series of SPI values for drought years. A drought was described as an SPI value less than zero. Drought-free years were given a zero value. The number of years for which

SPI values are available is used to calculate the duration of the sequence. Only non-zero values were used in the drought frequency calculation. To account for the number of zero values, a correction was made using nonexceedance likelihood (F') according to the following expression [55,61]:

$$F' = q + (1-q)F \tag{18}$$

where F is the non exceedance probability value obtained by using frequency analysis on the non zero values and q is the probability of zero values which can be calculated as the ratio of the number of time intervals without drought occurrences to the total number of time intervals in the recording period [55,61].

To estimate the return period of drought severity that may go beyond the values observed over the 40-year period for which we have data, we fitted a probability distribution to the derived AMS. In this case, the drought event time series were fitted with gamma distributions. The return period of drought with particular severity was then calculated as:

$$F'(s) = \frac{1}{1 - F'(x)} \tag{19}$$

4. Results and Discussion

4.1. Temporal Variability

The SPI was used to provide an indicator of drought severity in this study. The temporal characteristics of droughts in Wadi Mina basin was analyzed based on the 12-month timescale water-year SPI computed for each station (Figure 3). Analysis of the computed SPI series shows the basin has experienced droughts of high severity and duration in the 1980s and 1990s. A drought is defined whenever the SPI reaches a value of 0.00 and continues until the SPI becomes positive again.

The main historical droughts observed in the study area were in 1982/83, 1983/84, 1989/90, 1992/93, 1993/94, 1996/97, 1998/99, 1999/00, 2004/05 and 2006/07. Wet years were observed in 1971/72, 1972/73, 1995/96, 2008/09 and 2009/10. A decreasing trend of SPI, implying a likely increased frequency and intensity of drought, was observed on 13 of 16 rain gauge stations. Most of the stations with the strongest decreasing SPI trend are observed in the lower part of the Wadi Mina basin where are observed relatively lower sums of precipitation (Table 2). Increase of trend of SPI and likely decreased intensity of drought is observed on three rain gauge stations located mainly in upper part of the basin, in the Wadi Abd tributary. Spatio-temporal changes of SPI is caused by change of precipitation. Elouissi et al. [62] detected similar decreasing trends of precipitation in the northern part of the Macta basin (Algeria), close to the Mediterranean coast, and increasing trends in the southern part. The changes of precipitation and SPI can be affected by geographical position of the area in relation to the Atlantic Ocean, the Mediterranean Sea and the Atlas mountain ranges [63]. We can also see from Figure 3 that dry periods have tendency to cluster over long stretches of years. Clustering is especially visible in station S8 during 1975–1993, S5 (1981–1999), S6 (1981–1999) and S13 (1996–2007). Figure 3 also shows that at station S3 located in the upper part of the Wadi Taht subbasin, and S6 and S5 in the upper Wadi Mina, intensity of meteorological drought since 2000, expressed by SPI, was small, with wet years being more common.

Trend analysis determines whether the measured values of a variable show a significant increase or decrease during a time period. In this study, we used a simple method for evaluating trends, Mann-Kendall test and annual and seasonal Sen's slopes of trend values are expressed as rate of change per 40 years (1970–2010) in mm. The result of this analysis is shown in Table 5. At 7 of 16 rain gauge stations, or 44% all stations, there was a significant negative trend ($p < 0.1$). The significance level of trend in 6 cases was $p < 0.05$, and in the case of S13 station the p value was under 0.01. At most stations, from 1996 onward there were mainly severe droughts. Significant decreasing trends were observed at stations located in the upper part of the Wadi Mina and middle part of the Wadi Taht tributary. A significant increasing trend was not detected at any of the Wadi Mina basin stations.

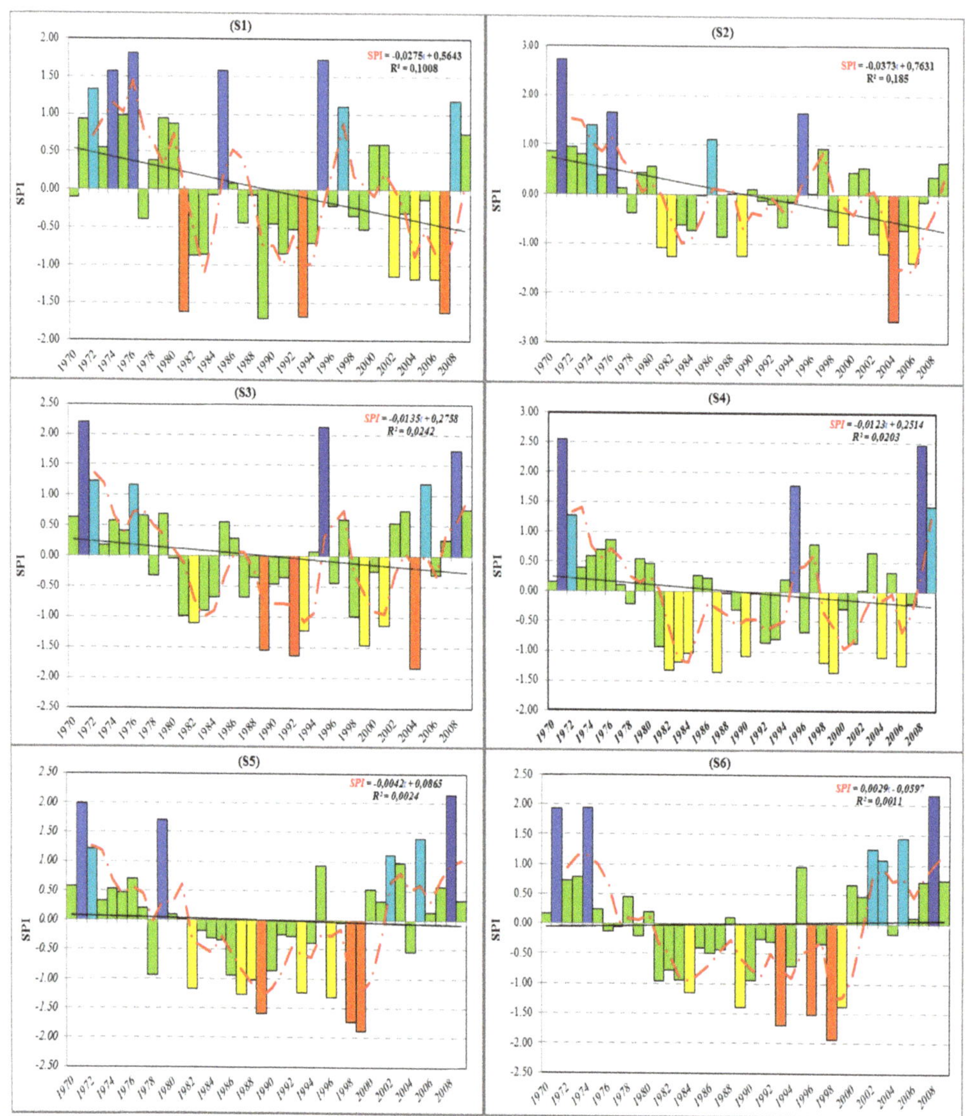

Figure 3. *Cont.*

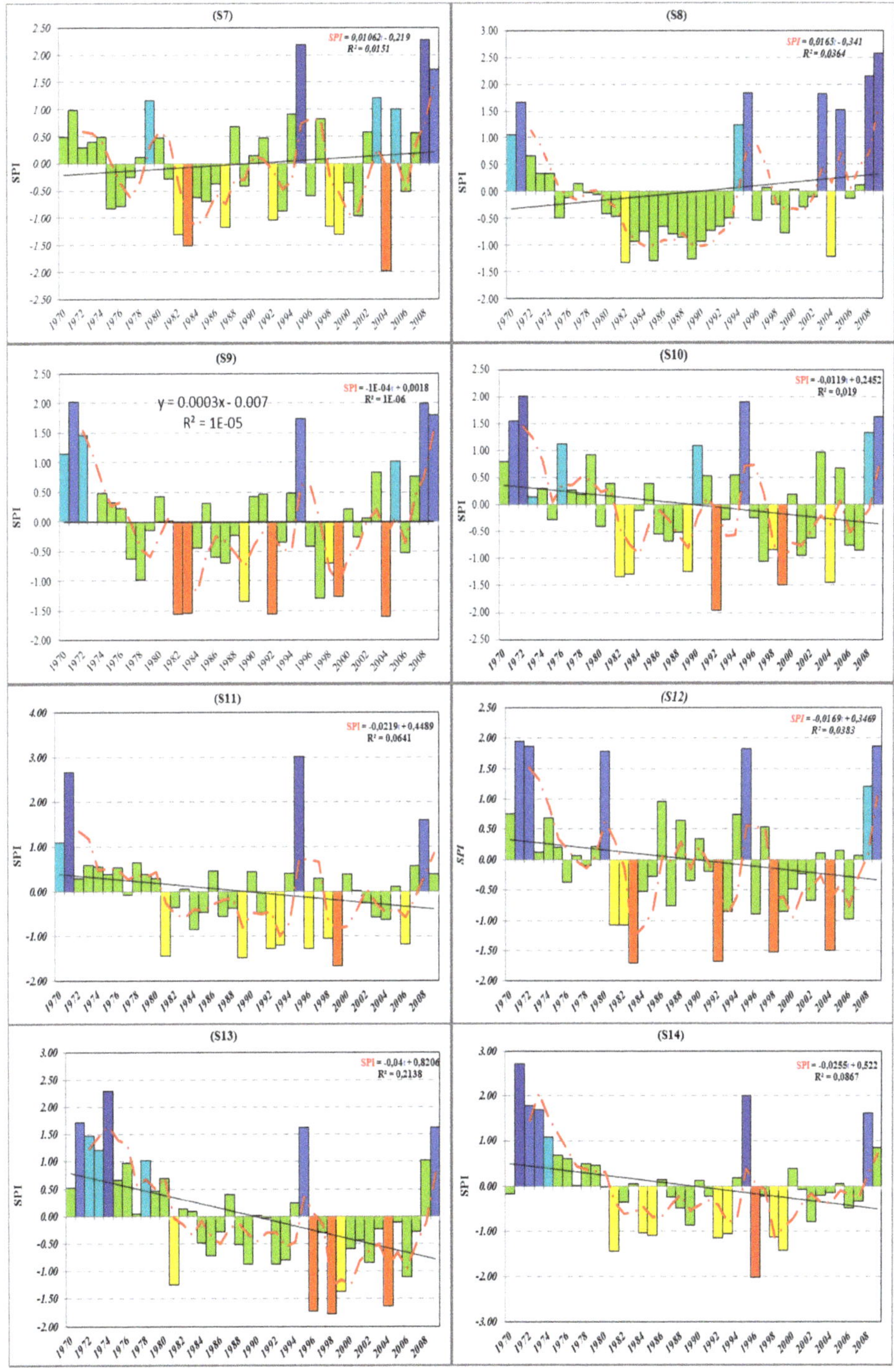

Figure 3. *Cont.*

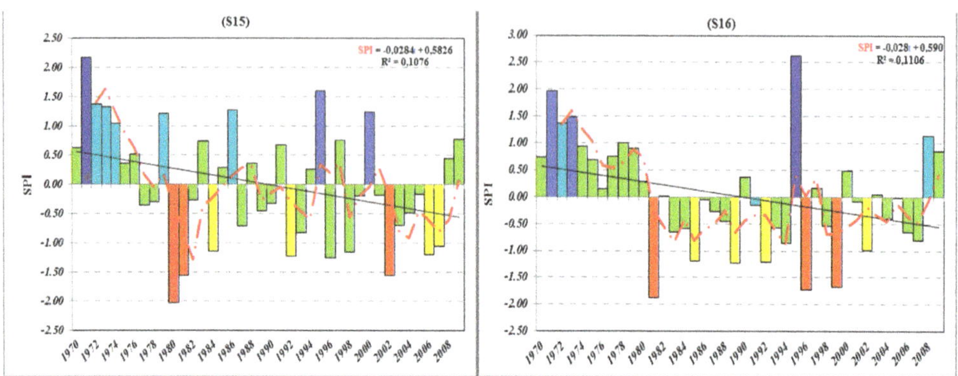

Figure 3. Annual SPI time series and its linear trend and 3 year moving average of the pluviometric stations of Wadi Mina basin. Note that colors are linked with drought classification based on Table 4. Number in branches above figures is number of station.

Table 5. Values of statistics b, Z of the Mann–Kendall test for the annual SPI series (1970–2010).

Stations	Area (km^2)	Z	Sen's Slope	Describtion
S1	57.40	−1.969	−0.031	Significant at 95% level of confidence or $p = 0.05$
S2	413.10	−2.540	−0.038	Significant at 95% level of confidence or $p = 0.05$
S3	160.60	−1.037	−0.015	
S4	560.20	−1.340	−0.02	
S5	150.30	−0.722	−0.012	
S6	254.60	0.011	0.001	
S	165.20	0.163	0.002	
S8	607.40	0.524	0.063	
S9	398.90	0.000	0.080	
S10	534.40	−1.002	0.334	
S11	261.70	−2.005	0.455	Significant at 95% level of confidence or $p = 0.05$
S12	568.90	−1.270	−0.019	
S13	193.30	−3.251	−0.906	Significant at 99% level of confidence or $p = 0.01$
S14	205.40	−1.969	−0.554	Significant at 95% level of confidence or $p = 0.05$
S15	300.40	−1.899	−0.653	Significant at 90% level of confidence or $p = 0.1$
S16	68.30	−2.237	−0.650	Significant at 95% level of confidence or $p = 0.05$

Table 1 presents drought classification for the 16 rain gauge stations in each year. The most common SPI category overall was near normal (NN). For several years (1971, 1972, 1995, 2008 and 2009) most stations were in wet categories (EW, VW and MW). For 1971, 1995 and 2008 only 2–5 out of 16 stations were dry, and no severe or extreme drought was observed. The highest number of stations with severe or extreme drought (SD and ED) was observed in the years 1981, 1983, 1989, 1992, 1996, 1998, 1999 and 2004. The highest number of years with unusually wet conditions (MW, VW and EW) were observed on stations S13 and S15-8 cases. These stations were located in the lower part of the Wadi Mina. The most cases of intense drought (ED, SD and ED) were observed at station S9-9 cases, and the highest number of years with severe and extreme drought were observed at stations S1, S9 and S12-4 cases.

4.2. Spatial Variability

To visualize the distribution of droughts in the basin, the study area is divided using Theissen Polygon tool in Arc GIS 10.2 into 16 polygons corresponding to the 16 rainfall stations. Stations that are closely spaced are assigned less area and vice versa (Figure 4). Lee et al. [64] showed that the spatial distribution of the rain gauge networks and the density have a significant influence on accurately calculating areal precipitation and Thiessen method gave good results when the spatial distribution of the rain gauge networks was even, as was the case here. Moreover, the weights assigned to the different stations do not vary with time, and thus it is easy to map the precipitation falling during each period. Geostatistical methods offer more sophisticated approaches to making maps based on station data, but the uncertainty of areal precipitation is in any case high if there are relatively few stations, like in this basin [65]. Even though some stations may show drought conditions, a regional drought is acknowledged only when some major portion of the total study area is under drought. Regional drought is determined by the intra-annual precipitation distribution, which can be affected by teleconnection patterns [66,67], and the North Atlantic Oscillation indices [68]. Moreover, regional-scale influence on the rainfall conditions in North Africa could result from the response of the African summer monsoon to oceanic forcing, amplified by land-atmosphere interaction [69].

The spatial distribution of drought intensity is shown (Figure 4) in each analyzed year. In 1971, 1995, and 2008, wet conditions prevailed over almost all the Wadi Mina basin (SPI $\geq$ 1.0). Less widespread wet conditions were seen in 1972 (east and central part of basin in wet condition) and 2009 (upper and middle part of basin). No droughts were seen between 1970 and 1979 in the region. The year 1980 is an example of intra-basin variability: almost all area of basin had near normal conditions, but particular areas had either very wet conditions (middle part of the Wadi Abd catchment) or extreme drought (upper part of the Wadi Haddad tributary). The years where a large part of the Wadi Mina basin was in drought were 1982, 1989, 1999, 2004 and 2006, but the worst situation was in 2004, where all the upper and middle parts of the basin had moderate to extreme drought. Spatial patterns of drought within the basin varied unpredictably during the study period, which could be due to the complex interaction of storm tracks with orographic features.

4.3. Drought Evaluation Indicators

4.3.1. Frequency Analysis

Drought frequency calculated for all analyzed stations is presented in Table 6. Near normal (NN) conditions occurred most frequently at all stations (57.5% to 72.5% of the time, depending on station). The extreme categories—extreme wetness (EW) and extreme drought (ED)—were the least frequently observed. Extreme drought only occurred at 3 of 16 rainfall gauge stations over the 40-year observation period. However, all but 2 rainfall gauge stations (S4 and S8) observed either ED and SD.

Table 2 shows drought duration, magnitude and intensity, as well as average, maximum and minimum SPI at annual time scale for the meteorological stations considered in this study area. All drought indicators have strong variability over the study area. Drought duration (DD) varied between 1 to 16 years, and most frequently was only 1 year. The highest DD of 12, 13 and 16 years were observed at stations S13, S6 and S5, respectively. The highest drought intensity was observed at S16 station and lowest at S8. The largest drought magnitude was observed at S5 and the smallest at S11. Extreme drought was observed in 1980/81, 1996/97 and 2004/05 in stations located in the lower part of basin.

Table 6. Frequency of each drought and wetness class for the considered stations, %.

	S1	S2	S3	S4	S5	S6	S7	S8	S9	S10	S11	S12	S13	S14	S15	S16
EW	0	2.5	5	5	0	2.5	5	5	2.5	2.5	5	0	2.5	5	2.5	2.5
VW	10	5	2.5	2.5	5	5	2.5	10	7.5	7.5	2.5	13	7.5	7.5	2.5	5
MW	7.5	5	7.5	5	7.5	7.5	7.5	5	7.5	7.5	2.5	2.5	10	2.5	15	5
NN	65	73	70	65	65	70	68	70	65	65	70	70	65	65	58	73
MD	7.5	13	7.5	23	13	7.5	13	10	7.5	13	18	5	7.5	18	15	7.5
SD	10	0	7.5	0	7.5	7.5	5	0	10	5	2.5	10	7.5	2.5	5	7.5
ED	0	2.5	0	0	2.5	0	0	0	0	0	0	0	0	0	2.5	0
Sums	100	100	100	100	100	100	100	100	100	100	100	100	100	100	100	100

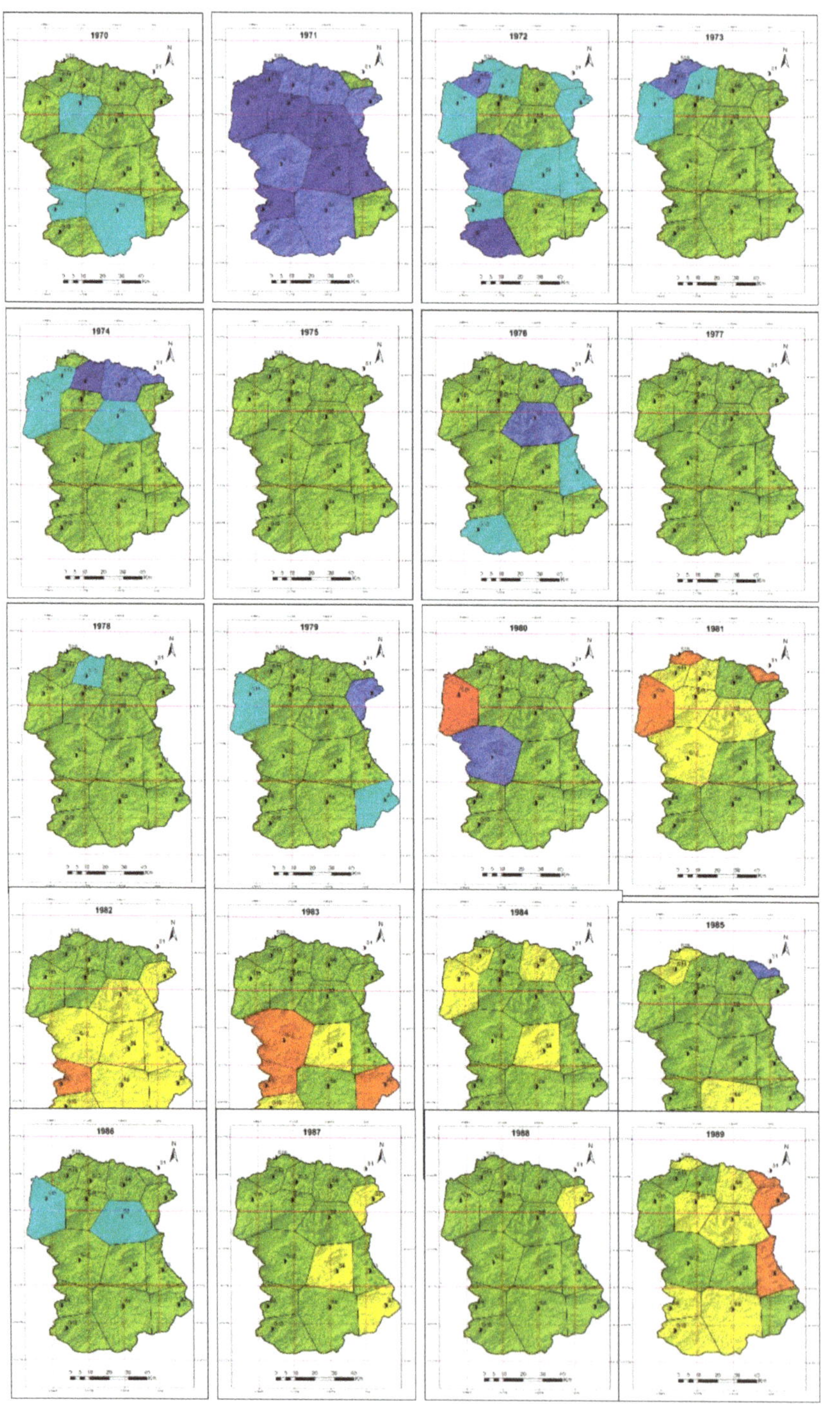

Figure 4. *Cont.*

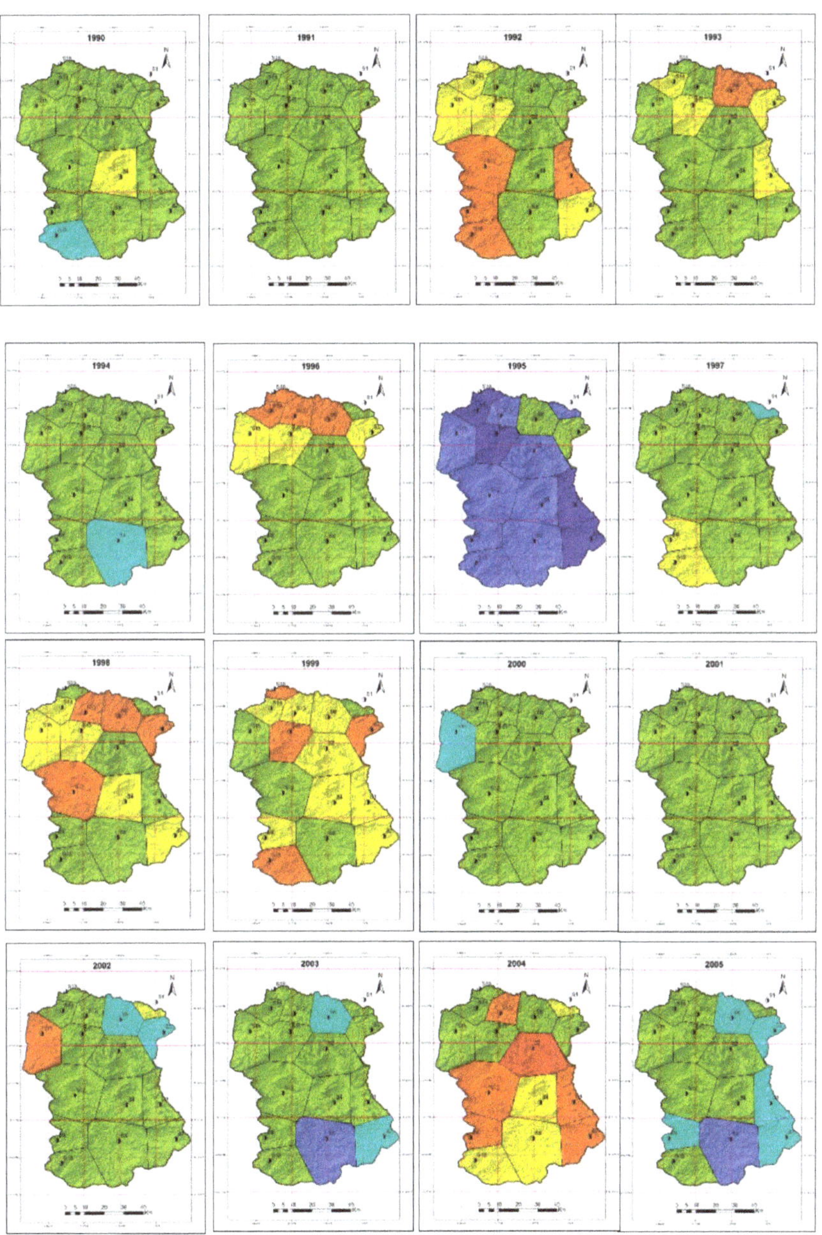

Figure 4. *Cont.*

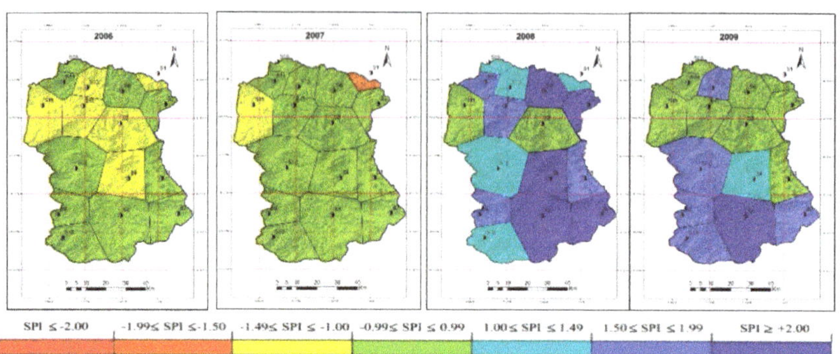

Figure 4. Spatial variability of meteorological drought in the wadi Mina basin.

4.3.2. Calculation of Return Period and the Severity-Area-Frequency (SAF) Curve

The Severity-Area-Frequency (SAF) curve is a very useful method for showing the spatial extent of different types of drought in a given area. This technique has been undertaken by several researchers around the world, for example in India [70]; China [71]; Southern Africa [72] and in Iraq [39]. Figure 5 illustrates the drought Severity-Area-Frequency curves of SPI annual scale time for 10-, 25-, 50-, 100-year exceedance periods, along with the curves for the four most severe drought years of 1984/85, 1993/94, 1998/99 and 2004/05 that affected the region.

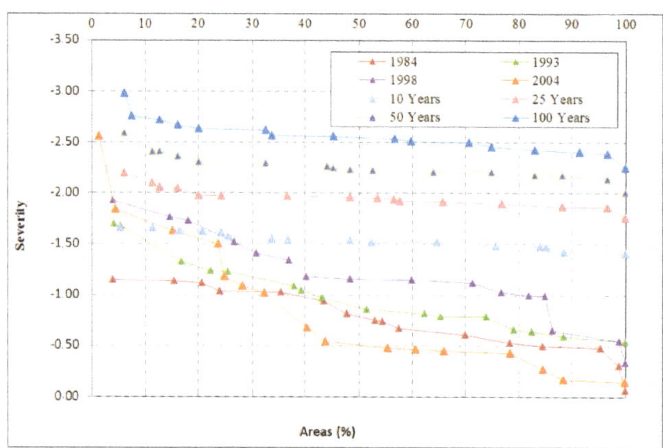

Figure 5. Estimated drought severity—area—frequency curves for the annual SPI values for the Wadi Mina basin, as compared those seen in historical droughts.

The severity analysis shows that all selected droughts have smaller severity than for return periods 10, 25, 50 and 100 years. Moreover, the high drought severity occurred on relatively small areas, less than 20% of analyzed basin and is observed mainly on the north part of the basin—Figure 4. Moderate or near normal years are observed on the most parts of the basin.

4.3.3. Spatial Pattern of Return Periods of Droughts

The return periods of moderate, severe and extreme droughts at all stations were calculated and the values were then used to prepare the corresponding maps by using inverse distance weighted (IDW) interpolation method analysis tool of ArcMap (Figure 6). The presentation of these maps shows spatial variability of the drought for the different classes, with extreme drought more likely (shorter return period) in the north and east, apparently modulated by the high heterogeneity of the spatial distribution of the rainfall [45,73]. Assessing vulnerability to drought across the Wadi Mina basin is important, considering that, as shown by Henchiri et al. [74], grasslands and croplands in the northern region of the Africa are highly vulnerable to drought risk.

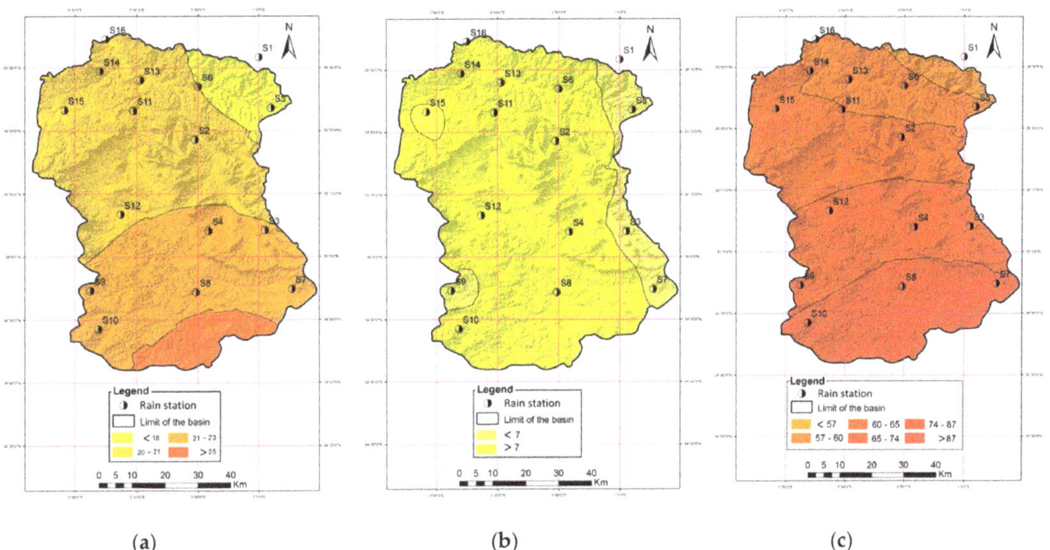

Figure 6. Return periods of meteorological droughts in Wadi Mina with (a) moderate; (b) severe; and (c) extreme severities.

To mitigate risks of drought, proper water management techniques must be adopted. One of these techniques is supplemental irrigation, which is an efficient practice used for increasing agricultural production under limited water resources in areas affected by drought [75].

5. Conclusions

This study was focused on analyzing temporal and spatial extents of droughts in the Wadi Mina basin, Algeria, using SPI as an indicator of drought severity. The aim of this study was to investigate spatial and temporal dimensions of meteorological droughts in the Wadi Mina basin. Meteorological drought was expressed by the Standardized Precipitation Index (SPI) method and GIS was used to detail temporal and geographical variations in the drought vulnerability based on severity of drought events at annual time steps. This study is applied to rainfall monthly records for the period 1970–2010 at 16 rainfall stations located in the Wadi Mina basin.

The results showed that the SPI was able to detect historical droughts of 1982/83, 1983/84, 1989/90, 1992/93, 1993/94, 1996/97, 1998/99, 1999/00, 2004/05 and 2006/07. Wet years were observed in 1971/72, 1972/73, 1995/96, 2008/09 and 2009/10. Decreasing SPI was observed on 13 of 16 rain gauge stations, with six showing statistically significant ($p < 0.05$) decreases. Most of the stations with the greatest decreasing trend were observed in the lower part of the Wadi Mina basin, where average precipitation is already low. As expected given the process used to construct SPI, near normal conditions dominated at all stations, and severe and extreme drought categories were uncommon. The spatial variability of the drought showed that extreme drought is more likely (shorter return period) in the north and east.

Severity-Area-Frequency curves that can aid the development of a drought preparedness plan were developed for Wadi Mina basin, so as to ensure sustainable water resource planning within the basin.

One limitation of the study is that we used only SPI to detect drought intensities. In the future, we plan to add evapotranspiration and calculate the SPEI indicator, which can give more complex information about meteorological conditions influencing drought events, particularly for agricultural and forestry applications. Moreover, in this study only annual sum of precipitation was used and thus seasonal variability of drought was not detected. While this is to some extent justified for this region given that precipitation is concentrated in only a few months per year, in a future study monthly and seasonal precipitation variations could also be explored. Moreover, as a future study, we plan to compare drought analyses based on different sources of rainfall data, including the Soil Moisture to Rainfall (SM2RAIN) [76] algorithm to estimate rainfall based on soil moisture time series.

The SM2RAIN is based on the inversion of the hydrological water balance, for estimation of rainfall from soil moisture observations. In this approach the soil is assumed as reservoir used for measuring the amount of rainfall [77]. This method gives independent rainfall product with a different error structure and allows integration with other satellite-based rainfall products. According to [76], the SM2RAIN method can be useful in regions for which satellite rainfall data are affected by higher errors or not available. Because Northwest Algeria is the region where water scarcity is high, we will perform analysis that can show potential use of SM2RAIN as indirect source of rainfall to detect meteorological drought, including seasonal variations.

Author Contributions: Conceptualization, M.A.; methodology, M.A., A.W., H.M. and N.K.; software, M.A. and A.K.T.; formal analysis, M.A., A.W.; validation: M.A., A.W. and N.K.; investigation, M.A., A.K.T., A.W. and H.M.; data curation, M.A.; writing—original draft preparation, M.A., A.W. and N.K.; writing—review and editing, M.A., A.W. and N.K.; visualization, A.W.; supervision, M.A., A.W. and N.K. All authors have read and agreed to the published version of the manuscript.

Funding: This research received no external funding. The APC was funded by MDPI Editor.

Data Availability Statement: The data presented in this study are available on request from the Corresponding authors.

Acknowledgments: Thanks to peer reviewers who improved this manuscript. We also thank the General Directorate of Scientific Research and Technological Development of Algeria (DGRSDT) for their financial support.

Conflicts of Interest: The authors declare no conflict of interest.

Appendix A

Table 1. Drought classification in the Wadi Mina basin.

Years	S1	S2	S3	S4	S5	S6	S7	S8	S9	S10	S11	S12	S13	S14	S15	S16
1970	NN	NN	NN	NN	NN	NN	NN	MW	MW	NN	MW	NN	NN	NN	NN	NN
1971	NN	EW	EW	EW	VW	VW	NN	VW	EW	VW	EW	VW	VW	EW	EW	VW
1972	MW	NN	MW	MW	MW	NN	NN	NN	MW	EW	NN	VW	MW	VW	MW	MW
1973	NN	NN	NN	NN	NN	NN	NN	NN	NN	NN	NN	NN	MW	VW	MW	VW
1974	VW	MW	NN	NN	NN	VW	NN	NN	NN	NN	NN	NN	EW	MW	MW	NN
1975	NN	NN	NN	NN	NN	NN	NN	NN	NN	NN	NN	NN	NN	NN	NN	NN
1976	VW	VW	MW	NN	NN	NN	NN	NN	NN	MW	NN	NN	NN	NN	NN	NN
1977	NN	NN	NN	NN	NN	NN	NN	NN	NN	NN	NN	NN	NN	NN	NN	NN
1978	NN	NN	NN	NN	NN	NN	NN	NN	NN	NN	NN	NN	MW	NN	NN	NN
1979	NN	NN	NN	NN	VW	NN	MW	NN	NN	NN	NN	NN	NN	NN	MW	NN
1980	NN	NN	NN	NN	NN	NN	NN	NN	NN	NN	NN	VW	NN	NN	ED	NN
1981	SD	MD	NN	NN	NN	NN	NN	NN	NN	NN	MD	MD	MD	MD	SD	SD
1982	NN	MD	MD	MD	MD	NN	MD	MD	SD	MD	NN	MD	NN	NN	NN	NN
1983	NN	NN	NN	MD	NN	NN	SD	NN	SD	MD	NN	SD	NN	NN	NN	NN
1984	NN	NN	NN	MD	NN	MD	NN	NN	NN	NN	NN	NN	NN	MD	MD	NN
1985	VW	NN	NN	NN	NN	NN	NN	MD	NN	NN	NN	NN	NN	MD	NN	MD
1986	NN	MW	NN	NN	NN	NN	NN	NN	NN	NN	NN	NN	NN	NN	MW	NN
1987	NN	NN	NN	MD	MD	NN	MD	NN	NN	NN	NN	NN	NN	NN	NN	NN
1988	NN	NN	NN	NN	MD	NN	NN	NN	NN	NN	NN	NN	NN	NN	NN	NN
1989	SD	MD	SD	NN	SD	MD	NN	MD	MD	MD	MD	NN	NN	NN	NN	MD
1990	NN	NN	NN	MD	NN	NN	NN	NN	NN	MW	NN	NN	NN	NN	NN	NN
1991	NN	NN	NN	NN	NN	NN	NN	NN	NN	NN	NN	NN	NN	NN	NN	NN
1992	NN	NN	SD	NN	NN	NN	MD	NN	SD	SD	MD	SD	NN	MD	MD	MD
1993	SD	NN	MD	NN	MD	SD	NN	NN	NN	NN	MD	NN	NN	MD	NN	NN
1994	NN	NN	NN	NN	NN	NN	NN	MW	NN	NN	NN	NN	NN	NN	NN	NN
1995	VW	VW	EW	VW	NN	NN	EW	VW	VW	VW	EW	VW	VW	EW	VW	EW
1996	NN	NN	NN	NN	MD	SD	NN	NN	NN	NN	MD	NN	SD	SD	MD	SD
1997	MW	NN	NN	NN	NN	NN	NN	NN	MD	MD	NN	NN	NN	NN	NN	NN
1998	NN	NN	NN	MD	SD	SD	MD	NN	NN	NN	MD	SD	SD	MD	MD	NN
1999	NN	MD	MD	MD	SD	MD	MD	NN	MD	SD	SD	NN	MD	MD	NN	SD
2000	NN	NN	NN	NN	NN	NN	NN	NN	NN	NN	NN	NN	NN	NN	MW	NN
2001	NN	NN	NN	NN	NN	NN	NN	NN	NN	NN	NN	NN	NN	NN	NN	NN
2002	MD	NN	NN	NN	MW	MW	NN	NN	NN	NN	NN	NN	NN	NN	SD	NN
2003	NN	NN	NN	NN	NN	MW	MW	VW	NN	NN	NN	NN	NN	NN	NN	NN
2004	MD	ED	SD	MD	NN	NN	SD	MD	SD	MD	NN	SD	SD	NN	NN	NN
2005	NN	NN	MW	NN	MW	MW	MW	VW	MW	NN	NN	NN	NN	NN	NN	NN
2006	MD	MD	NN	MD	NN	NN	NN	NN	NN	NN	MD	NN	MD	NN	MD	NN
2007	SD	NN	NN	NN	NN	NN	NN	NN	NN	NN	NN	NN	NN	NN	MD	NN
2008	MW	NN	VW	EW	EW	EW	EW	EW	VW	MW	VW	MW	MW	VW	NN	MW
2009	NN	NN	NN	MW	NN	NN	VW	EW	VW	VW	NN	VW	VW	NN	NN	NN

EW—extremely wet, VW—very wet, MW—moderately wet, NN—near normal, MD—moderate drought, SD—severe drought, ED—extreme drought.

Table 2. Average and maximum annual SPI values during drought years for the meteorological stations considered.

Stations	Events	Duration DD (Years)	Intensity DI	Magnitude DM	SPI Values				
					Average	Maximum	Year	Minimum	Year
S1	1	1	−0.11	−0.11	−0.74	−1.71	1989/90	−0.07	1984/85
	2	1	−0.39	−0.39					
	3	4	−0.86	−3.42					
	4	8	−0.80	−6.42					
	5	1	−0.21	−0.21					
	6	2	−0.43	−0.86					
	7	6	−0.93	−5.58					
S2	1	1	−0.39	−0.39	−0.79	−2.56	2004/05	−0.01	1985/86
	2	5	−0.74	−3.71					
	3	1	−0.86	−0.86					
	4	1	−1.25	−1.25					
	5	4	−0.29	−1.14					
	6	2	−0.83	−1.66					
	7	6	−1.13	−6.76					
S3	1	1	−0.32	−0.32	−0.84	−1.84	2004/05	−0.04	1980/81
	2	5	−0.74	−3.71					
	3	7	−0.89	−6.24					
	4	1	−0.44	−0.44					
	5	4	−0.96	−3.84					
	6	1	−1.84	−1.84					
	7	1	−0.31	−0.31					
S4	1	1	−0.21	−0.21	−0.80	−1.36	1999/00	−0.02	1991/92
	2	4	−1.12	−4.48					
	3	7	−0.63	−4.43					
	4	1	−0.67	−0.67					
	5	4	−0.92	−3.67					
	6	1	−1.09	−1.09					
	7	2	−0.71	−1.42					
S5	1	1	−0.94	−0.94	−0.86	−1.88	1999/00	−0.05	1997/98
	2	13	−0.76	−9.84					
	3	4	−1.25	−4.98					
	4	1	−0.54	−0.54					
S6	1	2	−0.10	−0.19	−0.77	−1.93	1989/99	−0.06	1977/78
	2	1	−0.21	−0.21					
	3	7	−0.74	−5.16					
	4	6	−0.88	−5.30					
	5	4	−1.29	−5.15					
	6	1	−0.17	−0.17					

Table 2. *Cont.*

Stations	Events	Duration DD (Years)	Intensity DI	Magnitude DM	SPI Values				
					Average	Maximum	Year	Minimum	Year
S7	1	3	−0.62	−1.87	−0.97	−1.98	2004/05	−0.26	1977/78
	2	7	−0.85	−5.97					
	3	1	−0.41	−0.41					
	4	2	−0.95	−1.90					
	5	1	−0.59	−0.59					
	6	4	−0.94	−3.77					
	7	1	−1.98	−1.98					
	8	1	−0.52	−0.52					
S8	1	2	−0.31	−0.62	−0.62	−1.33	1983/84	−0.02	1978/79
	2	16	−0.73	−11.67					
	3	1	−0.55	−0.55					
	4	2	−0.51	−1.02					
	5	2	−0.20	−0.40					
	6	1	−1.22	−1.22					
	7	1	−0.14	−0.14					
S9	1	1	−0.01	−0.01	−0.81	−1.60	2004/05	−0.01	1973/74
	2	3	−0.59	−1.77					
	3	3	−1.19	−3.56					
	4	8	−0.49	−3.91					
	5	4	−0.92	−3.69					
	6	1	−0.27	−0.27					
	7	1	−1.60	−1.6					
	8	1	−0.53	−0.53					
S10	1	1	−0.28	−0.28	−0.85	−1.96	1992/93	−0.11	1984/85
	2	1	−0.41	−0.41					
	3	3	−0.91	−2.74					
	4	4	−0.75	−2.98					
	5	2	−1.12	−2.24					
	6	4	−0.91	−3.65					
	7	2	−0.79	−1.57					
	8	1	−1.45	−1.45					
	9	2	−0.81	−1.61					

Table 2. *Cont.*

Stations	Events	Duration DD (Years)	Intensity DI	Magnitude DM	SPI Values				
					Average	Maximum	Year	Minimum	Year
S11	1	1	−0.07	−0.07	−0.85	−1.67	1999/00	−0.07	1977/78
	2	2	−0.90	−1.80					
	3	2	−0.66	−1.32					
	4	3	−0.81	−2.42					
	5	3	−1.00	−2.99					
	6	1	−1.29	−1.29					
	7	2	−1.37	−2.73					
	8	3	−0.49	−1.47					
	9	1	−0.19	−0.19					
S12	1	1	−0.37	−0.37	−0.8	−1.70	1983/84	−0.10	1978/79
	2	1	−0.10	−0.10					
	3	5	−0.93	−4.65					
	4	1	−0.76	−0.76					
	5	1	−0.34	−0.34					
	6	3	−0.91	−2.73					
	7	1	−0.90	−0.90					
	8	5	−0.75	−3.76					
	9	1	−1.50	−1.50					
	10	1	−0.98	−0.98					
S13	1	1	−1.25	−1.25	−0.77	−1.77	1998/99	−0.04	1991/92
	2	3	−0.49	−1.48					
	3	2	−0.69	−1.38					
	4	3	−0.57	−1.70					
	5	12	−0.87	−10.42					
S14	1	1	−0.17	−0.17	−0.68	−2.02	1996/97	−0.02	1980/81
	2	3	−0.61	−1.83					
	3	2	−1.07	−2.13					
	4	3	−0.54	−1.62					
	5	3	−0.81	−2.42					
	6	5	−0.56	−2.79					
	7	4	−0.31	−1.22					
	8	2	−0.42	−0.83					

Table 2. Cont.

Stations	Events	Duration DD (Years)	Intensity DI	Magnitude DM	SPI Values				
					Average	Maximum	Year	Minimum	Year
S15	1	2	−0.32	−0.64	−0.81	−2.03	1980/81	−0.17	2005/06
	2	3	−1.29	−3.86					
	3	1	−1.14	−1.14					
	4	1	−0.71	−0.71					
	5	2	−0.39	−0.78					
	6	2	−1.03	−2.05					
	7	1	−1.25	−1.25					
	8	2	−0.68	−1.35					
	9	7	−0.76	−5.32					
S16	1	1	−1.88	−1.88	−0.76	−1.88	1981/82	−0.01	2005/06
	2	7	−0.63	−4.43					
	3	4	−0.70	−2.80					
	4	1	−1.73	−1.73					
	5	2	−1.11	−2.21					
	6	2	−0.54	−1.07					
	7	4	−0.47	−1.87					

References

1. Rossi, G. Drought mitigation measures: A comprehensive frame work. In *Drought and Drought Migation in Europe*; Voght, J.V., Somma, F., Eds.; Kluwer: Dordrecht, The Netherlands, 2000.
2. Mishra, A.K.; Singh, V.P. A review of drought concepts. *J. Hydrol.* **2010**, *391*, 202–216. [CrossRef]
3. Achite, M.; Buttafuoco, G.; Toubal, K.A.; Lucà, F. Precipitation spatial variability and dry areas temporal stability for different elevation classes in the Macta basin (Algeria). *Environ. Earth Sci.* **2017**, *76*, 458. [CrossRef]
4. Hossard, L.; Fadlaoui, A.; Ricote, E.; Belhouchette, H. Assessing the resilience of farming systems on the Saïs plain, Morocco. *Reg. Environ. Chang.* **2021**, *21*, 36. [CrossRef]
5. Lionello, P.; Bhend, J.; Buzzi, A.; Della-Marta, P.M.; Krichak, S.O.; Jansa, A.; Maheras, P.; Sanna, A.; Trigo, I.F.; Trigo, R. Cyclones in the Mediterranean region: Climatology and effects on the environment. In *Mediterranean Climate Variability Developments in Earth and Environmental Sciences*; Lionello, P., Malanotte-Rizzoli, P., Boscolo, R., Eds.; Elsevier: Amsterdam, The Netherlands, 2006; Volume 4, pp. 325–372. [CrossRef]
6. Norrant, C.; Douguedroit, A. Tendances recentes des precipitations et des pressions de surface dans le Bassin mediterraneen. *Ann. Geogr.* **2003**, *631*, 298–305. [CrossRef]
7. Buttafuoco, G.; Caloiero, T.; Coscarelli, R. Spatial and temporal patterns of the mean annual precipitation at decadal time scale in southern Italy (Calabria region). *Appl. Clim.* **2011**, *105*, 431–444. [CrossRef]
8. Rossi, G. Drought Mitigation Measures: A Comprehensive Framework, in: Advances in Natural and Technological Hazards Research. *Adv. Nat. Technol. Hazards Res.* **2000**, *14*, 233–246. [CrossRef]
9. Ghouil, H.; Sancho-Knapik, D.; Ben Mna, A.; Amimi, N.; Ammari, Y.; Escribano, R.; Alonso-Forn, D.; Ferrio, J.P.; Peguero-Pina, J.J.; Gil-Pelegrín, E. Southeastern Rear Edge Populations of Quercus suber L. Showed Two Alternative Strategies to Cope with Water Stress. *Forests* **2020**, *11*, 1344. [CrossRef]
10. Akbas, A.; Freer, J.; Ozdemir, H.; Bates, P.D.; Turp, M.T. What about reservoirs? Questioning anthropogenic and climatic interferences on water availability. *Hydrol. Process.* **2020**, *34*, 5441–5455. [CrossRef]
11. Lyon, B. The strength of El Niño and the spatial extent of tropical drought. *Geophys. Res. Lett.* **2004**, *31*, 21204. [CrossRef]
12. Palmer, W.C. Keeping track of crop moisture conditions, nationwide: The Crop Moisture Index. *Weatherwise* **1968**, *21*, 156–161. [CrossRef]
13. Tsakiris, G.; Vangelis, H. Establishing a drought index incorporating evapotranspiration. *Eur. Water* **2005**, *9/10*, 3–11.
14. Byun, H.R.; Wilhite, D.A. Daily quantification of drought severity and duration. *J. Clim.* **1996**, *5*, 1181–1201.
15. Bokwa, A.; Klimek, M.; Krzaklewski, P.; Kukułka, W. Drought Trends in the Polish Carpathian Mts. in the Years 1991–2020. *Atmosphere* **2021**, *12*, 1259. [CrossRef]
16. Strommen, N.D.; Motha, R.P. An operational early warning agricultural weather system. In *Planning for Drought: Toward a Reduction of Societal Vulnerability*; Wilhite, D.A., Easterling, W.E., Wood, D.A., Eds.; Westview Press: Boulder, CO, USA, 1987.

17. Alley, W.M. The Palmer Drought Severity Index: Limitations and assumptions. *J. Appl. Meteorol.* **1984**, *23*, 1100–1109. [CrossRef]
18. McKee, T.; Doesken Kleist, J. The relationship of drought frequency and duration to time scales. In Proceedings of the Eighth Conference on Applied Climatology, Anaheim, CA, USA, 17–22 January 1993; American Meteorological Society: Boston, MA, USA, 1993; pp. 179–184.
19. Bordi, I.; Fraedrich, K.; Jiang, J.-M.; Sutera, A. Spatio-temporal variability of dry and wet periods in eastern China. *Theor. Appl. Climatol.* **2004**, *79*, 81–91. [CrossRef]
20. Lloyd-Hughes, B.; Saunders, M.A. A drought climatology for Europe. *Int. J. Climatol.* **2002**, *22*, 1571–1592. [CrossRef]
21. Hayes, M.J.; Svoboda, M.D.; Wilhite, D.A.; Vanyarkho, O.V. Monitoring the 1996 drought using the Standardized Precipitation Index. *Bull. Am. Meteorol. Soc.* **1999**, *80*, 429–438. [CrossRef]
22. Umran Komuscu, A. "Using the SPI to Analyze Spatial and Temporal Patterns of Drought in Turkey" Drought Network News (1994–2001). 1999. Available online: https://digitalcommons.unl.edu/droughtnetnews/49 (accessed on 2 November 2021).
23. Kemal Sonmez, K.; Umran Komuscu, A.; Erkan, A.; Turgu, E. An Analysis of Spatial and Temporal Dimension of Drought Vulnerability in Turkey Using the Standardized Precipitation Index. *Nat. Hazards* **2005**, *35*, 243–264. [CrossRef]
24. Lana, X.; Serra, C.; Burgueno, A. Patterns of monthly rainfall shortage and execess In terms of standardized precipitation index for Catalonia (NE Spain). *Int. J. Climatol.* **2001**, *21*, 1669–1691. [CrossRef]
25. Vicente-Serrano, S.M.; González-Hidalgo, J.C.; de Luis, M.; Raventós, J. Drought patterns in the Mediterranean area: The Valencia region (eastern Spain). *Clim. Res.* **2004**, *26*, 5–15. [CrossRef]
26. Bonaccorso, B.; Cancelliere, A.; Rossi, G.; Sutera, A. Spatial Variability of Drought: An Analysis of the SPI in Sicily. *Water Resour. Manag.* **2003**, *17*, 273–296. [CrossRef]
27. Piccarreta, M.; Capolongo, D.; Boenzi, F. Trend analysis of precipitation and drought in Basilicata from 1923 to 2000 within a Southern Italy context. *Int. J. Climatol.* **2004**, *24*, 907–922. [CrossRef]
28. Capra, A.; Consoli, S.; Scicolone, B.; Calafiore, G. Precipitation and drought variability in the last century in Calabria (Italy). In Proceedings of the XXXIII CIOSTA-CIGR V Conference Technology and Management to Ensure Sustainable Agriculture, Agro-Systems, Forestry and Safety, Reggio Calabria, Italy, 17–19 June 2009; pp. 1655–1659.
29. Vergni, L.; Todisco, F. Spatio-temporal variability of precipitation, temperature and agricultural drought indices in central Italy. *Agric. Meteorol.* **2010**, *151*, 301–313. [CrossRef]
30. Buttafuoco, G.; Caloiero, T.; Coscarelli, R. Analyses of Drought Events in Calabria (Southern Italy) Using Standardized Precipitation Index. *Water Resour. Manag.* **2015**, *29*, 557–573. [CrossRef]
31. Raziei, T.; Saghafian, B.; Paulo, A.A.; Pereira, L.S.; Bordi, I. Spatial Patterns and Temporal Variability of Drought in Western Iran. *Water Resour. Manag.* **2009**, *23*, 439. [CrossRef]
32. Moradi, H.R.; Rajabi, M.; Faragzadeh, M. Investigation of meteorological drought characteristics in Fars province, Iran. *Catena* **2011**, *84*, 35–46. [CrossRef]
33. Nafarzadegana, A.R.; Zadeha, M.R.; Kherada, M.; Ahania, H.; Gharehkhania, A.; Karampoora, M.A.; Kousari, M.R. Drought area monitoring during the past three decades in Fars Province, Iran. *Quat. Int.* **2012**, *250*, 27–36. [CrossRef]
34. Livada, I.; Assimakopoulos, V.D. Spatial and temporal analysis of drought in greece using the standardized precipitation index (SPI). *Theor. Appl. Climatol.* **2007**, *89*, 143–153. [CrossRef]
35. Nalbantis, I.; Tsakiris, G. Assessment of Hydrological Drought Revisited. *Water Resour. Manag.* **2009**, *23*, 881–897. [CrossRef]
36. Karavitis, C.A.; Alexandris, S.; Tsesmelis, D.E.; Athanasopoulos, G. Application of the Standardized Precipitation Index (SPI) in Greece. *Water* **2013**, *3*, 787–805. [CrossRef]
37. AL-Timimi, Y.K.; Loay, G.E.; AL-Jiboori Monim, H. Drought Risk Assessment In Iraq Using Remote Sensing And GIS Techniques. *Iraqi. J. Sci.* **2012**, *53*, 1078–1082.
38. Awchi, T.A.; Kalyana, M.M. Meteorological drought analysis in northern Iraq using SPI and GIS. Sustain. *Water Resour. Manag.* **2017**, *3*, 451–463. [CrossRef]
39. Jasim Ansam, I.; Awchi Taymoor, A. Regional meteorological drought assessment in Iraq. *Arab. J. Geosci.* **2020**, *13*, 284. [CrossRef]
40. Shadeed, S. Spatio-temporal Drought Analysis in Arid and Semi-arid Regions: A Case Study from Palestine. *Arab. J. Sci Eng.* **2013**, *38*, 2303–2313. [CrossRef]
41. Djellouli, F.; Abderrazak, B.; Baba-Hamed, K. Efficiency of some meteorological drought indices in different time scales, case study: Wadi Louza basin (NW-Algeria). *J. Water Land Dev.* **2016**, *31*, 33–41. [CrossRef]
42. Khezazna, A.; Amarchi, H.; Derdous, O.; Bousakhria, F. Drought monitoring in the Seybouse basin (Algeria) over the last decades. *J. Water Land Dev.* **2017**, *33*, 79–88. [CrossRef]
43. Brahim, H.; Meddi, M.; Torfs, P.J.J.F.; Remaoun, M.; Van Lanen, H.A.J. Characterisation and prediction of meteorological drought using stochastic models in the semi-arid Chéli–Zahrez basin (Algeria). *J. Hydrol. Reg. Stud.* **2018**, *16*, 15–31. [CrossRef]
44. Fellag, M.; Achite, M.; Wałęga, A. Spatial-temporal characterization of meteorological drought using the Standardized precipitation index. Case study in Algeria. *Acta Sci. Polonorum. Form. Circumiectus* **2021**, *20*, 19–31.
45. Achite, M.; Krakauer, N.Y.; Wałęga, A.; Caloiero, T. Spatial and Temporal Analysis of Dry and Wet Spells in the Wadi Cheliff Basin, Algeria. *Atmosphere* **2021**, *12*, 798. [CrossRef]
46. Ouatiki, H.; Boudhar, A.; Ouhinou, A.; Arioua, A.; Hssaisoune, M.; Bouamri, H.; Benabdelouahab, T. Trend analysis of rainfall and drought over the Oum Er-Rbia River Basin in Morocco during 1970–2010. *Arab. J. Geosci.* **2019**, *12*, 128. [CrossRef]

47. Jemai, S.; Ellouze, M.; Agoubi, B.; Abida, H. Drought intensity and spatial variability in Gabes Watershed, south-eastern Tunisia. *J. Water Land Dev.* **2016**, *31*, 63–72. [CrossRef]

48. Jemai, H.; Ellouze, M.; Abida, H.; Laignel, B. Spatial and temporal variability of rainfall: Case of Bizerte-Ichkeul Basin (Northern Tunisia). *Arab. J. Geosci.* **2018**, *11*, 177. [CrossRef]

49. Ben Abdelmalek Nouiri, I. Study of trends and mapping of drought events in Tunisia and their impacts on agricultural production. *Sci. Total Environ.* **2020**, *734*, 139311. [CrossRef]

50. Rapport d'Information Déposé en Application de L'article 145 du Règlement par la Commission des Affaires Étrangères de France en Conclusion des Travaux D'une Mission d'Information Constituée le 5 Octobre 2010 sur La Géopolitique de L'eau. 2011. Available online: https://www.assemblee-nationale.fr/dyn/15/rapports/cion_afetr/l15b3581_rapport-information (accessed on 30 September 2021).

51. Hamiche, A.; Stambouli, A.; Flazi, S. A review on the water and energy sectors in Algeria: Current forecasts, scenario and sustainability issues'. *Renew. Sustain. Energy Rev.* **2016**, *41*, 261–276. [CrossRef]

52. Mensah, C.; Šigut, L.; Fischer, M.; Foltýnová, L.; Jocher, G.; Acosta, M.; Kowalska, N.; Kokrda, L.; Pavelka, M.; Marshall, J.D.; et al. Assessing the Contrasting Effects of the Exceptional 2015 Drought on the Carbon Dynamics in Two Norway Spruce Forest Ecosystems. *Atmosphere* **2021**, *12*, 988. [CrossRef]

53. Achite, M. Sécheresse et Gestion des Ressources en Eau Dans le Bassin Versant de la Mina. Algérie. 2ème Colloque International Sur L'eau et L'Environnement. 30 et 31 Janvier 2007. Sidi Fredj. Alger (Algérie). Available online: https://www.worldwatercouncil.org/ (accessed on 5 September 2021).

54. Ng, C.K.; Ng, J.L.; Huang, Y.F.; Tan, Y.X.; Mirzaei, M. Tropical rainfall trend and stationarity analysis. *Water Supply* **2020**, *20*, 2471–2483. [CrossRef]

55. Santos, J.F.; Portela, M.M.; Pulido-Calvo, I. Regional frequency analysis of droughts in Portugal. *Water Resour. Manag.* **2011**, *25*, 3537–3558. [CrossRef]

56. Thom, H.C.S. A note on the gamma distribution. *Mon. Weather Rev.* **1958**, *86*, 117–122. [CrossRef]

57. Edwards, D.C.; McKee, T.B. Characteristics of 20th century drought in the United States at multiple scales. *Atmos. Sci. Pap.* **1997**, *634*, 1–30.

58. Abramowitz, M. Stegun, I.A. *Handbook of Mathematical Functions*; Dover Publications: New York, NY, USA, 1965.

59. Młyński, D.; Wałęga, A.; Petroselli, A.; Tauro, F.; Cebulska, M. Estimating Maximum Daily Precipitation in the Upper Vistula Basin, Poland. *Atmosphere* **2019**, *10*, 43. [CrossRef]

60. Sen, P.K. Estimates of the regression coefficient based on Kendall's tau. *J. Am. Stat. Assoc.* **1968**, *63*, 1379–1389. [CrossRef]

61. Vicente-Serrano, S.M. Differences in spatial patterns of drought on different time scales: An analysis of the Iberian Peninsula. *Water Resour. Manag.* **2006**, *20*, 37–60. [CrossRef]

62. Elouissi, A.; Sen, Z.; Habi, M. Algerian rainfall innovative trend analysis and its implications to Macta watershed. *Arab. J. Geosci.* **2016**, *9*, 1–12. [CrossRef]

63. Caloiero, T.; Caloiero, P.; Frustaci, F. Long-term precipitation trend analysis in Europe and in the Mediterranean basin. *Water Environ. J.* **2018**, *32*, 433–445. [CrossRef]

64. Lee, J.; Kim, S.; Jun, H. A Study of the Influence of the Spatial Distribution of Rain Gauge Networks on Areal Average Rainfall Calculation. *Water* **2018**, *10*, 1635. [CrossRef]

65. Hingray, B.; Picouet, C.; Muy, A. Hydrology. In *A Science for Engineers*; CRC Press: Boca Raton, FL, USA, 2014.

66. Vicente-Serrano, S.M.; López-Moreno, J.I.; Lorenzo-Lacruz, J.; Kenawy, A.; Azorin-Molina, C.; Morán-Tejeda, E.; Pasho, E.; Zabalza, J.; Beguería, S.; Angulo-Martínez, M. The NAO Impact on Droughts in the Mediterranean Region. In *Advances in Global Change Research*; Springer Science and Business Media LLC: Berlin, Germany, 2011; pp. 23–40.

67. Hoerling, M.P.; Eischeid, J.K.; Perlwitz, J.; Quan, X.; Zhang, T.; Pegion, P.J. On the Increased Frequency of Mediterranean Drought. *J. Clim.* **2012**, *25*, 2146–2161. [CrossRef]

68. Meddi, H.; Meddi, M.; Assani, A.A. Study of Drought in Seven Algerian Plains. *Arab. J. Sci. Eng.* **2013**, *39*, 339–359. [CrossRef]

69. Giannini, A.; Saravanan, R.; Chang, P. Oceanic Forcing of Sahel Rainfall on Interannual to Interdecadal Time Scales. *Science* **2003**, *302*, 1027–1030. [CrossRef]

70. Mishra, A.K.; Desa, V.R. Spatial and temporal drought analysis in the Kansabati river basin. India. *Intl. J. River Basin Manag.* **2005**, *3*, 31–41. [CrossRef]

71. Cai, W.; Zhang, Y.; Yao, Y.; Chen, Q. Probabilistic Analysis of Drought Spatiotemporal Characteristics in the Beijing-Tianjin-Hebei Metropolitan Area in China. *Atmosphere* **2015**, *6*, 431–450. [CrossRef]

72. Alemaw Berhanu, F.; Kileshye-Onema, J.M.; Love, D. Regional Drought Severity Assessment at a Basin Scale in the Limpopo Drainage System. *J. Water Resour. Prot.* **2013**, *5*, 1110–1116. [CrossRef]

73. Meddi, H.; Meddi, M. Variabilité spatiale et temporelle des précipitations du Nord-Ouest de l'Algérie. *Géogr. Tech.* **2007**, *2*, 49–55.

74. Henchiri, M.; Liu, Q.; Essifi, B.; Javed, T.; Zhang, S.; Bai, Y.; Zhang, J. Spatio-Temporal Patterns of Drought and Impact on Vegetation in North and West Africa Based on Multi-Satellite Data. *Remote Sens.* **2020**, *12*, 3869. [CrossRef]

75. Attia, A.; El-Hendawy, S.; Al-Suhaibani, N.; Alotaibi, M.; Tahir, M.U.; Kamal, K.Y. Evaluating deficit irrigation scheduling strategies to improve yield and water productivity of maize in arid environment using simulation. *Agric. Water Manag.* **2021**, *249*, 106812. [CrossRef]

76. Brocca, L.; Ciabatta, L.; Massari, C.; Moramarco, T.; Hahn, S.; Hasenauer, S.; Kidd, R.; Dorigo, W.; Wagner, W.; Levizzani, V. Soil as a natural rain gauge: Estimating global rainfall from satellite soil moisture data. *J. Geophys. Res. Atmos.* **2014**, *119*, 5128–5141. [CrossRef]
77. Brocca, L.; Melone, F.; Moramarco, T.; Wagner, W. A new method for rainfall estimation through soil moisture observations. *Geophys. Res. Lett.* **2013**, *40*, 853–858. [CrossRef]

Article

Hydrological Response of the Kunhar River Basin in Pakistan to Climate Change and Anthropogenic Impacts on Runoff Characteristics

Muhammad Saifullah [1,*], Muhammad Adnan [2], Muhammad Zaman [3], Andrzej Wałęga [4], Shiyin Liu [2,*], Muhammad Imran Khan [5], Alexandre S. Gagnon [6] and Sher Muhammad [7]

[1] Department of Agricultural Engineering, Muhammad Nawaz Shareef University of Agriculture, Multan 66000, Pakistan

[2] Institute of International Rivers and Eco-Security, Yunnan University, Kunming 650500, China; adnan@lzb.ac.cn

[3] Department of Irrigation and Drainage, University of Agriculture, Faisalabad 380000, Pakistan; muhammad.zaman@uaf.edu.pk

[4] Department of Sanitary Engineering and Water Management, University of Agriculture in Krakow, 31-120 Kraków, Poland; andrzej.walega@urk.edu.pl

[5] Research Center of Fluid Machinery Engineering and Technology, Jiangsu University, Zhenjiang 212013, China; imrankhan7792@yahoo.com

[6] School of Biological and Environmental Sciences, Liverpool John Moores University, Liverpool L3 3AF, UK; A.Gagnon@ljmu.ac.uk

[7] International Centre for Integrated Mountain Development (ICIMOD), Lalitpur 44700, Nepal; sher.muhammad@icimod.org

* Correspondence: muhammad.saifullah@mnsuam.edu.pk (M.S.); shiyin.liu@ynu.edu.cn (S.L.)

Citation: Saifullah, M.; Adnan, M.; Zaman, M.; Wałęga, A.; Liu, S.; Khan, M.I.; Gagnon, A.S.; Muhammad, S. Hydrological Response of the Kunhar River Basin in Pakistan to Climate Change and Anthropogenic Impacts on Runoff Characteristics. *Water* 2021, 13, 3163. https://doi.org/10.3390/w13223163

Academic Editor: Aizhong Ye

Received: 22 September 2021
Accepted: 2 November 2021
Published: 9 November 2021

Publisher's Note: MDPI stays neutral with regard to jurisdictional claims in published maps and institutional affiliations.

Abstract: Pakistan is amongst the most water-stressed countries in the world, with changes in the frequency of extreme events, notably droughts, under climate change expected to further increase water scarcity. This study examines the impacts of climate change and anthropogenic activities on the runoff of the Kunhar River Basin (KRB) in Pakistan. The Mann Kendall (MK) test detected statistically significant increasing trends in both precipitation and evapotranspiration during the period 1971–2010 over the basin, but with the lack of a statistically significant trend in runoff over the same time-period. Then, a change-point analysis identified changes in the temporal behavior of the annual runoff time series in 1996. Hence, the time series was divided into two time periods, i.e., prior to and after that change: 1971–1996 and 1997–2010, respectively. For the time-period prior to the change point, the analysis revealed a statistically significant increasing trend in precipitation, which is also reflected in the runoff time series, and a decreasing trend in evapotranspiration, albeit lacking statistical significance, was observed. After 1996, however, increasing trends in precipitation and runoff were detected, but the former lacked statistical significance, while no trend in evapotranspiration was noted. Through a hydrological modelling approach reconstructing the natural runoff of the KRB, a 16.1 m^3/s (or 15.3%) reduction in the mean flow in the KRB was simulated for the period 1997–2010 in comparison to the period 1971–1996. The trend analyses and modeling study suggest the importance of anthropogenic activities on the variability of runoff over KRB since 1996. The changes in streamflow caused by irrigation, urbanization, and recreational activities, in addition to climate change, have influenced the regional water resources, and there is consequently an urgent need to adapt existing practices for the water requirements of the domestic, agricultural and energy sector to continue being met in the future.

Keywords: climate change; Kunhar River Basin; streamflow; trend analysis; Soil and Water Assessment Tool (SWAT); anthropogenic impacts

179

1. Introduction

Human development has been increasing at a substantial rate since the Industrial Revolution [1], at the expense of anthropogenic activities causing land use changes, an increase in emissions of greenhouse gases (GHGs), and therefore climate change [2]. Both changes in climate and anthropogenic activities can affect hydrological processes of a river catchment [3,4], thus impacting on water resources, hydropower production [5,6], and crop yield [7]. According to the Intergovernmental Panel on Climate Change (IPCC) [8], GHG emission have caused a 0.85 °C increase in global mean temperature with associated changes in precipitation over the period from 1880 to 2012, thus impacting on the hydrological characteristics of the catchment [9], notably the volume of runoff [10] and peaks value [11], in addition to other runoff characteristics [12].

The identification of trends in precipitation and evapotranspiration time series (or more commonly temperature as a proxy for evapotranspiration) can help us understand the complex temporal variability of streamflow. Canchala et al. [10], for example, examined the variability of river flow of two Colombian rivers, which they corelated with various indices of atmospheric teleconnections as well as precipitation. Similarly, a trend analysis was performed on the runoff of the Athabasca River Basin in western Canada [11]. The author found that the decreasing trend in streamflow in recent decades was coherent with the temperature and precipitation trends, and that the trends in hydrological variables that the catchment have recently experienced are projected to continue under climate change.

The generation of runoff and its characteristics are not only affected by climatic changes, but also by anthropogenic activities [13]. For this reason, land-use/land-cover (LULC) types and changes in the latter are important to monitor, as the hydrological response of forested land, urbanized land [14], and cultivated land [15,16] to a precipitation event does vary. Applying a catchment-based approach with regard to managing the hydrology of a river basin, including water resources availability, for the benefits of all users is thus necessary [17].

There is a growing body of research on simulating the hydrological response and estimating changes in the characteristics of streamflow to the impacts of climate change and anthropogenic activities. The impacts of anthropogenic activities on streamflow were examined by [10,11,18,19], for instance, while other studies have focused on assessing the impacts of climate change on the hydrological response [20–24]. However, there is a limited number of studies that investigated the hydrological response to the impacts of both climate change and anthropogenic activities in river basin [2,9,25–30], those that have been performed to date have used various methods, including hydrological modeling [31,32], statistical techniques [33,34], empirical methods [35,36] and paired catchment techniques [37,38] and paired years methods [39,40], with each method having its limitations. For instance, in the case of hydrological modeling, the models require long-term and detailed data for their calibration, which are not always available. Moreover, the calibration of the model with limited input data can causes uncertainties and discrepancies in the model outputs. Empirical methods for their part, provide physical interpretation and have fewer data requirements, hence they are usually preferred in developing countries where there are fewer resources to monitor the hydrological conditions.

Many catchments have their source in mountainous regions and hence these regions play an important role in the water balance of areas located further downstream. The majority of inhabitants of the Kunhar River Basin (KRB), located in the western Himalaya of Pakistan, for instance, rely on agriculture for their livelihoods, with the water required for the latter originating from upstream mountain areas, and used for irrigation, power generation, recreation, and municipal use [41]. The region also has great potential for tourism and for further development in hydropower generation and agriculture, which may put additional pressure on water supply in the future.

The KRB has experienced changes in LULC types in recent decades. Woods burning to meet the domestic energy needs and a growing population have caused deforestation in the uphill areas of the basin [42], with forests being replaced by grasses and shrub [43].

Moreover, slope agriculture is becoming more and more common, causing further changes to the vegetation type over the basin [44]. A study conducted by Saifullah et.al. [45] determined the threshold levels and a climate-sensitivity model for the hydrological regime of the KRB. The current study is directly associated with the water-related ecosystem that is designed in accordance with the Sustainable Development Goals of the United Nations (SDGs 6.3), to integrate climate change measures and policies (SDGs 13.2), and anthropogenic impacts (SDGs 15.1.1).

The objectives of this study are: (1) To examine the temporal variability and trends in hydro-climatic variables over the KRB (2) to quantify of the relative contribution of climate change and anthropogenic activities on the variability of runoff in the KRB.

2. Materials and Methods

2.1. Study Area

This study was conducted in the KRB, a high-altitude catchment in Northern Pakistan (Figure 1). The Kunhar River is 171 km long; it originates in lake Lulusar in the Kaghan Valley of Khyber Pakhtunkhwa passing through the town of Jalkhand, Bata Kundi, Naran, Kaghan, Kwai, Balakot, and Garhi Habibullah and exiting into the Jhelum River at Rara [22]. The Kunhar River is an important source of water for the Mangla reservoir, which contributes nearly 11% of its water [46]. The drainage area of KRB is approximately 2600 km^2, it is mountainous [44], with elevation ranging from 672 to 5192 m above sea level [47].

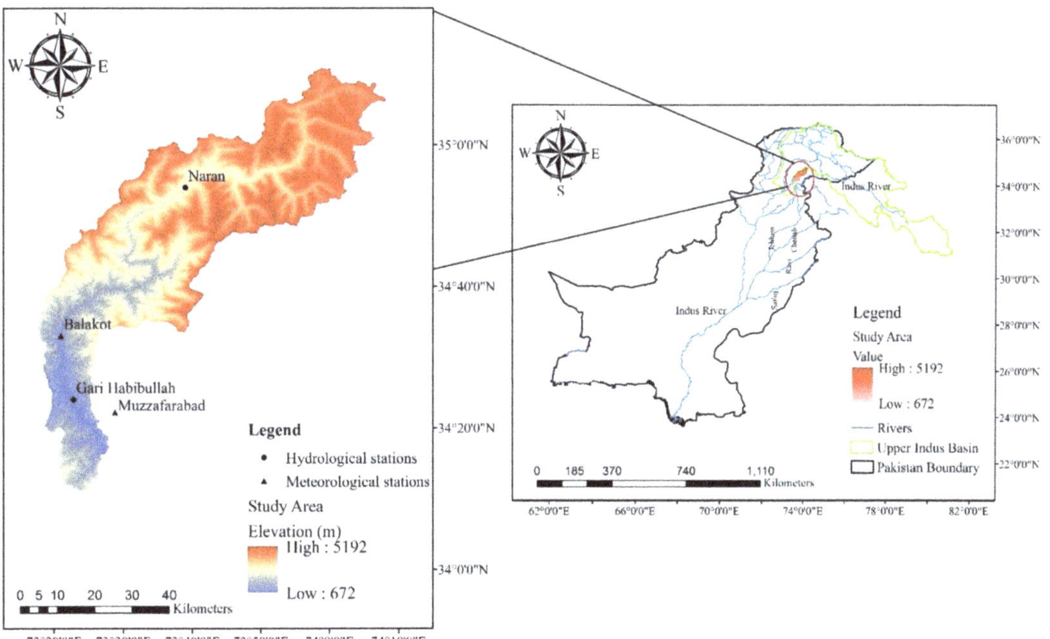

Figure 1. Geographical location of the KRB in Pakistan.

Mangla is the second-largest reservoir in Pakistan and its storage is used to irrigate nearly six million hectares of the country's agricultural land, in addition to producing nearly 1000 MW of hydroelectricity [41,48]. The region experiences mild summers and cold winters. Average annual maximum temperature at Naran, Balakot, and Muzaffarabad is 12.3, 24.9, and 28.4 °C, respectively, whereas the average annual minimum temperature is 3.2, 12.4, and 13.5 °C, respectively [22]. Annual rainfall at Muzaffarabad and Balakot is on average, 1351 and 1531 mm, respectively [43]. The basin also receives a substantial amount of precipitation in the form of snowfall in the winter (December-March) [44].

The vegetation of the KRB is divers, consisting of coniferous and broadleaf forests, grasses and shrubs, in addition to bare soil and snow and glaciers, and water bodies (Figure 2, Table 1). Three dominant land cover types dominate the basin: grasses/shrubs, bare soil/rocks, and dense coniferous forests, with crops covering only 4.35% of the basin (Table 1).

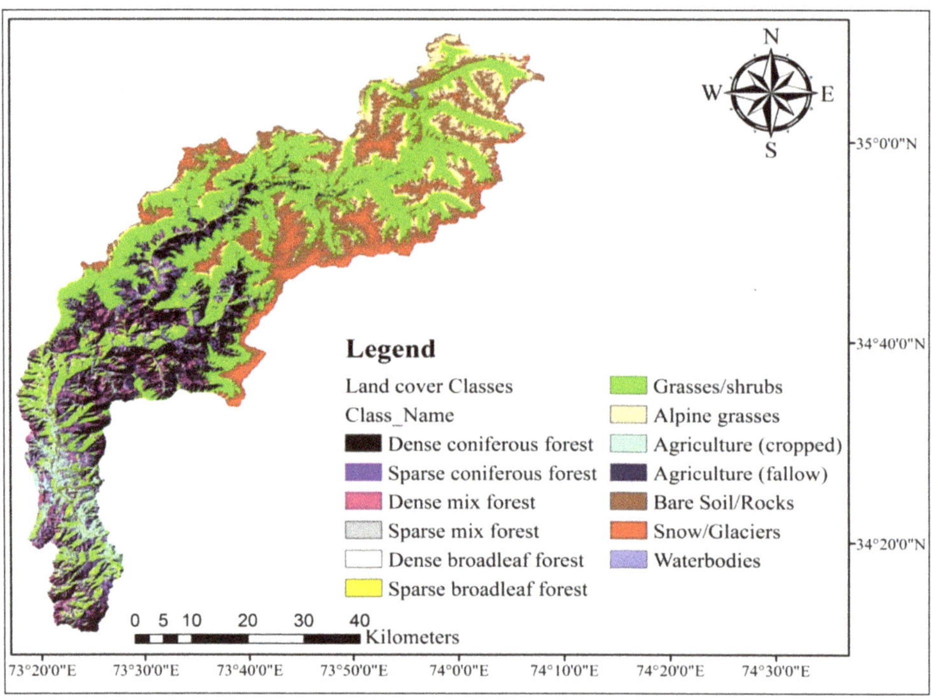

Figure 2. Land-cover classes of the KRB.

Table 1. Land-cover classes of the KRB.

ID	Class Name	Area (km^2)	Area (%)
1	Dense coniferous forest	359.9	13.70
2	Sparse coniferous forest	160.8	6.12
3	Dense mix forest	122.7	4.67
4	Sparse mix forest	61.6	2.35
5	Dense broadleaf forest	30.6	1.17
6	Sparse broadleaf forest	21.5	0.82
7	Grasses/shrubs	853.5	32.49
8	Alpine grasses	183.3	6.98
9	Agriculture (cropped)	114.2	4.35
10	Agriculture (fallow)	0.9	0.03
11	Bare Soil/Rocks	585.7	22.29
12	Snow/Glaciers	128.2	4.88
13	Water bodies	4.0	0.15

2.2. Datasets and Pre-Processing

A Digital Elevation Model (DEM), with a 30 m resolution, was downloaded from the website of the United States Geological Survey (USGS) [49] and used to delineate the boundaries of the KRB. The land-cover data over the basin, for their part, were downloaded from the website of the International Center for Integrated Mountain Development (ICIMOD) [50], while the Harmonized World Soil Database v 1.2 was acquired from the website of Food and Agriculture Organization (FAO) [51]. There are two gauging stations and three meteorological stations in the KRB (Table 2). The Water and Power Development Authority (WAPDA) provided daily streamflow data from the two gauging stations, while daily temperature and precipitation data were obtained from the Pakistan Meteorological Department (PMD). Wind speed, solar radiation, and relative humidity were extracted from the ERA5 reanalysis dataset at an hourly time scale [52]. The analyses were performed at the monthly and annual time scales, but some methods required daily data, e.g., the hydrological model, as described below.

Table 2. List of hydro-meteorological data of the KRB.

Station	Latitude (°)	Longitude (°)	Altitude (m)	Period of Record	Source of Data
		Hydrological stations			
Naran	34.9	73.65	2362	1971–2010	WAPDA
Gari Habibullah	34.40	73.38	810	1971–2010	WAPDA
		Meteorological stations			
Balakot	34.55	73.35	995	1971–2010	PMD
Muzaffarabad	34.37	73.48	702	1971–2010	PMD
Naran	34.9	73.65	2421	1971–2010	PMD

Note: WAPDA refer to the Water and Power Development Authority and PMD to Pakistan Meteorological department.

2.3. Methods

2.3.1. Mann Kendall Trend Test

Mann-Kendall (MK) trend test [53,54], calculated using Equations (1) and (2), is a commonly used non-parametric test to detect trends in a climatological and hydrological time series.

$$S = \sum_{i=1}^{n-1} \sum_{j=i+1}^{n} sgn(x_j - x_i) \tag{1}$$

$$sgn(x_j - x_i) = \begin{cases} +1, & x_j > x_i \\ 0, & x_j = x_i \\ -1, & x_j < x_i \end{cases} \tag{2}$$

where x_i and x_j denote the data values at times i and j, respectively, n specifies the length of the data set. A positive value of S indicates an increasing trend, while a negative value refers to a decreasing trend. The variance of S, $Var(S)$, is calculated using Equation (3), assuming a normally distributed time series with $n > 10$ and.

$$Var(S) = \frac{n(n-1)(2n+5) - \sum_{i=1}^{n} t_i i(i-1)(2i+5)}{18} \tag{3}$$

where t_i denotes the number of data ties. The test statistic, Z, is then calculated using Equation (4).

$$z = \begin{cases} \frac{S-1}{\sqrt{var(S)}} & S > 0 \\ 0 & S = 0 \\ \frac{S-1}{\sqrt{var(S)}} & S < 0 \end{cases} \tag{4}$$

In a two-tailed test, the standard Z value is compared with the standard normal distribution table at a 5% significance level (α). The null hypothesis (H_0) is rejected if $|Z| > |Z1-\alpha/2|$, meaning that the trend is statistically significant, otherwise H_0 is accepted, i.e., there is no presence of a statistically significant trend in the time series at the 95% confidence level.

2.3.2. Innovative Trend Analysis

The Innovative Trend Analysis (ITA) was proposed by Sen [55]. The technique consists of the dividing a time series into two subsets. Hence, the observed hydrological time series was divided into two sub-series after having re-arranged the time series in ascending order. Based on the Cartesian coordinate system, the first sub-series (X_i) is drawn on X-axis while the second (X_j) is on Y-axis as shown in Figure 3. Generally, the hydrological time series is described as trendless if it is located on the 1:1 (45°) straight line whereas it is a decreasing trend in time series if the data values are found below the triangular area of the straight line (45°), and an increasing trend if it is above it.

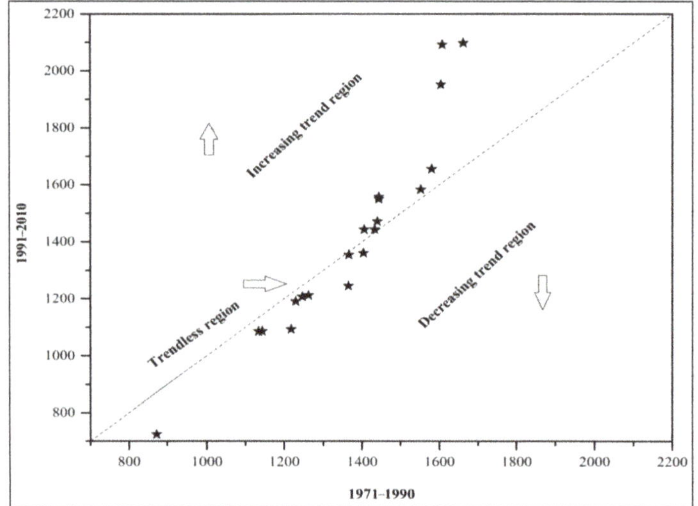

Figure 3. Interpretation of the ITA technique.

2.4. Change Point Analysis

A change-point analysis, as described in [11], was conducted on the time series with the purpose of determining the presence of abrupt changes in time series. Test determines the number of change points and estimate their time occurrence. This study used the Combinations of the Sum Boxes (CUSUM) plot to determine the change point in a time series. This procedure is described in [56] procedure. The accumulated sum of data points i.e., $X_1, X_2, \ldots, X_{24}$ is calculated and CUSUM graphs are produced. Moreover, in general, the average value displays periods that are greater than the average. Mostly, the values are below the average and the downward slope segment is displayed by the sudden changes in CUSUM direction, which points out an average or sudden change. The period of the CUSUM column is straightforward when the average or sudden change does not occur.

2.5. Double Mass Curve Analysis

The consistency of of hydrological time series is checked by comparing data for a single station using double mass curve analysis [57]. It can be used to correct the unstable precipitation data. It is basically a fraction of the accumulated image arithmetic figures in a variable different from the accumulated figures of another variable or the same variable occasionally. The relative mass ratios of these variables change concerning the interrelations between the variables. This may be due to changes in the data collection procedure or might be the physical changes that mark the relationship. It must not be thought that all discrepancies shown by a double mass curve were inconsistent due to changes in data collection methods or errors in data collection.

2.6. Flow Duration Curve

Flow Duration Curve (FDC) [58] is a graphical representation of the overall historical variability in river flow. FDC is a corresponding distribution function of daily flows. The probability that a certain value will be exceeded in a predefined future period is called the probability of exceedance. It is used to predict extreme events such as floods, hurricanes, and earthquakes. The exceedance probability P can be calculated using Equation (5).

$$P = \left[\frac{M}{(n+1)} \right] \times 100 \tag{5}$$

where M specifies the ranked position on the listing (dimensionless) and n denotes the number of events during the data record. The exceedance predicts the probability of a given streamflow [11]. Further detail about FDC can be found in a study conducted by [11].

2.7. Eco-Hydrological Framework

This framework focuses on studying the interactions between water and ecological systems. The annual hydrological budget for a catchment can be expressed by Equation (6):

$$P = ET + Q + D + \Delta S \tag{6}$$

where P stands for precipitation, ET for the evapotranspiration, Q for streamflow, D for deep groundwater losses, and ΔS for the changes in storage. Deep groundwater losses and changes in water storage can reasonably be assumed to be zero over a long time period. The available water and energy in an agricultural watershed can be assessed through the excess water (P_{Ex}) and excess energy (E_{Ex}), which can be estimated using the following equations:

$$P_{Ex} = \frac{(P - ET)}{P} \tag{7}$$

$$E_{Ex} = \frac{(PET - ET)}{PET} \tag{8}$$

where PET denotes potential evapotranspiration. The value of P_{Ex} and E_{Ex} can range from 0 to 1. PET denotes potential evapotranspiration. The eco-hydrological analysis represented by Figure 4 is an example of the conceptual model that is applicable for understanding the soil water conservations measures and climatic variability impacts on watershed hydrology [59].

2.8. Climate Elasticity Model

Climate change impacts can also be assessed using the climate elasticity model [35,60,61]. Schaake [60] presented the concept of climate elasticity to assess the sensitivity of streamflow to climate changes. It can be defined as the relative change in streamflow divided by the relative change in a climate variable, precipitation for instance. Schaake [60] defined the precipitation elasticity, ε_p, through Equation (9):

$$\varepsilon_p(P, Q) = \frac{\frac{dQ}{Q}}{\frac{dp}{P}} = \frac{dQ}{dP} \frac{P}{Q} \tag{9}$$

where P and Q represent precipitation and streamflow, respectively, and ε_p denotes the precipitation elasticity. The precipitation elasticity of streamflow is a random variable that depends on P and Q. A non-parametric estimator of precipitation elasticity was defined by [35]. Another similar study conducted by [62] defined an estimator of precipitation elasticity of streamflow by Equation (10) given below:

$$\frac{\Delta Q_i}{\overline{Q}} = \varepsilon_p \cdot \frac{\Delta P_i}{\overline{P}} \tag{10}$$

where ΔQ_i and ΔP_i represents a change in annual streamflow and precipitation in comparison to the long-term average of Q and P, respectively. The current study used historical data of precipitation and streamflow from 1971 to 1996 for the calculating the precipitation elasticity. It also considers the impacts of evapotranspiration in the climate elasticity model by adding evapotranspiration to Equation (10), resulting in Equation (11):

$$\frac{\Delta Q_i}{\overline{Q}} = \varepsilon_p \cdot \frac{\Delta P_i}{\overline{P}} + \varepsilon_{et} \cdot \frac{\Delta ET}{\overline{ET}} \qquad (11)$$

where ε_p and ε_{et} designate precipitation and evapotranspiration elasticity of streamflow, respectively, in multiple regression systems whereas ΔET represents a change in annual mean evapotranspiration in comparison to the long-term mean evapotranspiration (ET).

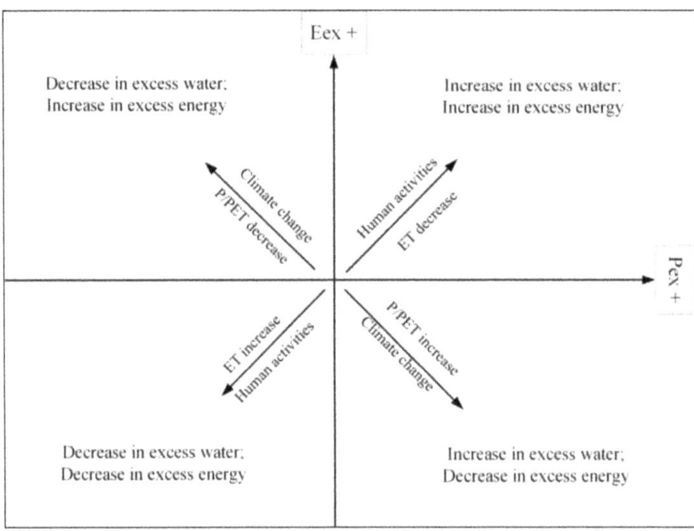

Figure 4. Tomer Schilling framework.

2.9. Statistical Model

The linear regression between averaged annual precipitation (P_{ref}) and annual runoff (Q_{ref}) can be represented by Equation (12):

$$Q_{ref} = aP_{ref} + bET + c \qquad (12)$$

where 'a and b' represents the regression equation constant and 'c' denotes the regression intercept. The coefficients of the equation are determined by the least square method. Climate variability is due to external factors such as a, change in LULC and climate variability as represented by Equation (13):

$$\Delta Q_{total} = \Delta Q_p + \Delta Q_L \qquad (13)$$

where ΔQ_{total} denotes a change in observed mean annual streamflow which is the result of a change in streamflow due to climate change (ΔQ_p) and anthropogenic activities (ΔQ_L). Equation (14) specifies the difference between streamflow of the reference period and change period and the output is a change in average annual streamflow ΔQ.

$$\Delta Q = Q_2 - Q_1 \qquad (14)$$

where ΔQ denotes the change in average annual streamflow, Q_1 denotes the streamflow of the reference period whereas Q_2 denotes the streamflow during the change period.

2.10. Hydrological Modeling

The Soil and Water Assessment Tool (SWAT) is a semi-distributed river basin scale model that can be used to compute the impacts of land-use changes by using climatic and streamflow data over a long period [32,63]. The SWAT model is based on the physical characteristics of the basin. In the Arc SWAT model, a watershed is divided into multiple subbasins and then each subbasin is further sub-divided into Hydrological Response Units (HRUs). The SWAT model simulates streamflow based on the following water balance Equation (15) [64]:

$$SW_t = SW_0 + \sum_{i=1}^{t} (R_{day} - Q_{surf} - E_a - W_{seep} - Q_{gw}) \tag{15}$$

where SW_t denotes the final soil water content (mm); SW_0 denotes the initial soil water content on the day i; 't' designates the time (days); R_{day} specifies the amount of precipitation on the day i (mm); Q_{surf} represents the amount of precipitation on the day i (mm); E_a denotes the amount of evapotranspiration on the day i (mm); W_{seep} specifies the amount of water entering the vadose zone from the soil profile on the day i (mm) and Q_{gw} represents the amount of return flow on the day i (mm).

The simulated runoff in the SWAT model can be achieved at the HRUs level by using the Soil Conservation Service (SCS) curve number method [65]. This method mainly depends on the soil, land cover, and hydrological characteristics. The basic equation of the SCS curve number method:

$$Q_{surf} = \frac{\left(R_{day} - 0.2S\right)^2}{\left(R_{day} + 0.8S\right)} \tag{16}$$

where Q_{surf} denotes the daily surface runoff (mm), R_{day} specifies the daily rainfall (mm) and S represents the retention parameter (mm). The parameter 'S' changes due to change in soil, land-use, water conservation measures, slope, and also with time due to variations in the soil water contents of the watershed. The retention parameter can be defined by Equation (17):

$$S = 25.4 \left(\frac{1000}{CN} - 10\right) \tag{17}$$

where S denotes the retention parameter which gives the value of the drainable volume of soil water per unit area of the saturated thickness (mm/day); CN represents the curve number. A more detailed explanation of the SWAT model can be found [65].

The current study divided the KRB into 15 subbasins and 223 HRUs were created based on the areas with homogeneous soil types, vegetation, and slope (Figure 5). The number of subbasins divisions depends on the stream networks of the watershed. Generally, the runoff values are not affected by the number of subbasins [66]. The land use, soil, and slope classification of the KRB in the Arc SWAT model are displayed in Figure 5. The major land-use type in the KRB is pasture whereas the major soil type is lithosols. The slope map shows that a major part of the basin lies at a slope greater than 60% (Figure 5).

2.10.1. Model Setup

The SWAT model requires the preparation of the forcing data such as soil type, land use, DEM, and climatic data. The key steps in the application of the SWAT model are displayed in Figure 6 and include; (a) watershed delineation and subbasin feature derivation; (b) HRU definition; (c) climatic inputs and weather generator; (d) simulation of the SWAT model; (e) sensitivity and uncertainty analysis, and (f) calibration and validation of the SWAT model.

The surface runoff volume was computed using the SCS curve number method [67]. The Penman-Monteith method was used for the calculation of evapotranspiration [68,69]. The SWAT model was run at the monthly time scale. The SWAT calibration and uncertainty program (SWAT-CUP) [70] was used for calibration, validation, sensitivity, and uncertainty analysis using the sequential uncertainty fitting algorithm (SUFI-2 algorithm). The SWAT model was calibrated for the period (1972–1981) and validated (1983–1996) against the observed flows at Gari Habibullah streamflow gauging station in SWAT-CUP.

The sensitivity analysis was performed in SWAT-CUP to determine the parameters that streamflow is most sensitive to. A total of 18 parameters of the SWAT model that relate to surface runoff, groundwater, and snowmelt were selected for model calibration in Table 3. The parameters related to snowmelt such as SMFMX, SMFMN, SFTMP and SMTMP, TLAPS, and others such as CN2, PLAPS, and ALPHA_BF were found to be more sensitive in the KRB as the basin is mainly snow-fed.

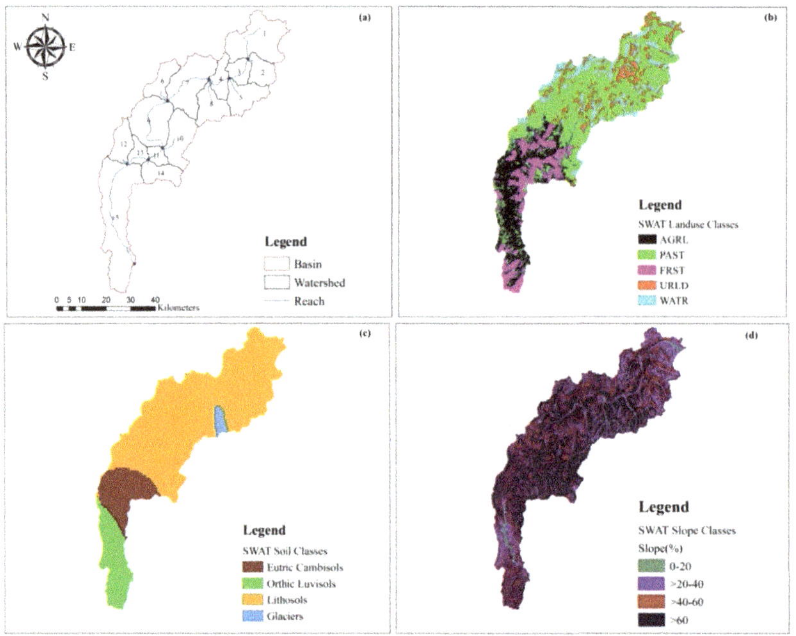

Figure 5. (**a**) Sub-basins; (**b**) Land-use classification; (**c**) Soil classification and (**d**) Slope classification of the KRB in the Arc SWAT model.

2.10.2. Model Performance and Evaluation Criteria

The performance of the SWAT model was assessed using the Nash-Sutcliffe model efficiency (NSE) and coefficient of determination (R^2):

$$\text{NSE} = 1 - \frac{\sum_{i=1}^{n} (Q_0 - Q_s)^2}{\sum_{i=1}^{n} (Q_0 - \overline{Q}_o)^2} \tag{18}$$

$$R^2 = \left[\frac{\sum_{i=1}^{n} (Q_o - \overline{Q}_o)(Q_s - \overline{Q}_s)}{\sqrt{\sum_{i=1}^{n} (Q_o - \overline{Q}_o)^2 \sum_{i=1}^{n} (Q_s - \overline{Q}_s)^2}} \right]^2 \tag{19}$$

where i denotes the time step and n specifies the total number of simulated time steps. Q_o and Q_s represent the observed and simulated streamflow values, respectively. NSE specifies how well the graph between observed and simulated data fits the 1:1 line. The value of NSE is considered satisfactory when NSE is higher than 0.5 [71]. The greater value of NSE, the better is the model accuracy. The model's simulations are considered perfectly fit if NSE = 1 [72].

Whereas, R^2 measures the strength of the linear relationship between the simulated and observed streamflow time series. R^2 value can range between 0 and 1. The model is considered ideal if the R^2 value is equal to 1, and satisfactory if $R^2 > 0.6$ [71].

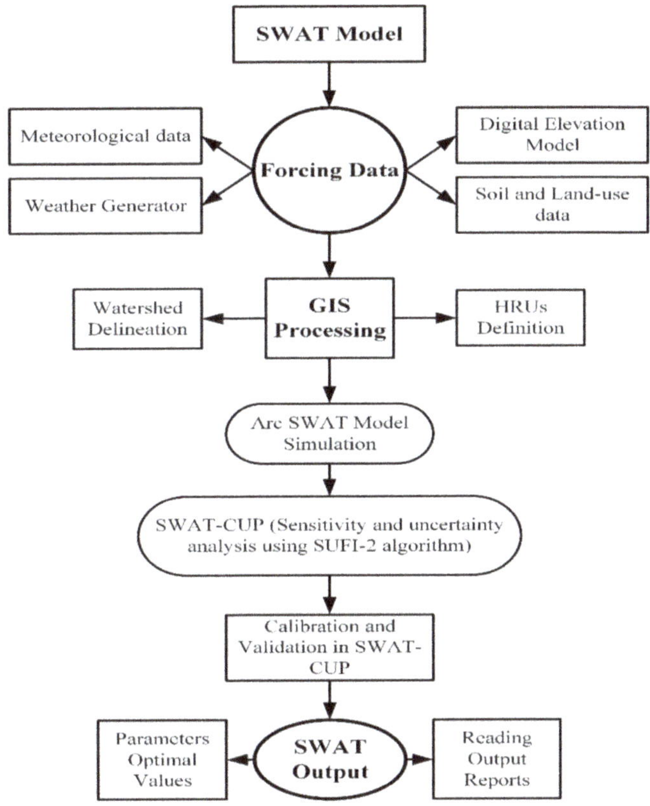

Figure 6. Flowsheet diagram of the SWAT model.

Table 3. List of parameters used for calibrating of the SWAT model.

Parameter	Description	Adjusted Value
r_CN2.mgt	Initial SCS runoff curve number for moisture condition II	−0.25
v__CH_N2.rte	Manning's "n" for the main channel	−0.01
r_SOL_AWC.sol	Soil available water capacity (mm H_2O/mm soil)	0.48
v__GWQMN.gw	Threshold depth in the shallow aquifer for return flow (mm)	1888
v__RCHRG_DP.gw	Fraction of root zone percolation that reaches the deep aquifer	0.62
v_ALPHA_BF.gw	Baseflow alpha-factor (days)	0.034
v_GW_DELAY.gw	Groundwater delay (days)	71
v_SURLAG.bsn	Surface Runoff lag coefficient	4.47
v_SMFMN.bsn	Minimum melt rate for snow during the year (occurs on winter solstice) H_2O/°C-day)	4.88
v_SMFMX.bsn	Maximum melt rate for snow during the year (occurs on the summer solstice). (mm H_2O/°C-day)	11.20
v_SMTMP.bsn	Snowmelt base temperature (°C)	−2.81

Table 3. *Cont.*

Parameter	Description	Adjusted Value
v_SFTMP.bsn	Snowfall temperature (°C)	4.91
v__TIMP.bsn	Snowpack temperature lag factor	0.045
v_SNOCOVMX.bsn	Minimum snow water content that corresponds to 100% snow cover	192.17
v__PLAPS.sub	Precipitation lapse rate (mm H_2O/km)	135.2
v_TLPAS.sub	Temperature lapse rate (°C/km)	−6.4
v__ESCO.hru	Soil evaporation compensation factor	0.42
v__EPCO.hru	Plant uptake consumption factor	0.80

Note: where 'v' designates that the parameter value is replaced by a given value, whereas 'r' specifies that the parameter value is multiplied by (1 + a given value).

3. Results and Discussion

3.1. Trends in Hydro-Climatic Variables

The statistical indices explaining the annual variations in observed precipitation, evapotranspiration, and runoff (from 1971 to 2010) of the KRB are listed in Table 4. Table 4 shows the combined annual trend analysis of all the variables. It was observed that the coefficient of variance (C_v) of precipitation was higher than that of runoff and evapotranspiration whereas the C_v of evapotranspiration was found the lowest i.e., 0.12 as compared to precipitation (0.22) and runoff (0.21). Moreover, the ratio of precipitation to runoff was found smaller than the ratio of precipitation to evapotranspiration. Overall, the mean value of precipitation was found to be 14% higher than runoff whereas runoff was found to be 70% higher than evapotranspiration. The skewness of evapotranspiration was observed to be 74% higher than runoff while 72% greater than precipitation. The skewness of precipitation was found as 10% higher than runoff (Table 4).

Table 4. Statistical indices for annual variations in precipitation, evapotranspiration, and runoff of the KRB.

Climatic Variables	1971–2010				
	Mean (mm)	Standard Deviation (mm)	Skewness	Coefficient of Variance (C_V)	Ratio of (C_V)
Precipitation	1636	365	0.611	0.22	0.96
Evapotranspiration	422	48	2.15	0.12	1.85
Runoff	1413	302	0.55	0.21	-

The results of the MK test on the precipitation, evapotranspiration, and runoff time series are presented in Table 4, and are also illustrated in Supplementary Figure S1 (for the entire data record), and Figure S2 (for the pre- and post-change time-periods separately). Precipitation and evapotranspiration exhibited statically significantly increasing trends, whereas no statistically significant trends were observed in the runoff time series.

The results of the ITA supports the trend analyses conducted using the MK test, particularly for precipitation, where all data points are above the 1:1 line (Table 5, Figure 7).

Table 5. Results of the MK trend test and ITA on the annual for the hydro-climatic time series of the KRB.

Variables	Z-Value	Trend	ITA
Precipitation	1.74 [+]	Sig. Increasing	↑
Evapotranspiration	1.67 [+]	Sig. Increasing	↑
Runoff	1.36	Non-significant	→

Note: 'Sig.' stands for statistically significant and '+.' Refers to the 0.1 level of significance.

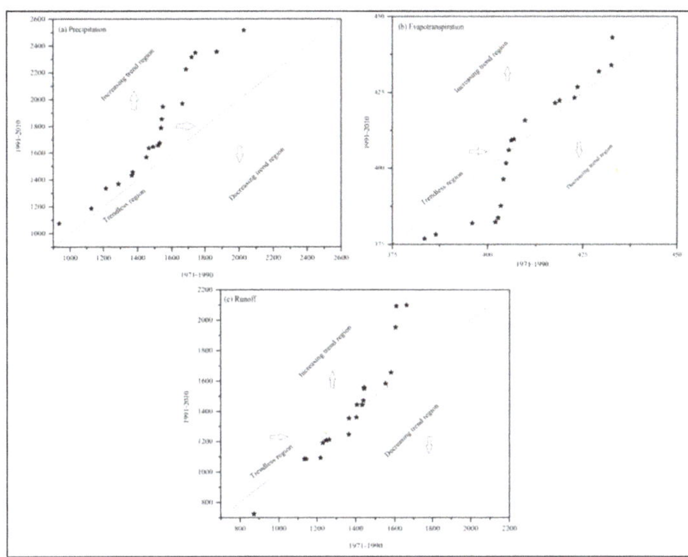

Figure 7. ITA of precipitation, evapotranspiration, and runoff time series of the KRB. (**a**) Precipitation; (**b**) Evapotranspiration; (**c**) Runoff.

The results of the recent trend analyses by Latif et al. [73] which was conducted over the Jhelum River Basin (JRB) also located in Northern Pakistan and extending into India, were found to be consistent with current study as they detected increasing trends in annual precipitation and evapotranspiration. Similarly, [74] examined the flow regime of the JRB and found trends similar to those in our study in the runoff of a few stations, whereas trends contrasting those of our study were identified at a few weather stations. Furthermore, [75] found an increase in the frequency of wet precipitation days from the northeast to the middle parts of the northern highlands of Pakistan. The trends in temperature and precipitation of our previous study i.e., [45] were found consistent with our current study but runoff in the upper part of the KRB was found to increase in comparison to the lower part (downstream) of the basin. Moreover, [3] also identified the lack of a trend in the runoff of the upper and middle stream of the Syr Darya River basin, and their results were found consistent with our current study.

3.2. Abrupt Changes in the Hydrological Time Series

The abrupt change or change point in the annual precipitation, evapotranspiration, and runoff of the KRB is displayed in Figure 8. The accumulative difference curve and double mass curve methods were used to identify the abrupt change in the runoff. An abrupt change in all three-time series was identified in 1996 (Figure 8). A study conducted by [34] used the accumulative anomaly method to determine the abrupt change in the runoff of the Huangfuchuan River, China, and their results were found to be consistent. Moreover, a study performed by [76] also observed an abrupt change in the runoff of the Wei He River basin in the year (i.e., 1993) as similar to our study. A study conducted by [74] also determined the change point in streamflow which was consistent with our results.

Table 6 shows the smaller C_v of during the post-change period in comparison to the pre-change period. The C_v of evapotranspiration, for its part, is higher during the post change than in the pre-change period, while for runoff, it is also higher during the later period.

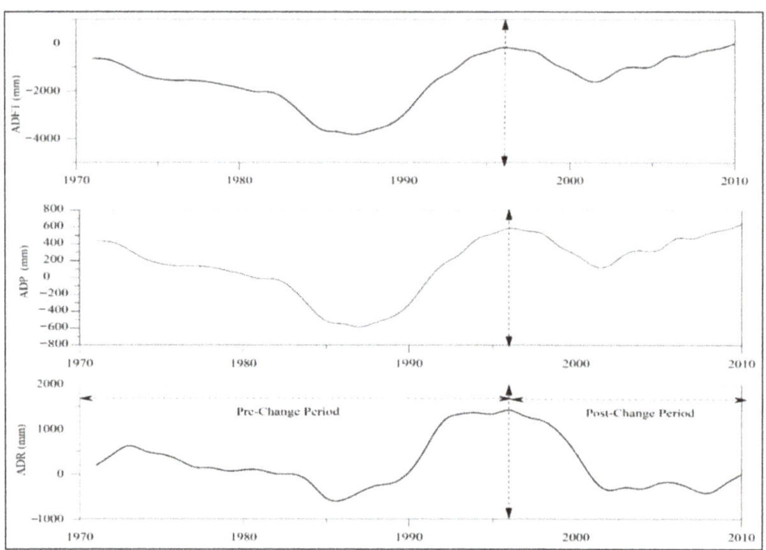

Figure 8. Change point analysis on the annual precipitation (ADP), evapotranspiration (ADET), and runoff (ADR) time series.

Table 6. Descriptive statistics of the time period before and after the abrupt change in time series.

Variables	Pre-Change (1971–1996)			Post-Change (1997–2010)			Relative Change	
	Mean (mm)	C_v	Ratio of C_v	Mean (mm)	C_v	Ratio of C_v	mm	%
Precipitation	1643	0.23	0.88	1624	0.22	1.10	−19	−1.0
Evapotranspiration	417	0.10	2.06	431	0.14	1.72	14	3.25
Runoff	1426	0.20		1390	0.25		−36	−2.59

Regarding the pre-change period, the precipitation (−1%) and runoff (−2.59%) decreased whereas evapotranspiration (+3.25%) increased during the post-change period (Table 6, Figure 9). The ratio of C_v for precipitation to runoff was observed to be 20% higher in the post-change period whereas the ratio of C_v for evapotranspiration to runoff was 18% less during the post-change period in relation to the pre-change period.

The abrupt change was also determined by plotting the data of cumulative precipitation and cumulative runoff in double mass curve analysis (Figure 10). Figure 10a shows the deviation of the double mass curve from the linear relation.

The MK test for the pre (1971–1996) and post-change (1997–2010) periods for precipitation, evapotranspiration, and runoff are displayed in Figure S3 (Supplementary Material) and Table 7 From the results of the MK trend test and ITA method, one can see increasing trends in precipitation and runoff but a decreasing trend, and lacking statistical significance, in evapotranspiration during the pre-change period. After the change point, both the MK test and ITA show statistically significant increasing trend for runoff and an insignificant increasing trend for precipitation and no trend for evapotranspiration (Table 7). The status of evapotranspiration changed from decreasing trend (pre-change) to a trendless (post-change) period which showed that evapotranspiration has increased during the post-change period in comparison with the pre-change period.

Liang [77] investigated the trends in precipitation, temperature, and runoff time series upstream of the Minjiang River Basin, China and observed an increasing trend in precipitation for a similar period to that of our study. They also observed an increase in runoff during the post-change period. Another study conducted by Latif [74] also identified increasing trends in several hydrological stations of the Indus, Jhelum, and Kabul River basins. They observed decreasing trends in precipitation in recent decades as compared to the reference period, which is consistent with the trends of precipitation for the post-change period of our study. Moreover, Khattak [78] observed increasing trends in temperature and runoff but noted an inconsistent pattern for precipitation trend in UIB, Pakistan. Their precipitation and temperature innovative trends were found consistent, whereas runoff innovative trends were found conflicting with the findings of [45].

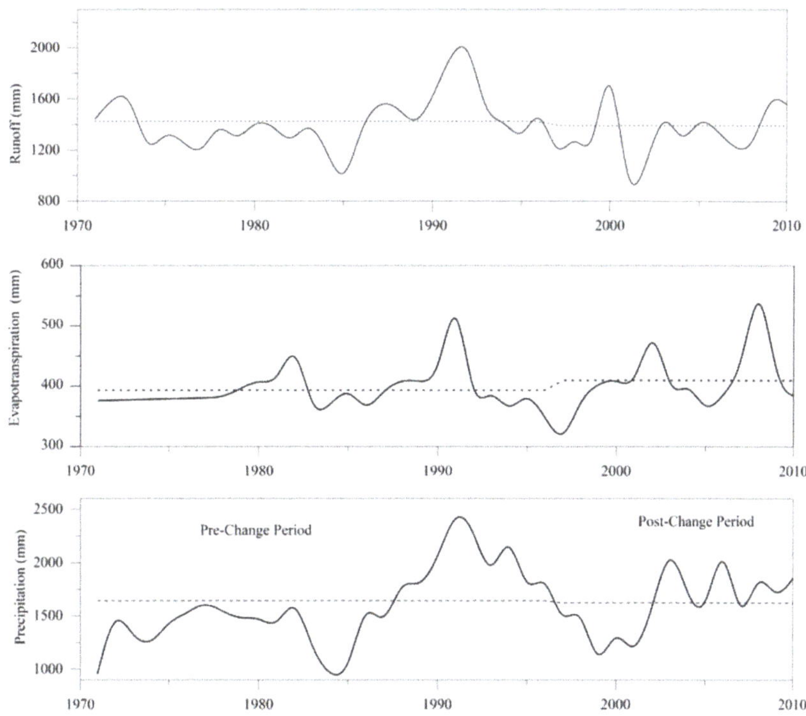

Figure 9. Temporal variation in precipitation, evapotranspiration, and runoff over the KRB.

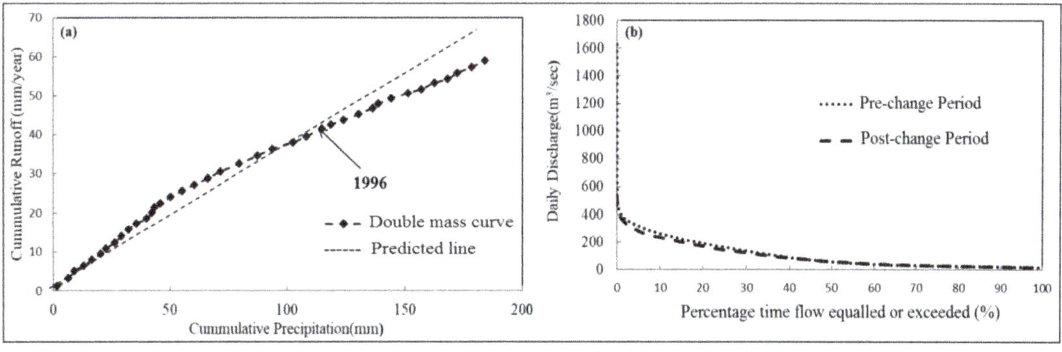

Figure 10. The determination of abrupt change in a time series using (**a**) a double mass curve and; (**b**) a flow duration curve analysis.

Table 7. Results of the MK trend test and ITA on the hydro-climatic variables of the KRB during the pre and post-change periods.

Variables	1971–1996			1997–2010		
	Z-Value	Trend	ITA	Z-Value	Trend	ITA
Precipitation	3.13 **	Sig. Increasing	⬆	1.53	Non-significant	⬆
ET_p	1.06	Non-significant	⬇	0.88	No-trend	➡
Runoff	2.07 *	Sig. Increasing	⬆	2.19 *	Sig. Increasing	⬆

Note: ** refers to the 0.01 significance level and * to the 0.05 significance level.

3.3. Runoff Response to Climate Change and Anthropogenic Activities using Different Methods

In the eco-hydrological framework, the variations in runoff are assessed by considering similar precipitation and evapotranspiration of the basin in different pairs of years. This framework was used to quantify the contribution of climate change and anthropogenic activities in response to variations in the runoff. In this analysis, the abrupt change was also observed in the year 1996 (Figure 11). Moreover, in this framework, P_{ex} was found to decrease from 0.88 to 0.84 while E_{ex} was found to decrease from 0.45 to 0.37 for pre and post change period as shown in Figure 11. The decrease in both excess energy and excess water implies that anthropogenic activities are pronounced in this region.

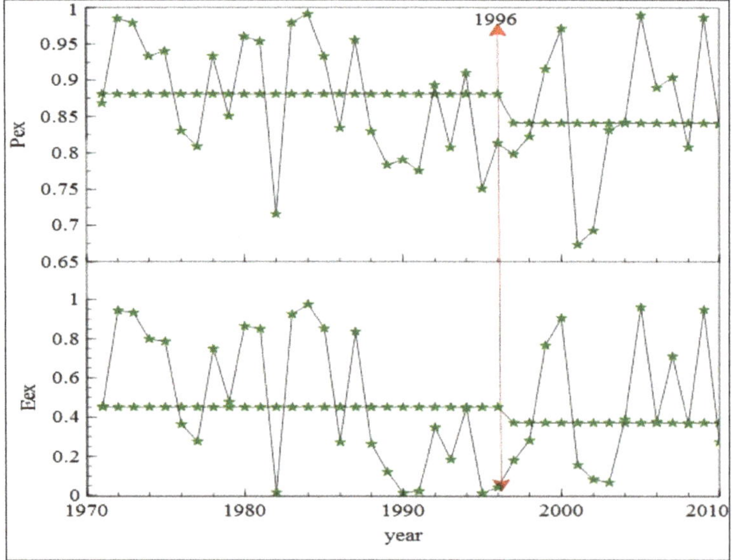

Figure 11. Eco-hydrological approach applied to the KRB.

In the modeling approach, the SWAT model was successfully calibrated (1972–1981) and validated (1983–1996) on the KRB on a monthly time scale as shown in Figures 12 and 13. Overall, the SWAT model simulated both low and high flows very well during both calibration and validation. Table 8 shows the results of the calibration and validation of the SWAT model, which confirm that the SWAT model can be used with confidence in simulating the streamflow of the KRB.

Moreover, several storage dams and small ponds have been built in the KRB in recent years. The different hydropower projects have been initiated under the United Nations (UN) goal of sustainable clean energy and the Paris agreement. Several hydropower projects have been started by the government of Khyber Pakhtunkhawa such as the Balakot project (190 MW), Kari-Muskhui (446 MW), Naran Dam (210 MW), Batakundi (65 MW), Laspur-Muri Gram (130 MW), Shushghai-Zhendoli (144 MW), Shogo Sin (132 MW), Torkum-Gudubar (409 MW) and Samshel Toren (260 MW) [87].

3.5. Comparison with Other River Basins

In the last fifty years, a number of studies have related abrupt changes in river runoff to climate change in different parts of the world [75,76] and anthropogenic activities [13]. Several researchers found that the watershed characteristics were the main driving factors that influenced the hydrological response [82]. Another study conducted by [88] determined the runoff characteristics in different environmental conditions of the United States. Moreover, [89] observed the impacts of climate change on the hydro-climatology of Lake Tana Basin, Ethiopia. Another study performed by [90] observed that climate change and anthropogenic activities are mainly responsible for influencing the runoff characteristics of the Tualatin River basin, Oregon. Similarly, [91] observed the remarkable changes in the hydrological response of the Swedish Rivers due to the impacts of anthropogenic activities and climate change. The grassland plains in Russia showed a greater delay in runoff due to watershed characteristics observed by [92].

Moreover, few Asian rivers are largely attributed to watershed characteristics, such as the Yellow River [39,93], Ganga River [94], Huifa River [32], Kofarnihon River [95], Bagmati River [2], and Pearl River [96], and the results of these studies were found consistent with our study. Another study conducted by [79] observed that just after the change point, reduction in Soan River runoff was due to climate change and land cover changes. Similarly, a study conducted by [80] on the Tarbela catchment found that watershed characteristics play a major role in changes in runoff and climate change relatively contributes less as compared to watershed characteristics (anthropogenic activities). Moreover, [97] observed that watershed characteristics were evident for the increased runoff in the Simly watershed. Under the different climatic zone, there is a need to plan watershed-scale water resources management to face global climate change and frequent extreme precipitation as well as land use degradation. The national self-regulation ability is not strong enough to obtain the key object of hydrologic protection and management for climate-sensitive areas.

This study examined the impact of climate change and anthropogenic activities on the hydrology of a river basin. It is necessary to determine the impact of individual anthropogenic activities on the hydrological response. This research also assumed that climate variability and anthropogenic activities are independent, which is not the case.

4. Conclusions

This study investigated the impacts of climate change and anthropogenic activities on the runoff characteristics of the KRB during the period 1971–2010 using empirical, statistical, and modeling techniques. Potential causes for the observed changes in streamflow were identified and, the following conclusions were drawn:

- The MK trend test and the ITA technique revealed statistically significant increases in both precipitation and evapotranspiration over the study region but no trend in runoff during the period 1971 to 2010.
- A change point analysis identified a change in the annual runoff time series in 1996. This abrupt change in 1996 was also observed using the double mass and flow duration curves
- The time series was divided into two time-periods: 1971–1996 and 1997–2010. The MK test applied over those two time-period revealed statistically significant increasing tends in precipitation and runoff, and a decreasing trend in evapotranspiration, although lacking statistical significance during the pre-change period. During the

post-change period, only an increasing trend in runoff was found to be statistically significant; no trend was seen in the evapotranspiration time series, while the increasing trend in precipitation was not found to be statistically significant.

- An eco-hydrological framework showed a decrease in both excess energy and excess water between two time periods, implying an evident contribution from anthropogenic activities to variations in runoff in the KRB.
- A climate elasticity model quantified the relative contribution of climate change and anthropogenic activities (25% and 75% respectively) to the variability of streamflow in the KRB.
- The statistical model developed in this paper estimated a 95% contribution from anthropogenic sources and a 5% contribution from climate change to streamflow variability in the KRB.
- The method used to reconstruct natural runoff estimated a 16.1 m^3/s (or 15.3%) reduction in the mean flow of the KRB during the post-change period in comparison to the pre-change period of which, -76% changes was calculated to be due to climate change and 176% to changes in anthropogenic activities over the catchment.
- Overall, it is concluded role of anthropogenic activities was evident in terms of runoff variability in the KRB in comparison to climate change, especially since 1996.
- This study quantified the impacts of climate change and anthropogenic activities on the streamflow of the KRB using the different techniques and identifies the areas that have experienced most change within the basin.
- The results of this study improved our understanding of the main cause of streamflow variability in the KRB, which will help with the planning of water management strategies.

Supplementary Materials: The following are available online at https://www.mdpi.com/article/10.3390/w13223163/s1, Figure S1: MK trend analysis of Runoff (R), Precipitation (Pr), and Evapotranspiration (Et) for whole period (1971–2010); Figure S2: MK trend analysis of Precipitation (Pr), Evapotranspiration (Et), and Runoff (R) during pre-change (1971–1996) and post-change (1997–2010) periods; Figure S3: ITA analysis of precipitation (a,b), evapotranspiration (c,d), and runoff (e,f) during pre-change (1971–1996), and post-change (1997–2010) periods.

Author Contributions: Conceptualization and data analysis, M.S.; Hydrological Modeling and abstract, M.S. & M.A.; Introduction, M.S., A.S.G., M.Z. & M.I.K.; S.M., Methodology, M.S. & A.W.; Idea, supervision and funding, S.L. & M.S; Results, M.S. & M.A.; Discussion, M.A., M.S. &; S.L. Paper formatting and conclusion M.S., M.A. & A.S.G. All the authors contributed to the finalization of this manuscript. All authors have read and agreed to the published version of the manuscript.

Funding: This study is supported by the NSFC-ICIMOD joint project (Grant no. 41761144075), Yunnan University grant (2018M643541 and 209071) and Higher Education Commission (HEC) of Pakistan (HEC/STR/279).

Institutional Review Board Statement: Not applicable.

Informed Consent Statement: Not applicable.

Data Availability Statement: The data presented in this study is available on request from the corresponding author.

Acknowledgments: We are thankful to the Surface Water Hydrology Project of the Water and Power Development Authority (SWHP-WAPDA) and the Pakistan Meteorological Department for providing the required hydro-meteorological data to conduct this study. We are thankful to editor and associate editor, as well as reviewers for their valuable comments and suggestions, which have improved the final version of this manuscript.

Conflicts of Interest: The authors declare that there is no conflict of interest.

Abbreviations

KRB	Kunhar River Basin
SDGs	Sustainable Development Goals
ITA	Innovative Trend Analysis
MK	Mann Kendall
KPK	Khyber Pakhtunkhwa
DEM	Digital Elevation Model
ICIMOD	International Center for Integrated Mountain Development
WAPDA	Water and Power Development Authority
PMD	Pakistan Meteorological Department
CUSUM	Combinations of the Sum Boxes
FDC	Flow Duration Curve
PET	Potential Evapotranspiration
ET	Evapotranspiration
SWAT	Soil and Water Assessment Tool
HRUs	Hydrological Response Units
SCS	Soil Conservation Service
CN	Curve Number
SUFI	Sequential Uncertainty Fitting
NSE	Nash-Sutcliffe model efficiency
C_v	Coefficient of Variance
UN	United Nations

References

1. Roser, M. *Human Development Index (HDI)*; Our World in Data: Oxford, UK, 2019.
2. Lamichhane, S.; Shakya, N.M. Integrated Assessment of Climate Change and Land Use Change Impacts on Hydrology in the Kathmandu Valley Watershed, Central Nepal. *Water* **2019**, *11*, 2059. [CrossRef]
3. Bissenbayeva, S.; Abuduwaili, J.; Saparova, A.; Ahmed, T. Long-term variations in runoff of the Syr Darya River Basin under climate change and human activities. *J. Arid. Land* **2021**, *13*, 56–70. [CrossRef]
4. AghaKouchak, A.; Mirchi, A.; Madani, K.; di Baldassarre, G.; Nazemi, A.; Alborzi, A.; Anjileli, H.; Azarderakhsh, M.; Chiang, F.; Hassanzadeh, E.; et al. Anthropogenic Drought: Definition, Challenges, and Opportunities. *Rev. Geophys.* **2021**, *59*, e2019RG000683. [CrossRef]
5. Dhaubanjar, S.; Lutz, A.F.; Gernaat, D.E.; Nepal, S.; Smolenaars, W.; Pradhananga, S.; Biemans, H.; Ludwig, F.; Shrestha, A.B.; Immerzeel, W.W. A systematic framework for the assessment of sustainable hydropower potential in a river basin—The case of the upper Indus. *Sci. Total Environ.* **2021**, *786*, 147142. [CrossRef] [PubMed]
6. Wu, Y.; Huang, L.; Zhao, C.; Chen, M.; Ouyang, W. Integrating hydrological, landscape ecological, and economic assessment during hydropower exploitation in the upper Yangtze River. *Sci. Total Environ.* **2021**, *767*, 145496. [CrossRef] [PubMed]
7. Liu, Q.; Niu, J.; Sivakumar, B.; Ding, R.; Li, S. Accessing future crop yield and crop water productivity over the Heihe River basin in northwest China under a changing climate. *Geosci. Lett.* **2021**, *8*, 2. [CrossRef]
8. Pachauri, R.K.; Allen, M.R.; Barros, V.R.; Broome, J.; Cramer, W.; Christ, R.; van Ypserle, J.P. *Climate Change: Synthesis Report. Contribution of Working Groups I, II and III to the Fifth Assessment Report of the Intergovernmental Panel on Climate Change*; IPCC: Geneva, Switzerland, 2014; p. 151.
9. Rahman, K.; da Silva, A.G.; Tejeda, E.M.; Gobiet, A.; Beniston, M.; Lehmann, A. An independent and combined effect analysis of land use and climate change in the upper Rhone River watershed, Switzerland. *Appl. Geogr.* **2015**, *63*, 264–272. [CrossRef]
10. Canchala, T.; Cerón, W.L.; Francés, F.; Carvajal-Escobar, Y.; Andreoli, R.; Kayano, M.; Alfonso-Morales, W.; Caicedo-Bravo, E.; de Souza, R.F. Streamflow Variability in Colombian Pacific Basins and Their Teleconnections with Climate Indices. *Water* **2020**, *12*, 526. [CrossRef]
11. Ghaderpour, E.; Vujadinovic, T.; Hassan, Q.K. Application of the Least-Squares Wavelet software in hydrology: Athabasca River Basin. *J. Hydrol. Reg. Stud.* **2021**, *36*, 100847. [CrossRef]
12. Alifujiang, Y.; Abuduwaili, J.; Groll, M.; Issanova, G.; Maihemuti, B. Changes in intra-annual runoff and its response to climate variability and anthropogenic activity in the Lake Issyk-Kul Basin, Kyrgyzstan. *Catena* **2021**, *198*, 104974. [CrossRef]
13. Jehanzaib, M.; Shah, S.A.; Kwon, H.-H.; Kim, T.-W. Investigating the influence of natural events and anthropogenic activities on hydrological drought in South Korea. *Terr. Atmospheric Ocean. Sci.* **2020**, *31*, 85–96. [CrossRef]
14. Chung, E.-S.; Park, K.; Lee, K.S. The relative impacts of climate change and urbanization on the hydrological response of a Korean urban watershed. *Hydrol. Process.* **2011**, *25*, 544–560. [CrossRef]
15. Saifullah, M.; Li, Z.; Li, Q.; Hashim, S.; Zaman, M. Quantifying the Hydrological Response to Water Conservation Measures and Climatic Variability in the Yihe River Basin, China. *Outlook Agric.* **2015**, *44*, 273–282. [CrossRef]

16. Zaman, M.; Anjum, M.N.; Usman, M.; Ahmad, I.; Saifullah, M.; Yuan, S.; Liu, S. Enumerating the Effects of Climate Change on Water Resources Using GCM Scenarios at the Xin'anjiang Watershed, China. *Water* **2018**, *10*, 1296. [CrossRef]

17. Wu, J.; Miao, C.; Zhang, X.; Yang, T.; Duan, Q. Detecting the quantitative hydrological response to changes in climate and human activities. *Sci. Total Environ.* **2017**, *586*, 328–337. [CrossRef]

18. Magritskii, D.V. Anthropogenic impact on the runoff of Russian rivers emptying into the Arctic Ocean. *Water Resour.* **2008**, *35*, 1–14. [CrossRef]

19. Yin, J.; Gentine, P.; Zhou, S.; Sullivan, S.C.; Wang, R.; Zhang, Y.; Guo, S. Large increase in global storm runoff extremes driven by climate and anthropogenic changes. *Nat. Commun.* **2018**, *9*, 4389. [CrossRef] [PubMed]

20. Abbasa, N.; Wasimia, S.A.; Al-Ansari, N. Assessment of Climate Change Impact on Water Resources of Lesser Zab, Kurdistan, Iraq Using SWAT Model. *Engineering* **2016**, *8*, 697–715. [CrossRef]

21. Ali, S.; Shah, S. Climate Change Impact on Flow Discharge of Kunhar River Catchment using Snowmelt Runoff Model. *J. Basic Appl. Sci.* **2015**, *11*, 184–192. [CrossRef]

22. Mahmood, R.; Jia, S.; Babel, M.S. Potential Impacts of Climate Change on Water Resources in the Kunhar River Basin, Pakistan. *Water* **2016**, *8*, 23. [CrossRef]

23. Hirabayashi, Y.; Mahendran, R.; Koirala, S.; Konoshima, L.; Yamazaki, D.; Watanabe, S.; Kim, H.; Kanae, S. Global flood risk under climate change. *Nat. Clim. Chang.* **2013**, *3*, 816–821. [CrossRef]

24. Held, I.M.; Soden, B.J. Robust Responses of the Hydrological Cycle to Global Warming. *J. Clim.* **2006**, *19*, 5686–5699. [CrossRef]

25. Akbar, H.; Gheewala, S.H. Impact of Climate and Land Use Changes on flowrate in the Kunhar River Basin, Pakistan, for the Period (1992–2014). *Arab. J. Geosci.* **2021**, *14*, 1–11. [CrossRef]

26. El-Khoury, A.; Seidou, O.; Lapen, D.; Que, Z.; Mohammadian, A.; Sunohara, M.; Bahram, D. Combined impacts of future climate and land use changes on discharge, nitrogen and phosphorus loads for a Canadian river basin. *J. Environ. Manag.* **2015**, *151*, 76–86. [CrossRef] [PubMed]

27. He, Y.; Wang, F.; Mu, X.; Yan, H.; Zhao, G. An Assessment of Human versus Climatic Impacts on Jing River Basin, Loess Plateau, China. *Adv. Meteorol.* **2015**, *2015*, 478739. [CrossRef]

28. Mango, L.M.; Melesse, A.M.; McClain, M.E.; Gann, D.; Setegn, S.G. Land use and climate change impacts on the hydrology of the upper Mara River Basin, Kenya: Results of a modeling study to support better resource management. *Hydrol. Earth Syst. Sci.* **2011**, *15*, 2245–2258. [CrossRef]

29. Neupane, R.P.; Kumar, S. Estimating the effects of potential climate and land use changes on hydrologic processes of a large agriculture dominated watershed. *J. Hydrol.* **2015**, *529*, 418–429. [CrossRef]

30. Strasser, U.; Förster, K.; Formayer, H.; Hofmeister, F.; Marke, T.; Meißl, G.; Nadeem, I.; Stotten, R.; Schermer, M. Storylines of combined future land use and climate scenarios and their hydrological impacts in an Alpine catchment (Brixental/Austria). *Sci. Total Environ.* **2019**, *657*, 746–763. [CrossRef] [PubMed]

31. Wang, W.; Shao, Q.; Yang, T.; Peng, S.; Xing, W.; Sun, F.; Luo, Y. Quantitative assessment of the impact of climate variability and human activities on runoff changes: A case study in four catchments of the Haihe River basin, China. *Hydrol. Process.* **2013**, *27*, 1158–1174. [CrossRef]

32. Zhang, A.; Zhang, C.; Fu, G.; Wang, B.; Bao, Z.; Zheng, H. Assessments of Impacts of Climate Change and Human Activities on Runoff with SWAT for the Huifa River Basin, Northeast China. *Water Resour. Manag.* **2012**, *26*, 2199–2217. [CrossRef]

33. Jiang, S.; Ren, L.; Yong, B.; Singh, V.P.; Yang, X.; Yuan, F. Quantifying the effects of climate variability and human activities on runoff from the Laohahe basin in northern China using three different methods. *Hydrol. Process.* **2011**, *25*, 2492–2505. [CrossRef]

34. Wang, S.; Yan, Y.; Yan, M.; Zhao, X. Quantitative estimation of the impact of precipitation and human activities on runoff change of the Huangfuchuan River Basin. *J. Geogr. Sci.* **2012**, *22*, 906–918. [CrossRef]

35. Sankarasubramanian, A.; Vogel, R.M.; Limbrunner, J.F. Climate elasticity of streamflow in the United States. *Water Resour. Res.* **2001**, *37*, 1771–1781. [CrossRef]

36. Zhou, X.; Zhang, Y.; Yang, Y. Comparison of Two Approaches for Estimating Precipitation Elasticity of Streamflow in China's Main River Basins. *Adv. Meteorol.* **2015**, *2015*, 924572. [CrossRef]

37. Dey, P.; Mishra, A. Separating the impacts of climate change and human activities on streamflows: A review of methodologies and critical assumptions. *J. Hydrol.* **2017**, *548*, 278–290. [CrossRef]

38. Krajewski, A.; Sikorska-Senoner, A.E.; Hejduk, L.; Banasik, K. An Attempt to Decompose the Impact of Land Use and Climate Change on Annual Runoff in a Small Agricultural Catchment. *Water Resour. Manag.* **2021**, *35*, 881–896. [CrossRef]

39. Saifullah, M.; Li, Z.; Li, Q.; Zaman, M.; Hashim, S. Quantitative Estimation of the Impact of Precipitation and Land Surface Change on Hydrological Processes through Statistical Modeling. *Adv. Meteorol.* **2016**, *2016*, 6130179. [CrossRef]

40. Wang, F.; Hessel, R.; Mu, X.; Maroulis, J.; Zhao, G.; Geissen, V.; Ritsema, C. Distinguishing the impacts of human activities and climate variability on runoff and sediment load change based on paired periods with similar weather conditions: A case in the Yan River, China. *J. Hydrol.* **2015**, *527*, 884–893. [CrossRef]

41. De Scally, F.A. Relative importance of snow accumulation and monsoon rainfall data for estimating annual runoff, Jhelum basin, Pakistan. *Hydrol. Sci. J.* **1994**, *39*, 199–216. [CrossRef]

42. Schickhoff, U. Himalayan Forest-Cover Changes in Historical Perspective: A Case Study in the Kaghan Valley, Northern Pakistan. *Mt. Res. Dev.* **1995**, *15*, 3–18. [CrossRef]

43. Star Hydropower Limited. *Environmental Impact Assessment of Patrind Hydropower Project*; Star Hydropower Limited: Islamabad, Pakistan, 2010.

44. Soomro, S.-E.-H.; Hu, C.; Jian, S.; Wu, Q.; Boota, M.W.; Soomro, M.H.A.A. Precipitation changes and their relationships with vegetation responses during 1982–2015 in Kunhar River basin, Pakistan. *Water Supply* **2021**. [CrossRef]

45. Saifullah, M.; Liu, S.; Tahir, A.A.; Zaman, M.; Ahmad, S.; Adnan, M.; Chen, D.; Ashraf, M.; Mehmood, A. Development of Threshold Levels and a Climate-Sensitivity Model of the Hydrological Regime of the High-Altitude Catchment of the Western Himalayas, Pakistan. *Water* **2019**, *11*, 1454. [CrossRef]

46. Khan, M.A.; Stamm, J.; Haider, S. Simulating the Impact of Climate Change with Different Reservoir Operating Strategies on Sedimentation of the Mangla Reservoir, Northern Pakistan. *Water* **2020**, *12*, 2736. [CrossRef]

47. Akbar, H.; Gheewala, S.H. Changes in hydroclimatic trends in the Kunhar River Watershed. *J. Sustain. Energy Environ.* **2020**, *11*, 31–41.

48. Mahmood, R.; Babel, M.S. Evaluation of SDSM developed by annual and monthly sub-models for downscaling temperature and precipitation in the Jhelum basin, Pakistan and India. *Theor. Appl. Clim.* **2013**, *113*, 27–44. [CrossRef]

49. USGS Terrain Data. Available online: https://lpdaac.usgs.gov/tools/data-pool/ (accessed on 3 February 2021).

50. ICIMOD Data Archive. Available online: http://rds.icimod.org/Home/DataDetail?metadataId=28630 (accessed on 25 February 2021). [CrossRef]

51. The Harmonized World Soil Database v 1.2. Available online: http://www.fao.org/soils-portal/data-hub/soil-maps-and-databases/harmonized-world-soil-database-v12/en/ (accessed on 25 December 2020).

52. ERA5 Reanalysis Data. Available online: https://cds.climate.copernicus.eu (accessed on 3 February 2021).

53. Mann, H.B. Nonparametric Tests Against Trend. *Econometrica* **1945**, *13*, 245–259. [CrossRef]

54. Kendall, M. *Rank Correlation Measures*; Charles Griffin: London, UK, 1975; p. 15.

55. Şen, Z. Innovative Trend Analysis Methodology. *J. Hydrol. Eng.* **2012**, *17*, 1042–1046. [CrossRef]

56. Taylor, W.A. *Change-Point Analysis: A Powerful New Tool for Detecting Changes*; Variation: Libertyville, IL, USA, 2000; Available online: https://variation.com/wp-content/uploads/change-point-analyzer/change-point-analysis-a-powerful-new-tool-for-detecting-changes.pdf (accessed on 31 October 2021).

57. Searcy, J.K.; Hardison, C.H.; Langein, W.B. *Double-Mass Curves, with a Section Fitting Curves to Cyclic Data*; US Geological Survey: Reston, VA, USA, 1960.

58. Searcy, J.K. *Flow-Duration Curves (No. 1542)*; US Government Printing Office: Boston, MA, USA, 1959.

59. Tomer, M.D.; Schilling, K.E. A simple approach to distinguish land-use and climate-change effects on watershed hydrology. *J. Hydrol.* **2009**, *376*, 24–33. [CrossRef]

60. Schaake, J.C. From climate to flow. In *Climate Change and US Water Resources*; Wiley-Interscience: Hoboken, NJ, USA, 1990; pp. 177–206.

61. Fu, G.; Charles, S.P.; Chiew, F.H.S. A two-parameter climate elasticity of streamflow index to assess climate change effects on annual streamflow. *Water Resour. Res.* **2007**, *43*, 43. [CrossRef]

62. Zheng, H.; Zhang, L.; Zhu, R.; Liu, C.; Sato, Y.; Fukushima, Y. Responses of streamflow to climate and land surface change in the headwaters of the Yellow River Basin. *Water Resour. Res.* **2009**, *45*, 45. [CrossRef]

63. Babar, S.; Ramesh, H. Streamflow Response to Land Use–Land Cover Change over the Nethravathi River Basin, India. *J. Hydrol. Eng.* **2015**, *20*, 05015002. [CrossRef]

64. Arnold, J.G.; Srinivasan, R.; Muttiah, R.S.; Williams, J.R. Large Area Hydrologic Modeling and Assessment Part I: Model Development. *JAWRA J. Am. Water Resour. Assoc.* **1998**, *34*, 73–89. [CrossRef]

65. Neitsch, S.L. *Soil and Water Assessment Tool*; User's Manual Version; Texas Water Resources Institute: College Station, TX, USA, 2005; p. 476.

66. Abu-Allaban, M.; El-Naqa, A.; Jaber, M.; Hammouri, N. Water scarcity impact of climate change in semi-arid regions: A case study in Mujib basin, Jordan. *Arab. J. Geosci.* **2015**, *8*, 951–959. [CrossRef]

67. Neitsch, S.L.; Arnold, J.G.; Kiniry, J.R.; Williams, J.R. *Soil and Water Assessment Tool Theoretical Documentation Version 2009*; Texas Water Resources Institute: College Station, TX, USA, 2011.

68. Monteith, J.L. Evaporation and environment. In *Symposia of the Society for Experimental Biology*; Cambridge University Press: Cambridge, UK, 1965; Volume 19, pp. 205–234.

69. Penman, H.L. Natural evaporation from open water, bare soil and grass. *Proc. R. Soc. A* **1948**, *193*, 120–145.

70. Abbaspour, K.C. *SWAT-Cup 2012. SWAT Calibration and Uncertainty Program—A User Manual*; Eawag: Dübendorf, Switzerland, 2013.

71. Moriasi, D.N.; Arnold, J.G.; van Liew, M.W.; Bingner, R.L.; Harmel, R.D.; Veith, T.L. Model evaluation guidelines for systematic quantification of accuracy in watershed simulations. *Trans. ASABE* **2007**, *50*, 885–900. [CrossRef]

72. Nash, J.E.; Sutcliffe, J.V. River flow forecasting through conceptual models part I—A discussion of principles. *J. Hydrol.* **1970**, *10*, 282–290. [CrossRef]

73. Azmat, M.; Liaqat, U.W.; Qamar, M.U.; Awan, U.K. Impacts of changing climate and snow cover on the flow regime of Jhelum River, Western Himalayas. *Reg. Environ. Chang.* **2017**, *17*, 813–825. [CrossRef]

74. Latif, Y.; Ma, Y.; Ma, W. Climatic trends variability and concerning flow regime of Upper Indus Basin, Jehlum, and Kabul river basins Pakistan. *Theor. Appl. Clim.* **2021**, *144*, 447–468. [CrossRef]

75. Zaman, M.; Ahmad, I.; Usman, M.; Saifullah, M.; Anjum, M.N.; Khan, M.I.; Qamar, M.U. Event-Based Time Distribution Patterns, Return Levels, and Their Trends of Extreme Precipitation across Indus Basin. *Water* **2020**, *12*, 3373. [CrossRef]
76. Guo, Y.; Li, Z.; Amo-Boateng, M.; Deng, P.; Huang, P. Quantitative assessment of the impact of climate variability and human activities on runoff changes for the upper reaches of Weihe River. *Stoch. Environ. Res. Risk Assess.* **2014**, *28*, 333–346. [CrossRef]
77. Liang, S.; Wang, W.; Zhang, D.; Li, Y.; Wang, G. Quantifying the Impacts of Climate Change and Human Activities on Runoff Variation: Case Study of the Upstream of Minjiang River, China. *J. Hydrol. Eng.* **2020**, *25*, 05020025. [CrossRef]
78. Khattak, M.; Babel, M.; Sharif, M. Hydro-meteorological trends in the upper Indus River basin in Pakistan. *Clim. Res.* **2011**, *46*, 103–119. [CrossRef]
79. Shahid, M.; Cong, Z.; Zhang, D. Understanding the impacts of climate change and human activities on streamflow: A case study of the Soan River basin, Pakistan. *Theor. Appl. Clim.* **2017**, *134*, 205–219. [CrossRef]
80. Shaukat, R.S.; Khan, M.M.; Shahid, M.; Shoaib, M.; Khan, T.A.; Aslam, M.A. Quantitative Contribution of Climate Change and Land Use Change to Runoff in Tarbela Catchment, Pakistan. *Pol. J. Environ. Stud.* **2020**, *29*, 3295–3304. [CrossRef]
81. Zhou, Y.; Lai, C.; Wang, Z.; Chen, X.; Zeng, Z.; Chen, J.; Bai, X. Quantitative Evaluation of the Impact of Climate Change and Human Activity on Runoff Change in the Dongjiang River Basin, China. *Water* **2018**, *10*, 571. [CrossRef]
82. Gao, G.; Fu, B.; Wang, S.; Liang, W.; Jiang, X. Determining the hydrological responses to climate variability and land use/cover change in the Loess Plateau with the Budyko framework. *Sci. Total Environ.* **2016**, *557–558*, 331–342. [CrossRef]
83. Zhou, X.; Zhang, Y.; Wang, Y.; Zhang, H.; Vaze, J.; Zhang, L.; Yang, Y.; Zhou, Y. Benchmarking global land surface models against the observed mean annual runoff from 150 large basins. *J. Hydrol.* **2012**, *470–471*, 269–279. [CrossRef]
84. Ruggiu, D.; Viola, F. Linking Climate, Basin Morphology and Vegetation Characteristics to Fu's Parameter in Data Poor Conditions. *Water* **2019**, *11*, 2333. [CrossRef]
85. Yang, H.; Yang, D. Derivation of climate elasticity of runoff to assess the effects of climate change on annual runoff. *Water Resour. Res.* **2011**, *47*, 47. [CrossRef]
86. Ramírez, J.A.; Hobbins, M.T.; Brown, T.C. Observational evidence of the complementary relationship in regional evaporation lends strong support for Bouchet's hypothesis. *Geophys. Res. Lett.* **2005**, *32*, L15401. [CrossRef]
87. Power, P. *Hydro Power Resources of Pakistan*; Annual Report of Private Power and Infrastructure Board, Islamabad; Government of Pakistan: Islamabad, Pakistan, 2011; pp. 1–2.
88. Caracciolo, D.; Pumo, D.; Viola, F. Budyko's Based Method for Annual Runoff Characterization across Different Climatic Areas: An Application to United States. *Water Resour. Manag.* **2018**, *32*, 3189–3202. [CrossRef]
89. Setegn, S.G.; Rayner, D.; Melesse, A.; Dargahi, B.; Srinivasan, R. Impact of climate change on the hydroclimatology of Lake Tana Basin, Ethiopia. *Water Resour. Res.* **2011**, *47*, 47. [CrossRef]
90. Praskievicz, S.; Chang, H. Impacts of Climate Change and Urban Development on Water Resources in the Tualatin River Basin, Oregon. *Ann. Assoc. Am. Geogr.* **2011**, *101*, 249–271. [CrossRef]
91. Van der Velde, Y.; Vercauteren, N.; Jaramillo, F.; Dekker, S.C.; Destouni, G.; Lyon, S.W. Exploring hydroclimatic change disparity via the Budyko framework. *Hydrol. Process.* **2014**, *28*, 4110–4118. [CrossRef]
92. Yasinskii, S.V.; Kashutina, E.A. Effect of regional climate variations and economic activity on changes in the hydrological regime of watersheds and small-river runoff. *Water Resour.* **2012**, *39*, 272–293. [CrossRef]
93. Zheng, Y.; Huang, Y.; Zhou, S.; Wang, K.; Wang, G. Effect partition of climate and catchment changes on runoff variation at the headwater region of the Yellow River based on the Budyko complementary relationship. *Sci. Total Environ.* **2018**, *643*, 1166–1177. [CrossRef] [PubMed]
94. Sharma, C.; Shukla, A.K.; Zhang, Y. Climate change detection and attribution in the Ganga-Brahmaputra-Meghna river basins. *Geosci. Front.* **2021**, *12*, 101186. [CrossRef]
95. Gulahmadov, N.; Chen, Y.; Gulakhmadov, A.; Rakhimova, M.; Gulakhmadov, M. Quantifying the Relative Contribution of Climate Change and Anthropogenic Activities on Runoff Variations in the Central Part of Tajikistan in Central Asia. *Land* **2021**, *10*, 525. [CrossRef]
96. Wu, C.; Ji, C.; Shi, B.; Wang, Y.; Gao, J.; Yang, Y.; Mu, J. The impact of climate change and human activities on streamflow and sediment load in the Pearl River basin. *Int. J. Sediment Res.* **2019**, *34*, 307–321. [CrossRef]
97. Abbas, T.; Nabi, G.; Boota, M.W.; Faisal, M.; Azam, M.I.; Yaseen, M.; Hussain, F. Continuous hydrologic modeling of major flood hydrographs using semi-distributed model. *Sci. Int.* **2015**, *27*, 6231e7.

water

Article

Development of a New 8-Parameter Muskingum Flood Routing Model with Modified Inflows

Eui Hoon Lee

School of Civil Engineering, Chungbuk National University, Cheongju 28644, Korea;
hydrohydro@chungbuk.ac.kr; Tel.: +82-043-261-2407

Abstract: Flood routing can be subclassified into hydraulic and hydrologic flood routing; the former yields accurate values but requires a large amount of data and complex calculations. The latter, in contrast, requires only inflow and outflow data, and has a simpler calculation process than the hydraulic one. The Muskingum model is a representative hydrologic flood routing model, and various versions of Muskingum flood routing models have been studied. The new Muskingum flood routing model considers inflows at previous and next time during the calculation of the inflow and storage. The self-adaptive vision correction algorithm is used to calculate the parameters of the proposed model. The new model leads to a smaller error compared to the existing Muskingum flood routing models in various flood data. The sum of squares obtained by applying the new model to Wilson's flood data, Wang's flood data, the flood data of River Wye from December 1960, Sutculer flood data, and the flood data of River Wyre from October 1982 were 4.11, 759.79, 18,816.99, 217.73, 38.81 $(m^3/s)^2$, respectively. The magnitude of error for different types of flood data may be different, but the error may be large if the flow rate of the flood data is large.

Keywords: hydrologic flood routing; Muskingum flood routing model; meta-heuristic optimization; self-adaptive vision correction algorithm

Citation: Lee, E.H. Development of a New 8-Parameter Muskingum Flood Routing Model with Modified Inflows. *Water* **2021**, *13*, 3170. https://doi.org/10.3390/w13223170

Academic Editor: Fi-John Chang

Received: 4 September 2021
Accepted: 5 November 2021
Published: 10 November 2021

1. Introduction

Water resources from rivers are sources of hydroelectric power generation, agricultural water, and industrial water; however, owing to the large volumes of water, such rivers are prone to floods that have adverse impacts on life and property [1]. To reduce or prevent such damage, engineering measures, such as the construction of flood control dams or flood walls (levees), are necessary. Therefore, the evaluation of engineering measures for flood control is critical, and these measures are generally directly related to flood routing. Flood routing can be defined as a procedure for determining the flood hydrograph at a point downstream from the base flood hydrograph at an upstream point. In other words, flood routing is the process of determining the amount by which a flood wave is reduced and how long it takes for a flood wave to pass through an arbitrary section of a river based on the amount of storage in that section.

There are two types of flood routing methodologies: hydraulic and hydrologic [2]. Hydraulic flood routing is a method for solving the partial differential continuity and momentum equations, the governing equations of an unsteady nonuniform flow, which hydraulically represent the flow of the flood wave according to the initial and boundary conditions [3]. In contrast, the hydrologic flood routing method yields an approximate solution using the storage equation based on the continuity equation of the flood wave [2]. The hydrologic flood routing method can be divided into three categories: reservoir routing, channel routing, and watershed routing. Channel routing allows the measurement of the storage effect of natural rivers on flood waves by calculating how the discharge of a flood changes as it progresses downstream and provides a standard hydrologic quantity for river planning. The Muskingum flood routing model is a representative channel-routing model [2].

The first Muskingum flood routing model proposed was the linear Muskingum flood routing model (LMM) with two parameters [4]. However, the LMM did not include lateral inflow, and a new Muskingum flood routing model with three variables (LMM-L) was thus proposed [5]. Additionally, a study using the nonlinear Muskingum flood routing model (NLMM) considered the nonlinear relationship between storage and outflow as means of improving upon the LMM [6]. Two types of NLMMs determined by the location of nonlinear factor have been proposed to calculate the storage [7]. The Broyden–Fletcher–Goldfarb–Shanno technique based on a mathematical gradient was applied to the NLMM [8]. NLMM incorporating lateral flow (NLMM-L) was developed to complement the existing NLMM [9]. In 2018, a Muskingum flood routing model called the advanced NLMM (ANLMM-L) was used for calculating a continuous inflow [2]. ANLMM-L is a type of nonlinear Muskingum flood routing model that considers lateral inflow and continuous flow with time. Generalized storage equations for the NLMM have also been suggested to apply more degrees of freedom in the suggested model [10].

In addition to the aforementioned studies, various other investigations have focused on recalculating the error between the outflow from flood data and the calculated outflow. Various studies on Muskingum flood routing models were conducted before the 2000s. The two parameters of the LMM including nonlinear relation between the storage and weighted flow were determined using the least-squares method [11]. The least-squares method was used to adjust the two parameters of LMM, K and X. The Muskingum parameter estimation/flood routing system was developed for linear LMMs and NLMMs and their results have been compared [12]. The results of the two different Muskingum flood routing models were compared. The genetic algorithm was used to estimate the parameters of the NLMMs [13]. In order to overcome the limitations of traditional methods used for Muskingum flood routing models, the genetic algorithm, a well-known meta-heuristic optimization algorithm, was applied.

Since the 2000s, studies applying various meta-heuristic optimization algorithms to the Muskingum flood routing models have continued. An immune clonal selection algorithm was suggested to improve the convergence speed and it was applied to estimate parameters of the NLMM [14]. The Nelder–Mead simplex algorithm was introduced to improve the usability, and it was used to estimate parameters of the NLMM [15]. Furthermore, the simulated annealing and shuffled frog leaping algorithms were used to estimate the parameters of the Muskingum flood routing model in two benchmark/real case studies, and they were compared with the results of Tung's method [16]. The honeybee mating optimization algorithm with past convergence speed has also been applied for the parameter estimation of the NLMM [17]. The elitist-mutated particle swarm optimization and improved gravitational search algorithm were applied to estimate the parameters of LMMs and NLMMs [18]. Particle swarm optimization was applied to the parameter estimation of the NLMM with four parameters to fit the multiple-peak hydrographs [19], and various NLMMs with different storage calculations such as parameterized initial storage have been proposed using a weed optimization algorithm [20]. Various meta-heuristic optimization algorithms, such as the genetic algorithm, evolution, particle swarm, and a harmony search have been used for parameter estimations of the nonlinear Muskingum model and the variable parameter McCarthy–Muskingum model [21]. The adaptive genetic algorithm was used to estimate the various exponent parameters of the NLMM and it was applied to Wilson's flood data [22]. In addition, genetic expression programming with faster convergence speed than existing genetic programming was developed for parameter estimation in the Muskingum flood routing model [23]. The water cycle algorithm was applied to estimate the parameters of the NLMM and compared with the genetic algorithm, particle swarm optimization, harmony search, and imperialist competitive algorithm [24]. Although various meta-heuristic optimization algorithms have been tested, studies focusing on comparing the results of each algorithm have indicated limited improvements for the Muskingum flood routing model.

Studies have also been conducted on the application of hybrid meta-heuristic optimization algorithms, combining a charged system search and particle swarm optimization, for parameter estimation of the Muskingum flood routing models [25]. For example, a hybrid meta-heuristic optimization algorithm combining particle swarm optimization and the Nelder–Mead simplex method was used to estimate the parameters of the Muskingum flood routing model [26]. Parameter estimation was conducted using a hybrid meta-heuristic optimization algorithm combining the shuffled frog leaping algorithm and the Nelder–Mead simplex method [27]. The improved real-coded adaptive genetic algorithm and the Nelder–Mead simplex algorithm were combined for the parameter estimation of two improved NLMMs [28]. The hybrid meta-heuristic optimization algorithm applied in particle swarm optimization and the bat algorithm were used to reduce the computational time of the Muskingum flood routing model [29]. Although good results were obtained in some studies, it is difficult to accurately compare them with the results of other existing Muskingum flood routing models because they were calculated using additional variables.

General improvements of the Muskingum flood routing model have also been considered. For example, a modified Muskingum flood routing approach, in conjunction with the HEC-RAS model, was implemented to determine floodplain flows [30]. A new NLMM with four parameters has been suggested [31]. A parameter estimation method of the Muskingum flood routing model in ungagged channel reaches has also been suggested [32].

Studies have been conducted to apply the hybrid method to Muskingum flood routing models. A hybrid harmony search combined with local search algorithm such as Broyden–Fletcher–Goldfarb–Shanno technique was developed and was applied to estimate parameters in NLMM [33]. The new hybrid optimization technique was suggested by combining the modified honeybee mating optimization and generalized reduced gradient algorithm for the application of the new Muskingum model with six parameters [34]. The particle swarm optimization hybridized with Nelder–Mead simplex method was proposed to improve precision and convergence speed in Muskingum model [26]. The hybrid algorithm combining the shuffled frog leaping algorithm and Nelder–Mead simplex was applied to NLMM with four parameters and NLMM with five parameters, and it was compared with the genetic algorithm-generalized reduced gradient [27]. The improved NLMM was suggested for flood prediction using the hybrid algorithm of particle swarm optimization and bat algorithm and it was compared with particle swarm optimization and bat algorithm [29]. The parameters in the two types of NLMM were estimated to improve precision using the hybrid algorithm combining the improved real-coded adaptive genetic algorithm and the Nelder–Mead simplex [28]. Most of the previously proposed hybrid methods combine optimization algorithms, but the hybrid method of this study is a method combining the inflow at the previous time and the inflow at the next time in the Muskingum flood routing. The honey bee mating optimization algorithm was combined with the generalized reduced gradient algorithm to estimate parameters of improved Muskingum flood routing model and applied to the single and multi-peak flood hydrographs [35].

In this study, a new Muskingum flood routing model was suggested. The new Muskingum flood routing model, which considers continuous inflow at previous and next time in the storage and inflow calculations, can enable accurate flood routing in various flood data. The self-adaptive vision correction algorithm (SAVCA), a recently developed meta-heuristic optimization algorithm, was applied to calibrate various parameters in the new Muskingum flood routing model. SAVCA can overcome the disadvantages of the previously developed vision correction algorithm (VCA). When applied to mathematical benchmark functions and water distribution problems, SAVCA has previously displayed good performance [36]. Various meta-heuristic optimization algorithms as well as SAVCA can be applied to the new Muskingum flood routing model to show good results. In the previous study, the type of meta-heuristic optimization algorithm did not significantly affect the results of Muskingum flood routing models [37].

2. Materials and Methodologies

2.1. Overview

The two primary methods used in this study were the SAVCA and new Muskingum flood routing model. The error between the flood outflow data and the calculated outflow in the new Muskingum flood routing model was used as an objective function in the SAVCA, which applied an iterative calculation to minimize the error. The iterative calculation progresses as follows:

1. A group of initial solutions is generated by a random value determined between the lower and upper boundaries for each variable of the new Muskingum flood routing model.
2. One among a group of existing solutions is then selected, or a new solution is generated according to the selected probability.
3. The inflow, storage, and outflow are calculated according to the generated solution, and the error between the flood outflow data and calculated outflow is determined as the objective function.
4. The error is calculated using the sum of squares (SSQ), the Nash–Sutcliffe efficiency (NSE), and the root mean square error (RMSE).

The solution refers to the values of the parameters for Muskingum flood routing models. The calculation in the SAVCA is as follows: If all initial solutions are calculated according to the new Muskingum flood routing model process, the errors of the initial solutions are calculated and sorted in ascending order. SAVCA consists of two types of parameters: self-adaptive and fixed. Division rate 1 (DR1), division rate 2 (DR2), and the compression factor (CF) are self-adaptive parameters. The modulation transfer function rate (MR) and astigmatic rate (AR) are fixed parameters.

DR1 determines whether a new solution should be generated in the range of each variable (global search) or if one solution should be selected from the existing solution group (local search). If the generation of a new solution is determined in DR1, the positive and negative direction searches are determined by DR2. The decision variables of the new solution, generated by global search or selected by local search, are adjusted in detail by MR, CF, and AR. After generating a new solution, the calculation process of the new Muskingum flood routing model is applied. The error (SSQ) is calculated for the inflow, storage, and outflow during each time period. If the error of the new solution is lower than that of the worst solution among the existing solution groups, the new solution is included in the existing solution group. DR1 and DR2 are adjusted according to the calculation process for the new solution. All processes are repeated until a certain number of iterations of SAVCA.

The SSQ was used to calculate the first error value for the Muskingum flood routing models in this study. The SSQ between the observed and calculated outflows was used as the objective function in the optimization process. In the new Muskingum flood routing model, eight parameters were used as decision variables, and the objective function is shown in Equation (1).

$$\text{Minimize SSQ} = \sum (O_o - O_s)^2 \tag{1}$$

where O_o is the observed outflow (m^3/s), and O_s is the calculated outflow (m^3/s). The NSE was used to calculate the second error value for the Muskingum flood routing models in this study. The equation of NSE is shown in Equation (2).

$$\text{NSE} = 1 - \frac{\sum_{i=1}^{n}(O_o - O_s)^2}{\sum_{i=1}^{n}(O_o - \overline{O}_o)^2} \tag{2}$$

where $\overline{O}_o$ is the average of observed outflow (m^3/s), and n is the number of data. The RMSE was used to calculate the third error value for the Muskingum flood routing models in this study. The equation of RMSE is shown in Equation (3).

$$\text{RMSE} = \sqrt{\frac{\sum_{i=1}^{n}(O_o - O_s)^2}{n}} \tag{3}$$

2.2. New Muskingum Flood Routing Model

The initial LMM was calculated by assuming the amount of storage in the channel as the sum of prism storage and wedge storage. Prism storage is proportional to the outflow, and wedge storage is proportional to the difference between the inflow and outflow. In the LMM, storage is calculated using Equation (4).

$$S_t = K[XI_t + (1 - X)O_t] \tag{4}$$

where S_t, I_t, and O_t are the storage, inflow, and outflow at time t, respectively, and X is the weighted factor. In the NLMM, the nonlinear factor is added in the exponential form of Equation (4). Equation (5) represents the storage in an NLMM.

$$S_t = K[XI_t + (1 - X)O_t]^m \tag{5}$$

where m is a nonlinear factor, namely different from 1. The storage calculation in the new Muskingum flood routing model is applied by considering an existing generalized storage. The storage is calculated by considering not only the inflow at the current time (t) but also the inflow at the next time point ($t + 1$). The reason for considering the inflow at $t + 1$ instead of $t - 1$ in the storage calculation is as follows. In the nonlinear Muskingum flood routing models, it is assumed that the storage at time t depends on the upstream storage S_{in}, and the downstream storage S_{out}. Inflow (I), outflow (O), S_{in}, S_{out} were organized as follows according to water depth [38]. I and S_{in} are shown in Equations (6) and (7).

$$I = a_1 y^{c_1} \tag{6}$$

$$S_{in} = b_1 y^{d_1} \tag{7}$$

where y is the water depth and a_1, b_1, c_1, d_1 are coefficients. If c_1 and d_1 are equal, then S_{in} is shown in Equation (8).

$$S_{in} = b_1 \left(\frac{I}{a_1}\right) \tag{8}$$

O and S_{out} are shown in Equations (9) and (10).

$$O = a_2 y^{c_2} \tag{9}$$

$$S_{out} = b_2 y^{d_2} \tag{10}$$

where a_2, b_2, c_2 and d_2 are coefficients. If c_2 and d_2 are equal, then S_{out} is shown in Equation (11).

$$S_{out} = b_2 \left(\frac{O}{a_2}\right) \tag{11}$$

In NLMM, inflow, outflow and storage are assumed to be water depth related. The storage can be summarized in Equation (12) [38].

$$S = XS_{in} + (1 - X)S_{out} \tag{12}$$

The storage with $K = b_1/a_1 = b_2/a_2$ can be expressed as Equation (13).

$$S = KXI + K(1 - X)O \tag{13}$$

209

Nonlinear parameter m was applied to Equation (11) and it can be expressed as Equation (14).

$$S = K[XI + (1 - X)O]^m \tag{14}$$

Additionally, it is assumed that there is an interdependence between the storage at time t and storage at time $t + 1$ in generalized storage [10].

$$S_t = X_1 S_{in,\,t} + X_2 S_{in,t+1} + (1 - X_1 - X_2) S_{out,t} \tag{15}$$

Equation (16) can be rearranged as in the process from Equations (12)–(14) and it represents storage in the new Muskingum flood routing model.

$$S_t = K[X_1 I_t + X_2 I_{t+1} + (1 - X_1 - X_2) O_t]^m \tag{16}$$

where X_1 is the weighted factor at time t, and X_2 is the weighted factor at time $t + 1$. In addition, I_t is the inflow at time t, and I_{t+1} is the inflow at time $t + 1$. Based on Equation (4), the outflow calculation is summarized in Equation (17).

$$O_t = \frac{1}{(1 - X_1 - X_2)} \left(\frac{S_t}{K}\right)^{\frac{1}{m}} - \frac{X_1}{(1 - X_1 - X_2)} I_t - \frac{X_2}{(1 - X_1 - X_2)} I_{t+1} \tag{17}$$

A weighted inflow, including a continuous inflow has been proposed previously [2]. In this study, the inflow at time $t + 1$ was included to consider the additional continuous inflow, and the weighted inflow could be calculated as shown in Equation (18).

$$W_t = [(1 - \theta_1 - \theta_2 - \theta_3) I_t + \theta_1 I_{t-1} + \theta_2 I_{t-2} + \theta_3 I_{t+1}] \tag{18}$$

where W_t is the weighted inflow at time t, and θ_1 is the weighted factor of the previous inflow at time $t - 1$. In addition, θ_2 is the weighted factor of the previous inflow at time $t - 2$, and θ_3 is the weighted factor of the next inflow at time $t + 1$. In previous studies, the inflow at time $t - 1$ (I_{t-1}) and the inflow at time $t - 2$ (I_{t-2}) were considered [2,9]. In this study, all inflows before and after the current time were considered by including the inflow at time $t + 1$ (I_{t+1}). If the weighted inflow of Equation (18) is substituted into Equation (17), the outflow is calculated using Equation (19).

$$O_t = \frac{1}{(1 - X_1 - X_2)} \left(\frac{S_t}{K}\right)^{\frac{1}{m}} - \frac{X_1}{(1 - X_1 - X_2)} W_t - \frac{X_2}{(1 - X_1 - X_2)} W_{t+1} \tag{19}$$

where W_{t+1} is the weighted inflow at time $t + 1$. The storage at time $t + 1$ can be calculated from the outflow in Equation (19) and the observed inflow. The general storage equation is calculated as Equation (20).

$$\frac{dS}{dt} = I_t - O_t \tag{20}$$

Equation (21) shows the modified storage equation that considers a change in lateral flow.

$$\frac{dS}{dt} = \frac{\Delta S}{\Delta t} = (1 + \beta) I_t - O_t \tag{21}$$

where β is the parameter accounting for the lateral flow. The storage at time $t + 1$ is shown in Equation (22).

$$S_{t+1} = S_t + \Delta S \tag{22}$$

Equation (23) represents the storage at time $t + 1$, and it can be obtained by substituting Equation (21) into Equation (22).

$$S_{t+1} = S_t + [(1 + \beta) I_t - O_t] \Delta t \tag{23}$$

where S_{t+1} is the storage at time $t + 1$. The outflow in the new Muskingum flood routing model is calculated using eight variables, i.e., K, X_1, X_2, m, β, θ_1, θ_2, and θ_3. The initial Muskingum flood routing model is based on mass conservation. Therefore, the new Muskingum flood routing model was calculated based on the mass conservation.

2.3. Self-Adaptive Vision Correction Algorithm

SAVCA has a total of six parameters: DR1, DR2, MR, CF, AR, and AF. Among these, DR1, DR2, and CF are self-adaptive, and MR, AR, and AF are fixed. Table 1 shows the parameter types of SAVCA [36].

Table 1. Parameter types of SAVCA.

Parameters	DR1	DR2	MR	CF	AR	AF
Types	Self-adaptive	Self-adaptive	Fixed	Self-adaptive	Fixed	Fixed

In SAVCA, the initial decision variables and decision variables generated by the global search are randomly generated within the range between the upper and lower boundaries. Decision variables in the global search are between the current optimal value and the upper boundary or between the lower boundary and the current optimal value based on the probability of DR2. The decision variable generated between the current optimal decision variable and the upper boundary by a global search is shown in Equation (24).

$$x_n = x_b + random(0, 1) \times (b_u - x_b) \tag{24}$$

where x_n is the new decision variable, and x_b is the current optimal value. In addition, $random(0, 1)$ is a random value generated from 0 to 1, and b_u is the upper boundary. The decision variable generated between the lower boundary and the current optimal decision variable through a global search is shown in Equation (25).

$$x_n = b_l + random(0, 1) \times (x_b - b_l) \tag{25}$$

where b_l is the lower boundary. In SAVCA, each decision variable is adjusted by the parameters used in the local search, such as MR, CF, AR, and AF. Equation (26) shows the calculation of the new decision variable.

$$x_n = x_n \times \left\{ 1 + MTF \times random(-1, 1) \times \left(1 - \frac{current\ iteration}{total\ iteration} \right)^{CF} \right\} \tag{26}$$

where MTF is the calculated modulation transfer function value. The term $random(-1, 1)$ is a random value between -1 and 1. In addition, CF is a parameter for lens compression. The calculation of MTF is based on the distance (dx) between the current best decision variable and the selected decision variable. dx can be calculated using Equation (27).

$$dx = \frac{rank(x_s) - rank(x_b)}{rank(x_l) - rank(x_b)} \tag{27}$$

where dx is the relative distance between each decision variable, $rank(x_s)$ is the fitness rank of the selected decision variable (x_s), $rank(x_b)$ is the fitness rank of the best decision variable (x_b), and $rank(x_l)$ is the fitness rank of the worst decision variable (x_l). The fitness rank is the order in which the values of the objective function are sorted. The worst decision variable has the lowest fitness rank. Figure 1 shows the relative distance of dx.

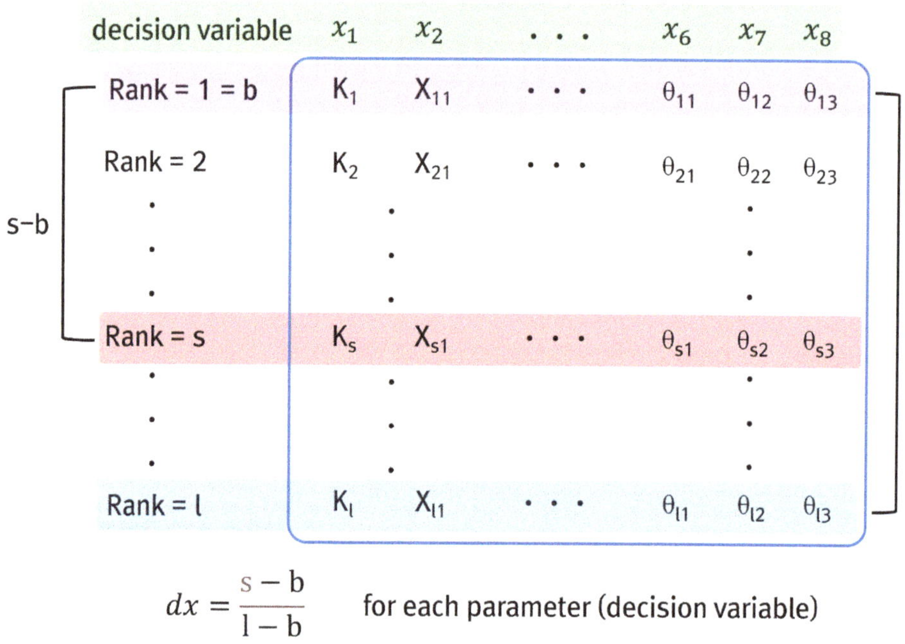

decision variable x_1 x_2 $\cdots$ x_6 x_7 x_8

Rank = 1 = b	K_1	X_{11}	$\cdots$	θ_{11}	θ_{12}	θ_{13}
Rank = 2	K_2	X_{21}	$\cdots$	θ_{21}	θ_{22}	θ_{23}
Rank = s	K_s	X_{s1}	$\cdots$	θ_{s1}	θ_{s2}	θ_{s3}
Rank = l	K_l	X_{l1}	$\cdots$	θ_{l1}	θ_{l2}	θ_{l3}

$$dx = \frac{s - b}{l - b} \quad \text{for each parameter (decision variable)}$$

Figure 1. Relative distance of dx.

The calculation of the MTF by applying dx is shown in Equation (28).

$$MTF_s = \left(\frac{dx_s}{\left(\sum_{i=1}^{k} dx_i^2 \right)^{0.5}} \right)^{0.5} \tag{28}$$

where $MTFs$ is the MTF value of the selected decision variable, and k is the total number of decision variables. In addition, dx_s is the relative distance of the selected decision variable, and dx_i is the relative distance of the i-th decision variable. The CF in SAVCA can be calculated as shown in Equation (29).

$$CF = 10 \times \left\{ \frac{standard\ deviation(x_i)}{average(x_i)} \right\} \tag{29}$$

where x_i is the i-th decision variable. The probability of applying the astigmatism correction process is determined using the AR. The new decision variable adjusted by the application of the astigmatism correction process on a local search is shown in Equation (30).

$$x_n = x_n \times \left\{ 1 + random(-1,\ 1) \times \sin^2(AF) \right\} \tag{30}$$

where AF is the angle of the astigmatic axis. The application process of SAVCA is summarized as follows: (1) generation of an initial solution group, (2) calculation of the fitness of the initial solution groups, (3) generation of a new solution, (4) application of MR and AR, (5) decision to replace after comparing the new solution with the worst solution in the existing solution group, and (6) repeating (2)–(5) until the total number of iterations is reached. The worst solution is the solution with the lowest fitness rank among the existing solution group. Figure 2 shows the application process.

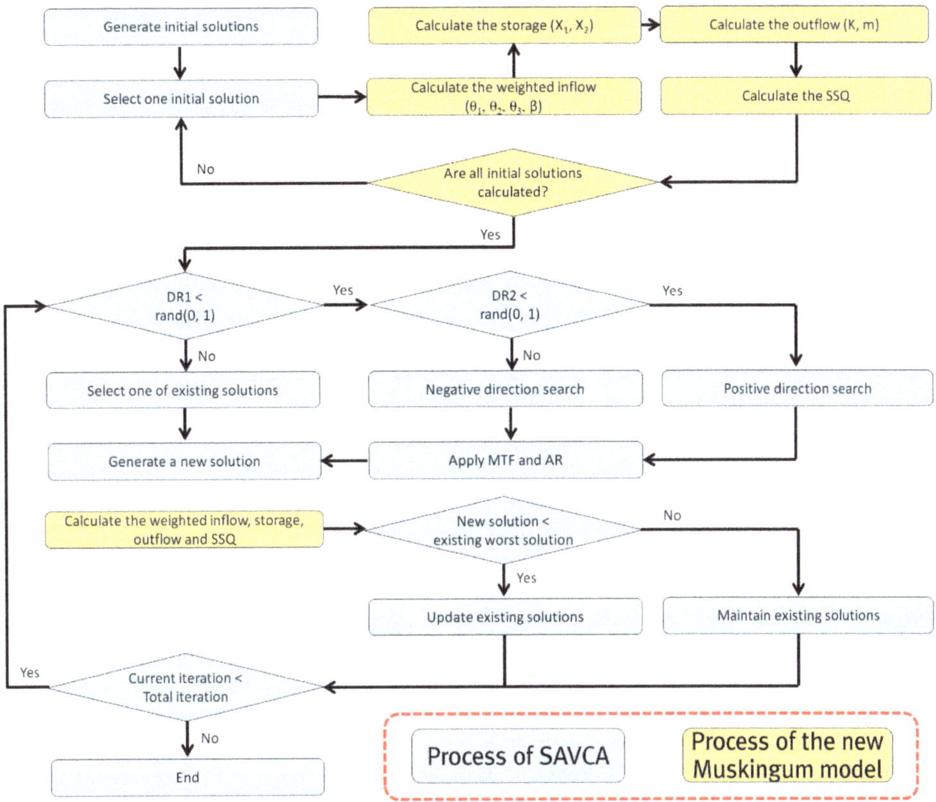

Figure 2. Application process.

An initial solution group was created to apply SAVCA to the new Muskingum flood routing model. The weighted inflow, storage, outflow, and SSQ were then calculated. According to the probability of DR1, a new solution was generated by a global search, or one of the existing solution groups was selected through a local search. When creating a new solution with a global search, a new decision variable was created in the positive and negative directions based on the current best decision variable. Each new decision variable was corrected using the MR and AR. In addition, the weighted inflow, storage, outflow, and SSQ were calculated using the new solution with new decision variables. Whether a new solution should be added to the existing solution group was determined by comparing the SSQ of the new solution with the SSQ of the worst solution among the existing group of solutions.

2.4. Flood Data

Five types of flood data were applied to the Muskingum flood routing models: Wilson's flood data, Wang's flood data, flood data for River Wye December in 1960, Sutculer flood data, and flood data for River Wyre October in 1982 [5,39–41]. All flood data used in this study have been applied in several Muskingum flood routing models in existing studies. The most important aspect of the Muskingum flood routing model is the range of each parameter. The range of each parameter in the new Muskingum flood routing model applied to the five flood datasets is presented in Table 2.

Table 2. Range of parameters in the new Muskingum flood routing model.

Parameters	Wilson's Flood Data	Wang's Flood Data	Flood Data for River Wye December in 1960	Sutculer Flood Data	Flood Data for River Wyre October in 1982
K	0.01–50.00	0.01–50.00	0.01–50.00	0.01–50.00	0.01–50.00
X_1	−0.50–0.50	−1.50–1.50	−0.50–0.50	−0.50–0.50	−0.50–0.50
X_2	−0.50–0.50	−1.50–1.50	−0.50–0.50	−0.50–0.50	−0.50–0.50
m	1.00–3.00	1.00–3.00	1.00–3.00	1.00–3.00	0.00–1.00
β	−0.10–0.10	−3.00–3.00	−0.10–0.10	−0.10–0.10	−3.00–3.00
θ_1	0.00–1.00	0.00–1.00	0.00–1.00	0.00–1.00	0.00–1.00
θ_2	0.00–1.00	0.00–1.00	0.00–1.00	0.00–1.00	0.00–1.00
θ_3	0.00–1.00	0.00–1.00	0.00–1.00	0.00–1.00	0.00–1.00

Various Muskingum flood routing models, i.e., LMM, LMM-L, NLMM, NLMM-L, ANLMM-L, and the new Muskingum flood routing model, were compared herein. The parameters used in each Muskingum flood routing model are listed in Table 3.

Table 3. Parameters used in each Muskingum flood routing model (○: applied, X: not applied).

Parameters	LMM	LMM-L	NLMM	NLMM-L	ANLMM-L	This Study
K	○	○	○	○	○	○
X_1	○	○	○	○	○	○
X_2	X	X	X	X	X	○
m	X	X	○	○	○	○
β	X	○	X	○	○	○
θ_1	X	X	X	○	○	○
θ_2	X	X	X	X	○	○
θ_3	X	X	X	X	X	○

Differences were observed in the results of the Muskingum flood routing models proposed in other studies. However, the simulation used in this study was conducted according to the parameters listed in Table 3. The data values from existing studies were considered only up to two decimal points when using them as an input for the models in this study. The parameters of Muskingum flood routing models without results from previous studies were obtained by applying SAVCA. However, the values of each parameter were all calculated differently.

3. Application and Results

The first flood dataset used for the application of all Muskingum flood routing models was Wilson's flood data. The parameters of the LMM for Wilson's flood data were determined to be 29.164640 for K, and 0.118200 for X_1. The parameters of the new Muskingum flood routing model for Wilson's flood data were determined to be 0.943442 for K, 0.340333 for X_1, −0.00102 for X_2, 1.744439 for m, −0.02166 for β, 0.758873 for θ_1, 0.230779 for θ_2, and 0.047773 for θ_3. The results, including those obtained using the new Muskingum flood routing model, are compared in Table 4.

Table 4. Results when using Wilson's flood data.

Time (h)	Inflow (m³/s)	Outflow (m³/s)	LMM (m³/s)	LMM-L (m³/s) [5]	NLMM (m³/s) [42]	NLMM-L (m³/s) [9]	ANLMM-L (m³/s) [2]	This Study (m³/s)
0	22	22	22.00	22.00	22.00	22.00	22.00	22.00
6	23	21	21.87	21.10	22.00	21.71	21.57	21.33
12	35	21	20.52	21.70	22.40	22.02	21.67	21.13
18	71	26	19.07	22.60	26.60	26.08	25.46	25.53
24	103	34	26.90	30.70	34.50	33.51	34.59	34.75
30	111	44	43.58	44.70	44.20	42.83	43.73	43.52
36	109	55	59.58	58.10	56.90	55.44	54.59	54.62
42	100	66	72.32	68.90	68.10	66.67	66.01	66.08
48	86	75	80.65	76.10	77.10	75.77	75.52	75.53
54	71	82	83.91	79.20	83.30	82.12	82.16	82.11
60	59	85	82.51	78.50	85.90	84.78	85.04	85.08
66	47	84	78.63	75.60	84.50	83.42	84.00	83.89
72	39	80	72.32	70.70	80.60	79.44	79.62	79.61
78	32	73	65.49	65.10	73.70	72.48	72.63	72.53
84	28	64	58.21	59.10	65.40	64.08	63.80	63.81
90	24	54	51.70	53.40	56.00	54.58	54.31	54.27
96	22	44	45.50	47.90	46.70	45.22	44.80	44.84
102	21	36	40.15	43.10	37.70	36.34	36.25	36.32
108	20	30	35.82	38.90	30.50	29.21	29.45	29.52
114	19	25	32.26	35.40	25.20	24.21	24.63	24.66
120	19	22	29.17	32.30	21.70	20.96	21.39	21.46
126	18	19	26.93	29.90	20.00	19.41	19.81	19.77
SSQ (m³/s)²	-	-	605.63	815.68	36.77	9.82	4.54	4.11
Squared root of SSQ (m³/s)	-	-	24.61	28.56	6.06	3.13	2.13	2.03
NSE	-	-	0.974322	0.974326	0.992412	0.999583	0.999808	0.999826
RMSE (m³/s)	-	-	5.310259	5.369885	2.919411	0.683993	0.464124	0.442254

Among the results in Table 4, those of the LMM-L, NLMM, NLMM-L, and ANLMM-L were calculated in previous studies [2,5,9,42]. The results of LMM and new Muskingum flood routing model were calculated using Wilson's flood data by applying SAVCA. Notably, the results of the LMM are better than those of the LMM-L. This is because the results of the LMM-L are the results of a previous study wherein the optimization method was not used, while the results of the LMM were obtained using SAVCA. It should be noted that the results of the new Muskingum flood routing model were better than those of the LMM, LMM-L, NLMM, NLMM-L, and ANLMM-L because the new Muskingum flood routing model showed the smallest error in the initial part from 0 to 24 h and showed the smallest error in the overall results. Because the errors of the NLMM-L and ANLMM-L were small, the new Muskingum flood routing model did not lead to a substantial improvement.

Among the existing Muskingum flood routing models, ANLMM-L showed the best results (smallest error). The difference in SSQ between the ANLMM-L and new Muskingum flood routing model was 0.43 (m³/s)². The differences between the results of the two models were most notable from 6 to 18 h and from 108 to 126 h. The new Muskingum flood routing model showed the closest outflow to the observed outflow from 6 to 18 h. Among other Muskingum flood routing models, the outflows obtained from LMM-L at 6 h, LMM at 12 h, and NLMM-L at 18 h were close to the observed outflow. The difference between the observed outflow and the outflow obtained from the new Muskingum flood routing model was smaller than that attained using other Muskingum flood routing models.

The second flood dataset was Wang's flood data. The parameters of the LMM-L for Wang's flood data were determined to be 1.075331 for K, -0.762101 for X_1, and -0.003024 for β. The parameters of the new Muskingum flood routing model for Wang's flood data were determined to be 0.079266 for K, -1.49742 for X_1, 0.011592 for X_2, 1.360300 for m, -0.000450 for β, 0.421275 for θ_1, 0.044483 for θ_2, and 0.261537 for θ_3. The results using Wang's flood data, including those for the new Muskingum flood routing model, are compared in Table 5.

The results of LMM, NLMM, NLMM-L, and ANLMM-L were calculated in previous studies [2,8,9,39]. The results of LMM-L and new Muskingum flood routing model were calculated by applying SAVCA. Notably, the results improved dramatically when the lateral inflow was considered, indicated by the difference between the results of the LMM and LMM-L and between those of the NLMM and NLMM-L. A difference between the results of the ANLMM-L and new Muskingum flood routing model was also observed, although it was not due to a lateral inflow but to differences in the calculation equations of the weighted inflow and storage. The results of the new Muskingum flood routing model were overwhelmingly better than those of other Muskingum flood routing models because the new Muskingum flood routing model showed the smallest error in the latter part from 19 to 29 h and showed the smallest error in the overall results.

Among the existing Muskingum flood routing models, the ANLMM-L showed the best results (smallest error). The difference in SSQ between the ANLMM-L and new Muskingum flood routing model was 149.56 $(m^3/s)^2$. The difference between the two results was clear from 228 (19) to 288 (24) h. The differences in the outflow obtained using the Muskingum flood routing models and the observed outflow was small. The outflow calculated by the new Muskingum flood routing model was closest to the observed outflow. Among other Muskingum flood routing models, the outflow obtained using NLMM at 21 h was close to the observed outflow. The difference between the observed outflow and the outflow obtained using the new Muskingum flood routing model was smaller than that when using other Muskingum flood routing models.

The third flood dataset was the flood data of River Wye December in 1960. The parameters for the LMM using these data were determined to be 23.877307 for K, and 0.153174 for X_1. The parameters for the new Muskingum flood routing model using these data were determined to be 2.318963 for K, 0.499999 for X_1, 0.000390 for X_2, 1.359406 for m, 0.057839 for β, 0.805567 for θ_1, 0.233550 for θ_2, and 6.88×10^{-11} for θ_3. The results of all the models for these data are compared in Table 6.

LMM-L, NLMM, NLMM-L, and ANLMM-L results were calculated in previous studies [2,5,9,42]. The results of LMM and new Muskingum flood routing model were calculated by applying SAVCA. Table 6 displays a notable difference between the results of the LMMs and NLMMs; the difference was large because the error in the flood data of River Wye from December 1960 was large. The new Muskingum flood routing model results were better than those of other Muskingum flood routing models because the new Muskingum flood routing model showed the smallest error from 102 to 198 h including the peak value and showed the smallest error in the overall result.

Among other Muskingum flood routing models, ANLMM-L showed the best results (smallest error). The difference in SSQ between the ANLMM-L and new Muskingum flood routing model was 1677.99 $(m^3/s)^2$. The greatest difference between the models occurred at 138–186 h. Among other Muskingum flood routing models, the outflow from ANLMM-L was close to the observed outflow at 180 and 186 h. The difference between the observed outflow and the outflow for the new Muskingum flood routing model was smaller than that for other Muskingum flood routing models.

Table 5. Results when using Wang's flood data.

Time (12 h)	Inflow (m^3/s)	Outflow (m^3/s)	LMM (m^3/s) [39]	LMM-L (m^3/s)	NLMM (m^3/s) [8]	NLMM-L (m^3/s) [9]	ANLMM-L (m^3/s) [2]	This Study (m^3/s)
1	261	228	228.00	228.00	228.00	228.00	228.00	228.00
2	389	300	305.19	300.19	303.80	299.74	300.92	301.75
3	462	382	382.00	377.92	382.30	382.57	381.51	382.38
4	505	444	442.70	440.10	442.40	442.76	443.15	442.81
5	525	490	483.60	482.17	482.40	482.16	482.69	483.63
6	543	513	513.00	511.70	511.2	509.89	510.09	510.15
7	556	528	534.29	532.96	532.30	530.72	530.66	530.75
8	567	543	550.44	548.97	548.50	546.77	546.62	546.79
9	577	553	563.53	561.89	561.70	559.96	559.77	559.53
10	583	564	573.16	571.53	571.60	569.94	569.80	569.75
11	587	573	580.02	578.38	578.70	577.07	576.95	577.89
12	595	581	587.32	585.44	586.20	584.39	584.22	584.03
13	597	588	592.14	590.40	591.20	589.68	589.60	589.77
14	597	594	594.59	592.93	593.90	592.34	592.30	591.61
15	589	592	592.02	590.68	591.80	590.33	590.34	586.67
16	556	584	574.89	574.62	575.70	574.68	574.86	576.15
17	538	566	556.85	556.15	558.50	556.41	556.23	556.07
18	516	550	536.93	536.22	539.00	537.43	537.13	536.33
19	486	520	512.18	511.79	514.80	513.47	513.35	521.23
20	505	504	507.96	505.60	509.60	507.07	506.51	502.72
21	477	483	493.22	492.40	484.90	494.86	494.95	492.05
22	429	461	462.34	462.82	464.80	464.39	464.94	463.80
23	379	420	421.87	422.73	425.10	423.97	424.15	422.09
24	320	368	372.34	373.60	376.10	375.05	375.07	374.32
25	263	318	318.97	320.23	322.40	321.35	321.35	322.59
26	220	271	270.39	271.06	272.50	271.42	271.40	271.68
27	182	234	226.99	227.38	227.50	226.94	227.09	229.70
28	167	193	197.20	196.67	195.70	194.92	195.13	194.64
29	152	178	174.87	174.28	172.60	172.46	172.76	174.61
SSQ (m^3/s)^2	-	-	1086.84	999.83	979.96	917.06	909.35	759.79
Squared root of SSQ (m^3/s)	-	-	32.97	31.62	31.30	30.28	30.16	27.56
NSE	-	-	0.998247	0.998326	0.998359	0.998464	0.998478	0.998728
RMSE (m^3/s)	-	-	6.008111	5.8711693	5.813054	5.623423	5.598762	5.118558

Table 6. Results when using flood data of River Wye December in 1960.

Time (h)	Inflow (m^3/s)	Outflow (m^3/s)	LMM (m^3/s)	LMM-L (m^3/s) [5]	NLMM (m^3/s) [42]	NLMM-L (m^3/s) [9]	ANLMM-L (m^3/s) [2]	This Study (m^3/s)
0	154	102	102.00	102.00	102.00	102.00	102.00	102.00
6	150	140	118.15	116.00	154.00	149.50	146.52	141.89
12	219	169	115.12	120.00	152.00	156.59	155.74	155.50
18	182	190	152.64	147.00	181.00	191.40	194.41	185.46
24	182	209	161.35	158.00	191.00	200.79	194.19	190.53
30	192	218	165.67	165.00	185.00	195.14	196.05	195.99
36	165	210	178.37	176.00	187.00	197.46	198.35	196.69
42	150	194	177.11	178.00	179.00	188.48	186.83	188.20
48	128	172	173.05	176.00	162.00	170.80	172.12	175.53
54	168	149	152.45	164.00	141.00	148.10	150.37	157.72
60	260	136	140.42	160.00	154.00	162.59	167.56	169.06
66	471	228	137.74	167.00	198.00	210.36	216.61	213.24
72	717	303	192.13	218.00	264.00	281.58	294.27	287.51
78	1092	366	280.05	303.00	344.00	367.75	378.29	378.89
84	1145	456	511.40	484.00	416.00	447.65	461.17	465.87
90	600	615	797.99	690.00	599.00	629.57	612.03	609.41
96	365	830	781.75	700.00	871.00	892.78	862.51	863.65
102	277	969	674.00	642.00	834.00	859.01	884.60	887.00
108	227	665	565.24	572.00	689.00	719.30	737.54	730.86
114	187	519	472.11	505.00	535.00	567.50	565.33	555.56
120	161	444	392.21	442.00	397.00	427.85	414.97	410.06
126	143	321	326.86	386.00	283.00	308.86	297.45	300.33
132	126	208	275.37	338.00	202.00	220.90	216.14	224.40
138	115	176	233.04	296.00	152.00	163.64	164.43	174.61
144	102	148	200.36	260.00	124.00	131.90	134.94	143.56
150	93	125	172.80	228.00	106.00	111.93	114.46	121.64
156	88	114	150.03	201.00	94.00	99.28	101.24	106.75
162	82	106	132.71	179.00	88.00	92.90	94.00	97.42
168	76	97	118.75	160.00	82.00	86.14	86.94	89.67
174	73	89	106.60	144.00	75.00	79.34	80.13	82.79
180	70	81	97.18	130.00	73.00	76.46	76.87	78.56
186	67	76	89.65	118.00	69.00	73.13	73.54	74.88
192	63	71	83.66	109.00	66.00	69.85	70.23	71.46
198	59	66	78.25	100.00	62.00	65.09	65.60	67.24
SSQ (m^3/s)^2	-	-	196,077.12	251,802.00	37,944.15	25,915.27	20,494.98	18,816.99
Squared root of SSQ (m^3/s)	-	-	442.81	501.80	194.79	160.98	143.16	137.18
NSE	-	-	0.916666	0.921600	0.959208	0.988986	0.991290	0.992003
RMSE (m^3/s)	-	-	77.082625	74.765750	53.930178	28.023612	24.921077	23.879109

The fourth flood dataset used was Sutculer flood data which is a flood data with a double-peak. The parameters of the LMM for Sutculer flood data were determined to be 1.0 for K, and -0.006097 for X_1. The parameters of the LMM-L for Sutculer flood data were determined to be 1.0 for K, -0.025914 for X_1, and -0.041042 for β. The parameters of the NLMM for Sutculer flood data were determined to be 1.0 for K, -0.053787 for X_1, and 1.002498 for m. The parameters of the new Muskingum flood routing model for Sutculer flood data were determined to be 0.931599 for K, -0.092988 for X_1, 0.009066 for X_2, 1.000013 for m, -0.036144 for β, 0.817639 for θ_1, 0.214801 for θ_2, and 0.745272 for θ_3. The results of all models are shown in Table 7.

Among the results in Table 7, those of NLMM-L and ANLMM-L were calculated in previous studies [2,9]. The results of LMM, LMM-L, NLMM, and new Muskingum flood routing model were calculated by applying the SAVCA. Notably, the models that consider the lateral inflow show good results. In addition, the results of LMM-L, NLMM-L, and ANLMM-L were better than those of LMM and NLMM because LMM-L, NLMM-L, and ANLMM-L showed relatively small errors in the overall results. Moreover, the difference between the results of the new Muskingum flood routing model and other Muskingum flood routing models was substantial.

The ANLMM-L showed the best results (smallest error) among the considered models. The difference in SSQ between the ANLMM-L and new Muskingum flood routing model was 63.22 $(m^3/s)^2$. The time required to show the difference between the two results ranged from 1 to 3 h. At 1 h, the outflow of most Muskingum flood routing models was close to the observed outflow. At 2 and 3 h, except for the new Muskingum flood routing model, the outflow of most Muskingum flood routing models showed a difference from the observed outflow.

The last flood dataset analyzed was the flood data of River Wyre October in 1982. The parameters of the LMM for the flood data of River Wyre from October 1982 were determined to be 3.950351 for K and 0.295668 for X_1. The parameters of the NLMM for the flood data of River Wyre from October 1982 were determined to be 8.248204 for K, 0.284338 for X_1, and 0.812821 for m. The parameters of the new Muskingum flood routing model for the flood data of River Wyre from October 1982 were determined to be 0.931599 for K, -0.092988 for X_1, 0.009066 for X_2, 1.000013 for m, -0.036144 for β, 0.817639 for θ_1, 0.214801 for θ_2, and 0.745272 for θ_3; the results, and their comparison with those of the other models, are shown in Table 8.

Of the results given in Table 8, the LMM-L, NLMM-L, and ANLMM-L results were calculated in previous studies [2,5,9], while those of the LMM and NLMM were calculated by applying SAVCA; the results of the latter two models were the same. The errors were the greatest for the results of both LMM and NLMM, and some of the calculated outflow values obtained were negative. The new Muskingum flood routing model results in Table 8 were also calculated by applying SAVCA.

Notably, the results obtained for the models considering the lateral inflow were good; those of the LMM-L, NLMM-L, and ANLMM-L were significantly better than those of the LMM and NLMM. The difference in the results of all Muskingum flood routing models occurs from 26 to 31 h. The outflows of NLMM-L, ANLMM-L, and new Muskingum flood routing model were the closest to the observed outflow from 26 to 31 h. Although the results of the NLMM-L and ANLMM-L were similar to the observed outflow at 26 h, these differed over time. However, the new Muskingum flood routing model results did not differ significantly from the observed outflow from 26 to 31 h. ANLMM-L showed the best results (smallest error) among the existing Muskingum flood routing models. The difference in SSQ between the ANLMM-L and new Muskingum flood routing model was 1.35 $(m^3/s)^2$. Although varying results were obtained when using different flood data for the various models, the overall results of new Muskingum flood routing model were better than those of other Muskingum flood routing models because the error of new Muskingum flood routing model was relatively small in the latter part from 16 to 31 h.

Table 7. Results when using Sutculer flood data.

Time (h)	Inflow (m³/s)	Outflow (m³/s)	LMM (m³/s)	LMM-L (m³/s)	NLMM (m³/s)	NLMM-L (m³/s) [9]	ANLMM-L (m³/s) [2]	This Study (m³/s)
0	7.53	7.00	7.00	7.00	7.00	7.00	7.00	7.00
1	9.06	8.00	7.59	7.25	7.58	7.24	7.26	8.14
2	28.00	23.00	10.06	9.11	9.94	9.00	9.01	11.97
3	79.80	25.00	29.95	27.66	29.56	27.35	27.35	25.63
4	64.30	75.00	76.04	74.92	75.86	74.84	74.81	73.93
5	38.20	60.00	63.47	61.36	63.70	61.57	61.59	62.61
6	41.40	40.00	39.84	37.33	39.96	37.40	37.41	37.54
7	41.30	41.00	41.30	39.64	41.31	39.63	39.62	39.37
8	33.80	41.00	40.87	39.42	40.92	39.47	39.47	39.72
9	32.00	32.00	34.10	32.55	34.15	32.57	32.58	32.56
10	29.00	30.00	31.95	30.66	31.98	30.68	30.68	31.10
11	35.00	34.00	29.51	28.03	29.49	28.00	28.00	29.48
12	63.10	35.00	36.30	34.10	36.10	33.93	33.93	36.09
13	110.00	60.00	64.26	60.98	63.81	60.62	60.62	63.47
14	170.00	105.00	110.82	105.81	110.12	105.25	105.25	108.08
15	216.00	160.00	169.24	162.69	168.46	162.06	162.07	157.14
16	131.00	206.00	208.43	203.95	208.54	204.11	204.11	205.42
17	101.00	128.00	133.73	126.88	134.55	127.58	127.61	126.32
18	65.00	97.00	100.81	96.74	101.33	97.14	97.10	98.08
19	62.40	61.00	66.91	63.14	67.19	63.33	63.32	63.18
20	53.80	60.00	62.16	59.71	62.26	59.78	59.76	59.18
21	36.30	50.00	53.27	51.37	53.44	51.51	51.50	51.71
22	29.60	33.00	36.89	35.07	37.03	35.16	35.16	35.19
23	25.00	27.00	29.75	28.44	29.82	28.49	28.48	28.49
24	21.30	23.00	25.06	24.00	25.11	24.03	24.03	24.13
25	19.60	19.00	21.42	20.47	21.44	20.49	20.49	20.53
26	18.00	18.00	19.61	18.80	19.63	18.81	18.81	18.90
27	17.30	17.00	18.05	17.28	18.06	17.29	17.29	17.38
28	17.00	17.00	17.33	16.60	17.33	16.60	16.60	16.63
29	16.00	17.00	16.96	16.29	16.97	16.29	16.29	16.53
SSQ (m³/s)²	-	-	512.87	282.89	510.18	281.11	280.95	217.73
Squared root of SSQ (m³/s)	-	-	22.65	16.82	22.59	16.77	16.76	14.76
NSE	-	-	0.992557	0.995895	0.992596	0.995921	0.995922	0.996840
RMSE (m³/s)	-	-	4.134694	3.070802	4.123823	3.061080	3.060593	2.694028

Table 8. Results when using flood data of River Wyre October in 1982.

Time (h)	Inflow (m³/s)	Outflow (m³/s)	LMM (m³/s)	LMM-L (m³/s) [5]	NLMM (m³/s)	NLMM-L (m³/s) [9]	ANLMM-L (m³/s) [2]	This Study (m³/s)
0	2.60	8.30	8.30	8.30	8.30	8.30	8.30	8.30
1	4.20	9.00	5.58	8.20	6.00	8.51	8.52	8.73
2	12.30	9.90	1.68	8.10	2.27	8.79	9.94	10.11
3	25.40	10.20	0.00	12.70	0.00	10.94	12.74	12.75
4	24.10	18.90	9.67	27.90	8.66	20.28	19.71	19.51
5	20.30	35.90	16.45	39.90	15.50	37.54	35.73	36.22
6	23.30	51.80	16.58	45.70	16.02	49.07	48.87	49.25
7	27.70	59.40	17.15	52.20	16.91	55.11	55.95	55.83
8	27.70	63.30	20.94	61.40	20.90	62.50	62.74	62.54
9	26.90	69.60	23.71	68.90	23.78	71.44	71.35	71.33
10	24.80	76.70	25.73	74.70	25.80	78.03	77.95	77.87
11	26.90	82.00	24.52	77.20	24.59	82.07	82.67	82.68
12	33.70	85.30	22.52	79.80	22.77	83.72	85.27	85.10
13	33.90	89.00	26.45	87.80	26.92	87.43	88.11	87.71
14	27.80	94.60	31.69	95.50	32.09	95.49	94.74	94.61
15	20.80	98.80	33.23	97.70	33.18	100.88	99.90	99.91
16	15.60	98.00	30.95	94.40	30.43	99.29	98.87	98.75
17	11.90	91.80	26.98	87.90	26.29	92.06	92.05	91.82
18	9.50	82.30	22.57	79.80	21.98	82.22	82.36	82.12
19	7.80	72.00	18.59	71.50	18.24	71.75	71.88	71.67
20	6.50	61.90	15.26	63.60	15.19	61.94	61.93	61.80
21	5.80	53.00	12.40	56.10	12.60	53.12	53.10	53.03
22	5.00	45.60	10.37	49.60	10.74	45.47	45.37	45.34
23	4.80	39.20	8.52	43.70	9.04	39.14	39.04	39.07
24	4.50	33.80	7.31	38.80	7.87	33.76	33.65	33.68
25	4.10	29.30	6.47	34.60	7.03	29.55	29.39	29.44
26	3.70	26.20	5.78	30.90	6.34	26.12	25.96	26.02
27	3.40	23.50	5.16	27.70	5.71	23.20	23.08	23.14
28	3.20	21.20	4.61	24.80	5.14	20.67	20.59	20.64
29	2.90	19.20	4.23	22.30	4.73	18.52	18.44	18.48
30	2.80	17.70	3.79	20.10	4.27	16.71	16.68	16.72
31	2.60	16.40	3.52	18.20	3.96	15.12	15.09	15.23
SSQ (m³/s)²	-	-	53,544.67	468.84	53,544.99	53.66	40.16	38.81
Squared root of SSQ (m³/s)	-	-	231.40	21.65	231.40	7.33	6.34	6.23
NSE	-	-	−0.213958	0.989570	−0.213965	0.998842	0.999090	0.999120
RMSE (m³/s)	-	-	40.905633	3.790780	40.905755	1.263563	1.120320	1.101288

The new flood data in Daechung were applied to calibrate and validate the new Muskingum flood routing model. The flood data in April, 2014 were used for calibration and the flood data in April, 2018 were used for validation. Among various Muskingum flood routing models, LMM-L considering lateral inflow to LMM, NLMM considering nonlinearity to LMM and new Muskingum flood routing model were applied to Daechung flood data and compared.

The parameters of LMM were 3.989981 for K and -0.034950 for X. The parameters of LMM-L were 3.865970 for K, -0.043293 for X, and -0.020080 for β. The parameters of NLMM were 2.225924 for K, -1.5 for X, and 1.0 for m. The parameters of NLMM-L were 2.225076 for K, -1.406061 for X, 1.000000 for m, -0.007072 for β, 1.000000 for θ. The parameters of ANLMM-L were 2.123526 for K, -1.500000 for X, 1.000000 for m, -0.010223 for β, 1.000000 for θ_1, and 0.017871 for θ_2. The parameters of new Muskingum flood routing model were 2.220910 for K, -1.498329 for X_1, 0.094832 for X_2, 1.000008 for m, -0.012616 for β, 0.999660 for θ_1, 0.000093 for θ_2, and 0.000288 for θ_3. Each parameter was applied equally for 2014 and 2018 data. A total of 100 simulations were conducted for each Muskingum flood routing model, yielding the best results. Table 9 shows the results of Daechung flood data in 2014.

Based on the variable values determined from the 2014 flood data, it was applied to the 2018 flood data. Table 10 shows the results of Daechung flood data in 2018.

Figure 3 shows the results of calibration and validation in Daechung flood data.

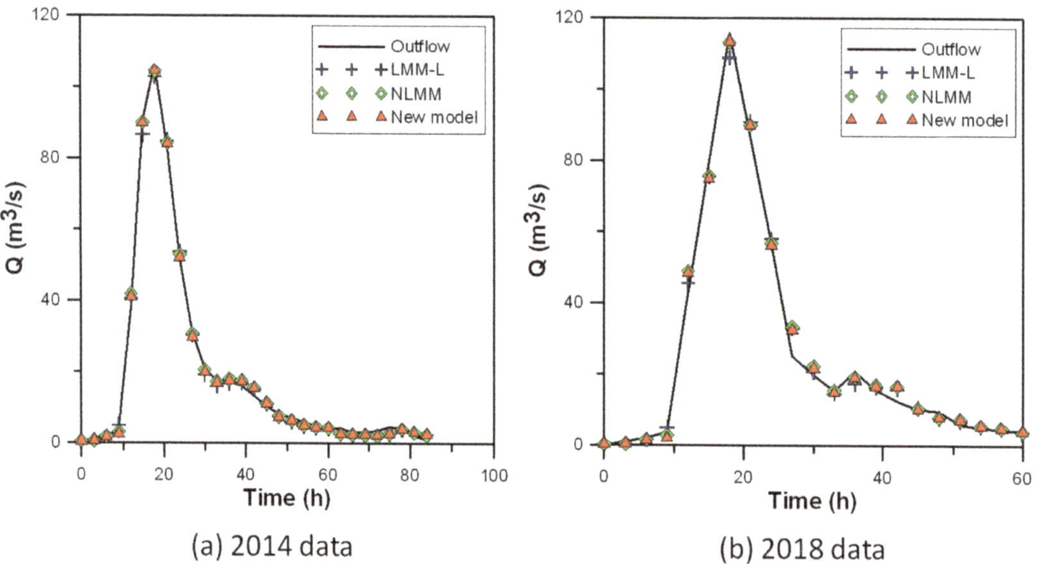

Figure 3. Results of calibration and validation in Daechung flood data (**a**) 2014 flood data; (**b**) 2018 flood data..

In the 2014 flood data, the SSQs of LMM, LMM-L, NLMM, NLMM-L, ANLMM-L and new Muskingum flood routing model were 88.23, 73.81, 43.79, 42.32, 40.16, and 39.55, respectively. In the 2014 flood data, the error of new Muskingum flood routing model was relatively small and the calibration of new Muskingum flood routing model was relatively accurate. The SSQs of new Muskingum flood routing model in the 2018 flood data was relatively small. In the 2018 flood data, the SSQs of LMM, LMM-L, NLMM, NLMM-L, ANLMM-L and new Muskingum flood routing model were 221.92, 180.41, 171.45, 161.99, 159.84, and 157.64, respectively. In the results of Figure 2, more accurate flood routing was performed by applying the new Muskingum flood routing model compared to LMM, LMM-L, NLMM, NLMM-L and ANLMM-L.

Table 9. Results of Daechung flood data in 2014.

Time (h)	Inflow (m³/s)	Outflow (m³/s)	LMM (m³/s)	LMM-L (m³/s)	NLMM (m³/s)	NLMM-L (m³/s)	ANLMM-L (m³/s)	This Study (m³/s)
0	0.79	0.47	0.47	0.47	0.47	0.47	0.47	0.47
3	2.12	1.04	0.75	0.75	0.64	0.65	0.65	0.62
6	3.54	2.11	1.79	1.80	1.81	1.80	1.78	1.79
9	50.75	3.19	4.66	4.96	3.14	3.12	2.62	2.34
12	103.90	38.32	39.94	40.42	41.86	41.65	40.88	41.01
15	112.33	90.68	86.69	86.43	89.99	89.68	89.43	89.79
18	82.41	106.18	104.31	102.80	104.36	104.08	104.41	104.44
21	45.06	82.32	87.14	84.88	84.25	83.98	84.39	84.04
24	23.12	51.59	55.83	53.70	52.80	52.46	52.58	52.12
27	15.76	31.21	31.82	30.31	30.74	30.33	30.17	29.87
30	15.03	20.82	20.13	19.22	20.63	20.23	19.98	19.86
33	17.22	17.24	16.50	15.97	17.41	17.07	16.86	16.80
36	17.27	17.65	17.02	16.64	17.91	17.64	17.51	17.50
39	15.04	16.09	17.13	16.76	17.58	17.38	17.32	17.30
42	9.97	13.02	15.44	15.05	15.59	15.44	15.44	15.41
45	6.38	10.39	11.34	10.98	11.16	11.04	11.04	10.98
48	5.67	8.17	7.71	7.43	7.59	7.49	7.45	7.39
51	4.45	7.07	6.19	5.99	6.36	6.27	6.23	6.21
54	4.23	5.99	4.92	4.77	4.99	4.92	4.89	4.87
57	4.18	4.91	4.42	4.30	4.52	4.46	4.43	4.42
60	2.25	4.20	4.18	4.07	4.32	4.27	4.27	4.27
63	2.33	4.02	2.78	2.69	2.67	2.64	2.63	2.60
66	2.24	3.01	2.45	2.38	2.51	2.48	2.46	2.45
69	2.11	3.13	2.29	2.24	2.34	2.31	2.30	2.30
72	2.83	3.49	2.18	2.14	2.18	2.16	2.14	2.13
75	4.25	4.55	2.70	2.67	2.73	2.71	2.68	2.67
78	2.83	4.20	3.78	3.72	3.94	3.92	3.92	3.94
81	2.15	2.08	3.07	2.99	2.95	2.93	2.94	2.92
84	2.13	1.04	2.40	2.33	2.33	2.31	2.30	2.37
SSQ (m³/s)²	-	-	88.23	73.81	43.79	42.32	40.16	39.55
Squared root of SSQ (m³/s)	-	-	1.78	1.62	1.25	1.23	1.20	1.19
NSE	-	-	0.996612	0.997166	0.998319	0.998375	0.998458	0.998482
RMSE (m³/s)	-	-	1.775135	1.623577	1.250501	1.229365	1.197584	1.188440

Table 10. Results of Daechung flood data in 2018.

Time (h)	Inflow (m^3/s)	Outflow (m^3/s)	LMM (m^3/s)	LMM-L (m^3/s)	NLMM (m^3/s)	NLMM-L (m^3/s)	ANLMM-L (m^3/s)	This Study (m^3/s)
0	0.53	0.32	0.32	0.32	0.32	0.32	0.32	0.32
3	1.86	1.02	0.52	0.52	0.43	0.44	0.44	0.41
6	3.30	2.09	1.54	1.55	1.57	1.56	1.54	1.55
9	59.34	3.79	4.71	5.08	2.91	2.90	2.30	1.95
12	85.37	42.46	45.27	45.61	48.82	48.58	48.05	48.45
15	124.77	79.88	75.73	75.51	75.72	75.50	75.05	74.93
18	88.73	115.68	110.14	108.82	113.04	112.65	112.85	113.28
21	48.98	87.42	93.24	90.93	89.96	89.69	90.22	89.81
24	25.23	57.05	60.29	58.03	56.88	56.55	56.70	56.21
27	17.09	25.19	34.54	32.93	33.26	32.85	32.68	32.36
30	12.43	19.64	21.71	20.70	22.29	21.88	21.66	21.54
33	19.03	15.39	15.19	14.63	15.69	15.35	15.09	14.96
36	16.39	20.74	17.89	17.51	19.31	19.02	18.88	18.94
39	16.44	15.69	16.80	16.43	17.01	16.81	16.74	16.68
42	8.97	12.45	16.28	15.89	16.71	16.54	16.55	16.56
45	6.89	10.17	10.90	10.52	10.48	10.36	10.36	10.26
48	6.76	9.50	7.98	7.71	7.97	7.86	7.80	7.76
51	4.90	5.93	7.03	6.83	7.28	7.19	7.15	7.15
54	4.55	4.98	5.47	5.31	5.49	5.41	5.39	5.36
57	3.51	4.50	4.76	4.63	4.88	4.82	4.80	4.79
60	2.45	4.05	3.82	3.70	3.85	3.80	3.80	3.88
SSQ (m^3/s)^2	-	-	221.92	180.41	171.45	161.99	159.84	157.64
Squared root of SSQ (m^3/s)	-	-	2.82	2.54	2.47	2.41	2.39	2.37
NSE	-	-	0.990792	0.992514	0.992886	0.993279	0.993368	0.993459
RMSE (m^3/s)	-	-	2.815292	2.538342	2.474522	2.405271	2.389282	2.372746

4. Discussion

Because the calculation process differs for each Muskingum flood routing model, the time required to find the parameters when applying a meta-heuristic optimization algorithm is different for each method. The time required to apply SAVCA was summarized to determine the parameters of each Muskingum flood routing model for Wilson's flood data. The parameters of SAVCA were set at a constant, and the simulation was conducted 10 times. In addition, the number of iterations was set to 100,000. Table 11 shows the time taken for SAVCA when using Wilson's flood data.

Table 11. Time required by Muskingum flood routing models when using Wilson's flood data.

Comparative Indicators	LMM	LMM-L	NLMM	NLMM-L	ANLMM-L	This Study
Time (s)	634	633	741	753	810	936

Depending on the parameters of the SAVCA and flood data, the results over time using the Muskingum flood routing models differ from the results in Table 9. As the number of parameters of the Muskingum flood routing models increased, the time required also increased. The time required by LMMs, LMM and LMM-L, was the shortest. NLMM and NLMM-L required a greater amount of time compared to the LMMs. The new Muskingum flood routing model required more time than the ANLMM-L, which in turn required more time than the NLMM-L. The time required by the new Muskingum flood routing model was approximately 1.5-times that required by LMM and LMM-L. In conclusion, the new Muskingum flood routing model produced more accurate results but took more time owing to the greater number of parameters and calculations.

An analysis was conducted on how each parameter of new Muskingum flood routing model affects the results. Daechung flood data in April, 2014 was applied to analyze the sensitivity of each parameter in the new Muskingum flood routing model. Figure 4 showed the results of the sensitivity analysis for the parameters in the new Muskingum flood routing model.

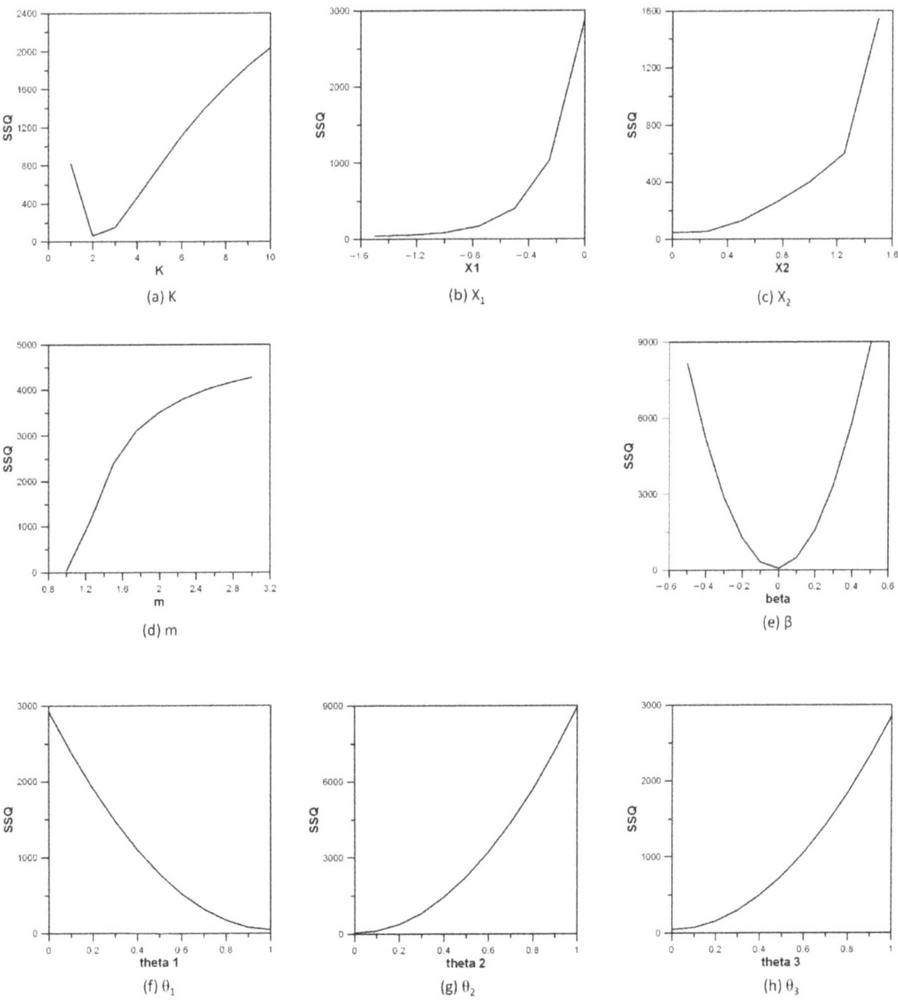

Figure 4. Results of sensitivity analysis for the parameters in the new Muskingum flood routing model.

SSQ decreases and then increases as parameter K increases, and SSQ increases as parameter X_1 increases. SSQ increases as parameter X_2 increases and SSQ increases as parameter m increases. SSQ decreases and then increases as parameter β increases, and SSQ decreases as the parameter θ_1 increases. SSQ increases as parameter θ_2 increases, and SSQ increases as parameter θ_3 increases. As each parameter changed, the change of the results was not constant. It is difficult to find a uniform pattern in all the results. However, what can be confirmed from the results of sensitivity analysis is that a meta-heuristic optimization algorithm such as SAVCA is required to produce results with a low SSQ.

5. Conclusions

The Muskingum flood routing model is a representative hydrologic flood routing model that is widely used owing to its easy applicability. The proposed Muskingum flood routing model in this study is a simple model that can be applied by researchers that use the existing Muskingum models for accurate flood routing.

In this study, the new Muskingum flood routing model was applied to various flood data, and the results obtained were compared with those of previously developed Muskingum flood routing models. As an index for comparison, the error was calculated between the observed and simulated outflows using the SSQ, NSE and RMSE. In addition, SAVCA, a meta-heuristic optimization algorithm, was applied to adjust the parameters of the new Muskingum flood routing model.

In the sensitivity analysis, the changes of the eight parameters in the new Muskingum flood routing model are different. There are parameters whose results are improved as the value (θ_1) increases, some parameters (m, θ_2, θ_3, x_1, x_2) whose results are improved as the value decreases, and some parameters (K, β) whose results are changed (improved and then deteriorated) as the value increases. The eight parameters of the new Muskingum flood routing model are decision variables of SAVCA and are calculated through the optimization process.

Muskingum flood routing models considering the lateral inflow are capable of relatively sophisticated simulations, which corroborates that the influence of lateral inflow on the outflow can be considered. Among the existing models, the ANLMM-L showed the highest accuracy, although the difference between its results and those of the other Muskingum flood routing models was insignificant. In the new Muskingum flood routing model, the improved calculation method of the inflow at previous time and next time reflected the trend of the observed outflow.

Since the Muskingum flood routing model proposed in this study has eight parameters, the calculation process is more complicated than the existing Muskingum flood routing models. Accurate flood prediction is possible due to the complicated calculation process, but the calculation time is long.

Many studies, including this study, have been performed by applying various meta-heuristic optimization algorithms to the Muskingum flood routing models. Since various optimization algorithms cannot show advantages in all problems, the results can be improved by appropriately selecting the operators of each meta-heuristic optimization algorithm. In addition, deep learning techniques have been widely used to apply various flood prediction methods including Muskingum flood routing models. By replacing the optimizer in deep learning techniques with meta-heuristic optimization algorithms, it would be possible to produce improved results in flood prediction.

Funding: This research was funded by the National Research Foundation (NRF) of Korea (NRF-2019R1I1A3A01059929).

Institutional Review Board Statement: Not applicable.

Informed Consent Statement: Not applicable.

Acknowledgments: This work was supported by a grant from The National Research Foundation (NRF) of Korea (NRF-2019R1I1A3A01059929).

Conflicts of Interest: The author declares no conflict of interest.

Abbreviations

ANLMM-L	Advanced nonlinear Muskingum flood routing model considering continuous inflow
NLMM-L	Nonlinear Muskingum flood routing model incorporating lateral flow
NLMM	Nonlinear Muskingum method
LMM-L	Linear Muskingum method incorporating lateral flow
LMM	Linear Muskingum method
SAVCA	Self-adaptive vision correction algorithm
DR1	Division rate 1
DR2	Division rate 2
MTF	Modulation transfer function
CF	Compression factor
AR	Astigmatic rate
AF	Astigmatic angle
SSQ	Sum of squares
NSE	Nash–Sutcliffe efficiency
RMSE	Root mean square error

References

1. Go, I.H. Development of watershed integrated water resource management technique. *J. Korea Water Resour. Assoc.* **2004**, *37*, 10–15.
2. Lee, E.H.; Lee, H.M.; Kim, J.H. Development and application of advanced Muskingum flood routing model considering continuous flow. *Water* **2018**, *10*, 760. [CrossRef]
3. Absi, R. Reinvestigating the parabolic-shaped eddy viscosity profile for free surface flows. *Hydrology* **2021**, *8*, 126. [CrossRef]
4. McCarthy, G.T. The unit hydrograph and flood routing. In *Proceedings of Conference of North Atlantic Division*; U.S. Army Corps of Engineers: New London, CT, USA, 1938.
5. O'donnell, T. A direct three-parameter Muskingum procedure incorporating lateral inflow. *Hydrol. Sci. J.* **1985**, *30*, 479–496. [CrossRef]
6. Tung, Y.K. River flood routing by nonlinear Muskingum method. *J. Hydraul. Eng.* **1985**, *111*, 1447–1460. [CrossRef]
7. Das, A. Parameter estimation for Muskingum models. *J. Irrig. Drain. Eng.* **2004**, *130*, 140–147. [CrossRef]
8. Geem, Z.W. Parameter estimation for the nonlinear Muskingum model using the BFGS technique. *J. Irrig. Drain. Eng.* **2006**, *132*, 474–478. [CrossRef]
9. Karahan, H.; Gurarslan, G.; Geem, Z.W. A new nonlinear Muskingum flood routing model incorporating lateral flow. *Eng. Optim.* **2015**, *47*, 737–749. [CrossRef]
10. Bozorg-Haddad, O.; Abdi-Dehkordi, M.; Hamedi, F.; Pazoki, M.; Loáiciga, H.A. Generalized storage equations for flood routing with nonlinear Muskingum models. *Water Resour. Manag.* **2019**, *33*, 2677–2691. [CrossRef]
11. Gill, M.A. Flood routing by the Muskingum method. *J. Hydrol.* **1978**, *36*, 353–363. [CrossRef]
12. Yoon, J.; Padmanabhan, G. Parameter estimation of linear and nonlinear Muskingum models. *J. Water Resour. Plann. Manag.* **1993**, *119*, 600–610. [CrossRef]
13. Mohan, S. Parameter estimation of nonlinear Muskingum models using genetic algorithm. *J. Hydraul. Eng.* **1997**, *123*, 137–142. [CrossRef]
14. Luo, J.; Xie, J. Parameter estimation for nonlinear Muskingum model based on immune clonal selection algorithm. *J. Hydrol. Eng.* **2010**, *15*, 844–851. [CrossRef]
15. Barati, R. Parameter estimation of nonlinear Muskingum models using Nelder-Mead simplex algorithm. *J. Hydrol. Eng.* **2011**, *16*, 946–954. [CrossRef]
16. Orouji, H.; Bozorg Haddad, O.; Fallah-Mehdipour, E.; Marino, M.A. Estimation of Muskingum parameter by meta-heuristic algorithms. In Proceedings of the Institution of Civil Engineers-Water Management; Thomas Telford Ltd.: London, UK, 2013; Volume 166, pp. 315–324.
17. Niazkar, M.; Afzali, S.H. Assessment of modified honey bee mating optimization for parameter estimation of nonlinear Muskingum models. *J. Hydrol. Eng.* **2015**, *20*, 04014055. [CrossRef]
18. Kang, L.; Zhang, S. Application of the elitist-mutated PSO and an improved GSA to estimate parameters of linear and nonlinear Muskingum flood routing models. *PLoS ONE* **2017**, *11*, e0147338. [CrossRef]
19. Moghaddam, A.; Behmanesh, J.; Farsijani, A. Parameters estimation for the new four-parameter nonlinear Muskingum model using the particle swarm optimization. *Water Resour. Manag.* **2016**, *30*, 2143–2160. [CrossRef]
20. Hamedi, F.; Bozorg-Haddad, O.; Pazoki, M.; Asgari, H.R.; Parsa, M.; Loáiciga, H.A. Parameter estimation of extended nonlinear Muskingum models with the weed optimization algorithm. *J. Irrig. Drain. Eng.* **2016**, *142*, 04016059. [CrossRef]

21. Perumal, M.; Tayfur, G.; Rao, C.M.; Gurarslan, G. Evaluation of a physically based quasi-linear and a conceptually based nonlinear Muskingum methods. *J. Hydrol.* **2017**, *546*, 437–449. [CrossRef]
22. Zhang, S.; Kang, L.; Zhou, B. Parameter estimation of nonlinear Muskingum model with variable exponent using adaptive genetic algorithm. In *Environmental Conservation, Clean Water, Air & Soil (CleanWAS)*; IWA Publishing: London, UK, 2017; pp. 231–237.
23. Bagatur, T.; Onen, F. Development of predictive model for flood routing using genetic expression programming. *J. Flood Risk Manag.* **2018**, *11*, S444–S454. [CrossRef]
24. Akbarifard, S.; Qaderi, K.; Alinnejad, M. Parameter estimation of the nonlinear Muskingum flood-routing model using water cycle algorithm. *J. Watershed Manag. Res.* **2018**, *8*, 34–43. [CrossRef]
25. Talatahari, S.; Sheikholeslami, R.; Farahmand Azar, B.; Daneshpajouh, H. Optimal parameter estimation for Muskingum model using a CSS-PSO method. *Adv. Mech. Eng.* **2013**, *5*, 480954. [CrossRef]
26. Ouyang, A.; Li, K.; Truong, T.K.; Sallam, A.; Sha, E.H.M. Hybrid particle swarm optimization for parameter estimation of Muskingum model. *Neural Comput. Appl.* **2014**, *25*, 1785–1799. [CrossRef]
27. Haddad, O.B.; Hamedi, F.; Fallah-Mehdipour, E.; Orouji, H.; Marino, M.A. Application of a hybrid optimization method in Muskingum parameter estimation. *J. Irrig. Drain. Eng.* **2015**, *141*, 04015026. [CrossRef]
28. Kang, L.; Zhou, L.; Zhang, S. Parameter estimation of two improved nonlinear Muskingum models considering the lateral flow using a hybrid algorithm. *Water Resour. Manag.* **2017**, *31*, 4449–4467. [CrossRef]
29. Ehteram, M.; Binti Othman, F.; Mundher Yaseen, Z.; Abdulmohsin Afan, H.; Falah Allawi, M.; Najah Ahmed, A.; Shaid, S.; Singh, V.; El-Shafie, A. Improving the Muskingum flood routing method using a hybrid of particle swarm optimization and bat algorithm. *Water* **2018**, *10*, 807. [CrossRef]
30. O'Sullivan, J.J.; Ahilan, S.; Bruen, M. A modified Muskingum routing approach for floodplain flows: Theory and practice. *J. Hydrol.* **2012**, *470*, 239–254. [CrossRef]
31. Easa, S.M. New and improved four-parameter non-linear Muskingum model. In *Proceedings of the Institution of Civil Engineers-Water Management*; Thomas Telford Ltd.: London, UK, 2014; Volume 167, pp. 288–298.
32. Yoo, C.; Lee, J.; Lee, M. Parameter estimation of the Muskingum channel flood-routing model in ungauged channel reaches. *J. Hydrol. Eng.* **2017**, *22*, 05017005. [CrossRef]
33. Karahan, H.; Gurarslan, G.; Geem, Z.W. Parameter estimation of the nonlinear Muskingum flood-routing model using a hybrid harmony search algorithm. *J. Hydrol. Eng.* **2013**, *18*, 352–360. [CrossRef]
34. Niazkar, M.; Afzali, S.H. Application of new hybrid optimization technique for parameter estimation of new improved version of Muskingum model. *Water Resour. Manag.* **2016**, *30*, 4713–4730. [CrossRef]
35. Niazkar, M.; Afzali, S.H. Parameter estimation of an improved nonlinear Muskingum model using a new hybrid method. *Hydrol. Res.* **2017**, *48*, 1253–1267. [CrossRef]
36. Lee, E.H. Application of self-adaptive vision-correction algorithm for water-distribution problem. *KSCE J. Civ. Eng.* **2021**, *25*, 1106–1115. [CrossRef]
37. Kim, Y.N. Development of Exponential Bandwidth Harmony Search with Centralized Global Search. Master's Thesis, Chungbuk National University, Cheongju, Korea, 2021. (In Korean)
38. Chow, V.T. *Open-Channel Hydraulics*; McGraw-Hill Civil Engineering Series; McGraw-Hill: New York, NY, USA, 1959.
39. Wang, W.; Xu, Z.; Qiu, L.; Xu, D. Hybrid chaotic genetic algorithms for optimal parameter estimation of Muskingum flood routing model. In Proceedings of the International Joint Conference on Computational Sciences and Optimization, Sanya, China, 24–26 April 2009; pp. 215–218.
40. Bajracharya, K.; Barry, D.A. Accuracy criteria for linearised diffusion wave flood routing. *J. Hydrol.* **1997**, *195*, 200–217. [CrossRef]
41. Ulke, A. Flood Routing Using Muskingum Method. Master's Thesis, Suleyman Demirel University, Kaskelen, Kazakhstan, 2003. (In Turkish)
42. Karahan, H.; Gurarslan, G. Modelling flood routing problems using kinematic wave approach: Sutculer example. In Proceedings of the 7th National Hydrology Congress, Isparta, Turkey, 26–27 September 2013.

Article

Selected Issues of Adaptive Water Management on the Example of the Białka River Basin

Monika Bryła *, Tomasz Walczykiewicz , Magdalena Skonieczna and Mateusz Żelazny

Institute of Meteorology and Water Management—National Research Institute, ul. Podleśna 61,
01-673 Warsaw, Poland; tomasz.walczykiewicz@imgw.pl (T.W.); magdalena.skonieczna@imgw.pl (M.S.);
mateusz.zelazny@imgw.pl (M.Ż.)
* Correspondence: monika.bryla@imgw.pl; Tel.: +48-(12)-63-98-219

Abstract: Water is a fundamental resource needed for human life and functioning and the environment. Water management requires a comprehensive, adaptive approach that also considers the dynamics of changes in the water management system. This is particularly important in areas where different groups of stakeholders intertwine, whose needs often contradict, which hampers effective water management, particularly in places of high natural value. This research aimed to analyze selected issues in water management in the Białka River Basin in Southern Poland. The analysis was based on a review of scientific publications, internet sources, and a survey on water management in the basin. Our research shows that the dominant issues in the study area are the flood risk and water pollution related to, among other factors, the intensive development of tourism. Moreover, the effective management of water resources is hampered by poor communication between the administration and stakeholders, which results in a low level of knowledge, negative attitudes towards nature protection, and the emergence of conflicts. The main conclusion of this paper indicates that, despite the existing social potential for implementing comprehensive water management methods, the lack of an appropriate legal framework prevents the implementation of concepts such as Adaptive Water Management.

Keywords: Adaptive Water Management; stakeholder engagement; legislation; survey; uncertainty in water management

Citation: Bryła, M.; Walczykiewicz, T.; Skonieczna, M.; Żelazny, M. Selected Issues of Adaptive Water Management on the Example of the Białka River Basin. *Water* **2021**, *13*, 3540. https://doi.org/10.3390/w13243540

Academic Editors: Tamara Tokarczyk and Andrzej Walega

Received: 10 November 2021
Accepted: 7 December 2021
Published: 10 December 2021

Publisher's Note: MDPI stays neutral with regard to jurisdictional claims in published maps and institutional affiliations.

1. Introduction

Water, as a fundamental resource needed for human life and the activities of daily living and the environment, requires the particular attention of all parties involved in the process of managing its resources. Contemporary social, economic, and climatic changes are causing growing problems related to water [1]. In many regions of the world, water resources are polluted, which negatively affects the aquatic ecosystems and reduces the availability of water to humans. On the one hand, developing urbanization reduces the resources of groundwater on a local scale, enlarging the runoff of waters from the catchment area, and, on the other hand, it increases the flood risk. On a global scale, the availability of water is uneven [2]. The progressive population growth, industrialization, and the lack of appropriate conservation practices of its quality and quantity make almost 80% of the world's population vulnerable to water stress [2,3]. Exposure to water stress depends on the Earth's climatic diversity and economic conditions—highly developed countries are able to allocate much greater financial resources to water resource management than less developed countries. An additional factor that affects the quantitative and spatial availability of water resources is climate change [4–7]. Similar to other regions of the world, Europe is affected by climate change [6]. Depending on latitude or other factors (e.g., high mountain areas), these changes can be positive or negative from a water resource perspective. The negative phenomena include an increase in temperature, a decrease in snowfall in winter, and an increase in evaporation. These changes may have a negative

229

impact on the availability of fresh water in countries such as Poland. The increased frequency of droughts [8–11] can affect both humans and the environment, causing, among others, the drying of forests [12], water shortages in agriculture [13,14], increased risk of disease, malnutrition, high infant mortality, a decrease in the efficiency of electricity production processes, and a negative impact on water quality [15]. The reason for water scarcity is not necessarily natural; for example, in certain areas of India, poor management practices and government policies exacerbate the water issues [16]. Another reason for water scarcity is also cross-border conflicts regarding access to water resources [17] or a lack of appropriate measures to protect the water quality [18–20]. Transboundary water management is one of the most difficult challenges. It requires cooperation at the level of a legal framework, communication between national authorities, a joint action plan, and compliance with agreements [21]. A further challenge in water management is the progressive urbanization and the increase in floods, including floods in cities, which are a growing threat to both people and the economy [22,23]. Extreme climatic phenomena associated with water resources can interact with many environmental and socio-economic sectors, including health, public safety, biodiversity, industry, shipping, and tourism [24].

Traditional management based on the provision of an adequate quantity of water of appropriate quality is insufficient in view of the increasing water-related issues. It requires a more comprehensive approach that considers the needs of the environment, society, and the economy in terms of access to water [25–29]. This approach is called Integrated Water Resource Management (IWRM) [30–32]. In this concept, stakeholder involvement is a key issue in water management. It should be based on the cooperation of various entities—representatives of the public administration, entrepreneurs, residents, nature protection associations, and other social groups that want to be involved in the water management process. This cooperation is a multi-stage process, from public consultations and meetings and the implementation of the agreed roadmaps to their evaluation [33–36]. Stakeholder involvement requires appropriate legal forms to develop tools to support the water management process [30]. IWRM points out that it is necessary to formalize this concept in water legislation as one of the stages leading to the decentralization of government management towards river basin management. Collaboration with stakeholders throughout the water management process is a key criterion for the success of IWRM. It is a comprehensive approach to the development of a water management policy in terms of both resources and services provided related to water [37]. On the other hand, decentralization can make it difficult to control the transparency and fairness of the process [38]. There are known cases around the world of newly created IWRM-related institutions becoming power-laden, gendered, and beset with conflict and factional divisions. According to critics, IWRM could be seen as a form of coercion as it imposes a set of principles and tools to be followed, or also as an idea of a hegemonic discourse that prevents any alternative [39]. The implementation of IWRM may encounter many problems, such as insufficient administrative structure, poor knowledge, and conflicts of interest in water needs [25,33,40]. As Michalak [40] indicates in his work, the lack of knowledge about the functioning of the environment and awareness of the negative effects of anthropopressure is one of the barriers to the implementation of effective water resource management. Furthermore, conflicting and often competing water needs require an appropriately integrated approach [33] that will seek to resolve water-related conflicts on a different scale—from local communities to national needs [25]. Measurement uncertainty is an additional obstacle to effective water management. According to McMillan et al. [41], this issue applies to all stages of data processing, from differences in the quality of measuring equipment and errors caused by individuals carrying out measurements to an incomplete/insufficient measurement network. This network in uncontrolled places requires empirical data interpolation with the selected method, also burdened with the uncertainty of the result. The measurement uncertainty or lack of measurement is significant in forecasting changes in the volume of water resources in view of climate change [42]. As a result, it also has a negative impact on the development of appropriate water management policies [43]. Therefore, water management requires the creation of a resilient management system capable of absorbing disruptions and adapting to changes while maintaining its functions, structure, and purpose [26,44]. Van der Keur et al. [37] indicate that Adaptive Integrated Water Resources Management (AWM) should therefore be used. Its main features are learning, reflection and adaptation capacity, co-management, the formal and informal involvement of decision makers [45], and the drive to decentralize management structures, which benefits stakeholder cooperation [28,34,46,47]. AWM provides added value to IWRM by taking into account uncertainty and adapting to changes (e.g., climate change, lack of complete hydrological data, changing water demand) in the system. It also emphasizes the education of the stakeholders involved [28,48–50]. One of the most significant elements of AWM is social learning, which aims to connect laypeople, enthusiasts, business representatives, and experts in the common goal of water resource management [27,35]. It can be defined as social interactions between stakeholders based on knowledge exchange and an understanding of the management of a water system, where knowledge is acquired from all sides—from legal persons and individuals to organizations [27]. According to Pahl-Wost [51], social learning should focus on learning the social entity as a whole to "learn management together". The framework for this process should be context-specific and involve multilateral cooperation, leading to particular outcomes. AWM also assumes the complexity of the managed systems and the limitations resulting from forecasting their behavior and the possibility of controlling them [37]. Water management will always have to proceed with an incomplete understanding of how the system operates and the effects of its management. Adaptation policy should therefore be planned, taking into account environmental and human behavior processes for measures implemented as part of water management.

Water management in Poland has been regulated by the provisions of the Water Law Act [52] since 2018. As Poland is a member state of the European Union, water management legislation is based on the assumptions of the Water Framework Directive [53]. The relatively small water resources of the country and the unfavorable climatic conditions related to, among others, the high dynamics of flow changes during the year make Poland a country that requires proper management of water resources [54]. The southern parts of Poland are mountainous areas—national and landscape park regions, which are also the main tourist and holiday regions [55]. It is an area where water management should be of particular concern to society due to its natural value and the broad group of stakeholders that benefit from the region's water resources. As a result of the research, this manuscript presents an analysis of selected issues related to water management in the Białka River Basin, located in the south of Poland—in a unique culturally and naturally significant area. The aim of the research was achieved by identifying the issues and the barriers involved in water resource management in the study area. The study was based on a query of scientific publications, press reports, and the results of a survey conducted in 2021. In essence, issues were identified based on query. The results of the survey were used as an additional source of information about problems in water resource management in the Białka River Basin and allowed for the identification of other management barriers, especially in the context of IWRM and AWM. The survey results additionally confirmed that the lack of legal solutions to stimulate cooperation between stakeholders at the lowest level is one of the main barriers to management.

2. Study Area

The Białka River is the right tributary of the Dunajec River, originating in the Tatra Mountains at an altitude of around 1075 m above sea level from the merging of the Rybi Potok, flowing from the Polish part of the Tatra Mountains, and Biała Woda, flowing from the Slovak part of the Tatra Mountains. The length of the Białka River is approximately 42 km [56], including the Biała Woda source stream flowing from the Kacza Valley at an altitude of around 1577 m above sea level. Moving northwards, the Białka River is run by numerous streams flowing from mountainous areas. Its major tributaries include, among others, Roztoka, Jaworowy Potok, Jurgowczyk, and Trybska Rzeka. Białka is a river with typical mountainous characteristics, characterized by a fast current and high dynamics of changes in the hydrological regime [57]. According to Wrzesiński [58], the Białka River is characterized by a pluvial–nival regime. Between September and February, lower than average flows are observed, and from March, snowmelt-related flow increases, which, in the longer term, are combined with summer floods. Data from the Institute of Meteorology and Water Management—National Research Institute (IMGW-PIB) water gauge network for the years 2001–2020 confirm this characteristic. The lowest flows on the Białka River in Trybsz and Łysa Polana occur from January to February (Figure 1), and the highest during the flood period, which is in May. The river flow variability coefficient in both water gauges exceeds 100% (Table 1). Essential flow characteristics in the catchment area are presented in Table 1. The water level in Białka is subject to significant fluctuations, caused by factors such as the intense melting of snow cover or heavy rainfall, especially in mountainous areas. In the upper flow, the river's bed fall reaches around 72‰, while, in the lower flow, it drops to below 20‰ [59]. Spring and summer floods also cause the river to change its bed frequently. The riverbank can be described as steep, rocky, and partially regulated, with sections of a natural character. The river's course is also distinguished by significant terrain height differences—from 530 to 883 m above sea level [60]. The Białka River is one of the few mountainous Carpathian rivers with a natural, anastomosing character [59]. The Białka Valley has been designated as Natura 2000 SOO site (under the Habitats Directive) no. PLH 120024, with an area of 716.03 ha, and as an area of community importance [61]. The site contains eleven natural habitats from Annex I of the Habitats Directive [62]. It is home to many unique habitat types and plant and animal species. At the Białka River, one can observe the region's largest resources of riverside habitats, rare on a national scale, linked to natural mountain rivers, e.g., German tamarisk thickets (Myricaria germanica) on stony river beds and willow thickets (Salix eleagnos). In the area of the Przełom Białki reserve in the vicinity of Krempachy, 457 species of vascular plants have been declared [59]. The fish that live here include brown trout (Salmo trutta m. fario) and barb (Barbus petenyi), while, on land, one can find deer (Cervus elaphus elaphus), wild boar (Sus scrofa), wolf (Canis lupus), European viper (Vipera berus), or bear (Ursus arctos) [63]. The Białka Valley also forms a significant ecological corridor on the north–south line, connecting the Tatra Mountains with the Gorce and Pieniny Mountains. The area of the Polish part of the Białka River Basin is approximately 123.27 km^2, which is 55% of the total area of the basin of around 225.33 km^2 [56] and includes, among others, the areas of the Tatra National Park, involving the whole area of the northern slopes of the High Tatras. The basin has varied physical and geographical locations. It is located within seven mesoregions. Listed from the south side, it belongs to the High Tatras, the Reglowe Tatras, the Podtatrzańska Bruzda, the Podtatrzański Foothills, the Magura Spiska, the Pieniny Mountains, and the Orawsko-Nowotarska Valley [64,65]. Various types of land cover can be observed in the basin. There are [66] anthropogenic areas, agricultural areas, forests, semi-natural ecosystems, and water areas (Table 2). The largest area in the basin is covered by coniferous forests, followed by meadows and pastures, arable land beyond the reach of irrigation equipment, exposed rocks, and land mainly occupied by agriculture, with a large share of natural vegetation. Other land cover classes account for less than 10% of the total catchment area. In the southern part of the catchment area, there are primarily forest and rocky areas (Figure 2). In contrast, the central part is dominated by agricultural areas, with discontinuous urban fabric along the Białka River and its tributaries. The northern part of the basin catchment area, covering a narrow strip of the river valley up to the river mouth, is mainly covered by forest and arable land.

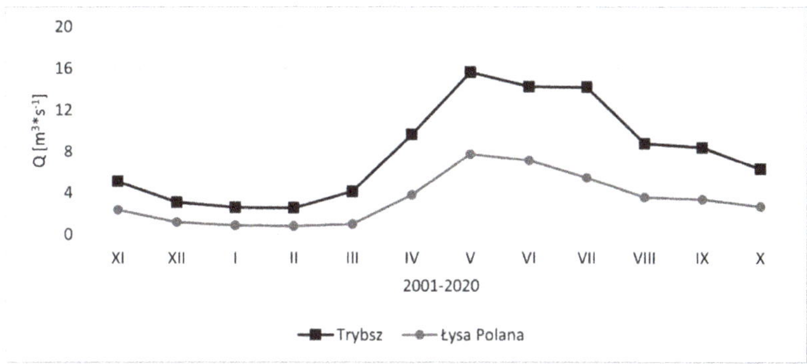

Figure 1. Average monthly water flow in 2001–2020.

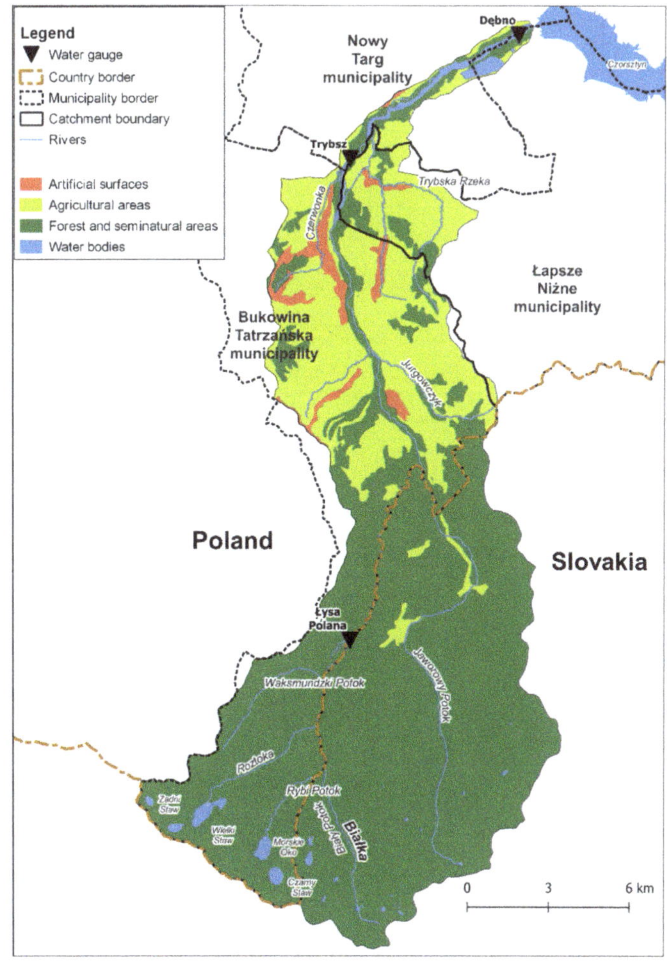

Figure 2. Study area based on Corine Land Cover [66].

Table 1. The Białka River flow.

2001–2020	σ	Min	$Q_{25\%}$	Me	Mean	Max	$Q_{75\%}$	Cv
				m^3/s				%
Łysa Polana	4.2	0.3	1.0	2.1	3.3	67.4	4.1	138
Trybsz	10.9	0.9	2.9	5.0	7.9	216.0	8.8	126

Table 2. Land cover classes in the Białka River Basin [66].

Type of Land Cover (Level 1)	Area	Area
	km^2	%
Artificial surfaces	5.9	4.8
Agricultural areas	52.2	42.3
Forest and semi-natural areas	62.1	50.4
Water bodies	3.1	2.5
Total	123.3	100

3. Materials and Methods

The study was carried out in two stages. In the first one, an inventory of the issues of the Białka River Basin was made based on a query of available knowledge—scientific publications, press reports, and information posted on the websites of public institutions (Table 3). The Statistics Poland (SP) data on population, tourism, nature protection, and water and sewage management from the years 2014–2020 [67] were used. Land cover classes were determined using the CORINE Land Cover 2018 database [66]. The database containing information on the water-legal permits in the selected river basin was obtained from the National Water Holding Polish Waters. Based on the available materials (literature, analysis of materials published by local media), the issues in the basin in the context of water management were reviewed. For the hydrological characterization of the area, data from the IMGW-PIB for two water gauges—Łysa Polana and Trybsz—were used. The analysis of the current legal situation regarding water management in the Białka River Basin was carried out on the basis of the Water Law Act [52].

In the second stage, a survey was conducted on water management in the Białka River Basin. The study took place from 15 June to 16 July 2021. The questionnaire was addressed to residents, entrepreneurs, and other people associated with the Białka River Basin. The survey was distributed via e-mail and social media (it was made available through the Facebook portal by a person associated with the catchment area). The sources of contact information were: municipal websites, schools, fire brigades, national parks, and other local government organizations. The booking.com portal and the Google search engine were also used. The online query focused on identifying small guesthouses and agrotourism, with the assumption that a significant portion of the people employed there were local residents. The questionnaire consisted of single and multiple-choice questions and, in selected cases, also included the possibility for the respondent's own answers. A total of 22 questions were formulated.

The first part of the questionnaire concerned the respondent's characteristics: age, gender, education, commune of residence, and the type of relationship with the Białka River Basin. In the next part, attempts were made to identify society's awareness related to the river basin in terms of water management. Questions were formulated based on the issues of water management, sources of information on this subject, knowledge of planning documents, institutions responsible for water management, forms of nature protection in the river basin, and the functioning of the river basin over time. The next part aimed to identify the perception of water-related issues in the river basin. The respondents were asked to select from a list of problems their causes and give them a rank. They were allowed to formulate their own statements on selected questions. The last part of the survey was to identify the potential of public involvement in water management. The questions concerned the interest of stakeholder groups in caring for the quality and quantity of water in the river basin, the respondent's direct interest in involvement in the water management process, and the willingness of both individuals and entrepreneurs to cooperate.

Table 3. Categories of identified issues with data sources division.

Category	The Type of Source	References
All	Survey research	-
Urbanization	Database	[66]
Tourism developmnent	Scientific publication	[68–73]
	Website	[74–76]
	Database	[67]
Water pollution from point sources	Scientific publication	[77–80]
	Other	[81]
	Website	[82–89]
	Databese	[67]
Use of water	Legal act	[52]
	Water-legal permits	-
	IMGW-PIB hydrological database	-
Flood risk	Scientific publication	[73,90]
	Other	[91,92]
	Website	[63,93–108]
Context of climate change	Scientific publication	[6,109–116]
Impact of Polish legal forms	Legal act	[52,117]
	Website	[118]

4. Results

The Białka River Basin is one of the most valuable landscape systems in the Polish Carpathians. It is characterized by high natural value—despite the settlement, it has preserved its natural mountainous character [68]. Due to environmental conditions and high economic activity, it is a zone where various issues and interest groups related to water resource management intertwine.

4.1. Tourism Development

One of the main aspects of the economic development of the region is tourism. The development of tourism in the Białka River Basin has a long history. The high natural value of the valley and the surrounding area makes these lands very attractive for tourists. The first ski lifts in Białka Tatrzańska were built in the early 1960s. In the following years, the ski infrastructure was expanded [69]. There are many ski resorts in the catchment area, including Jurgów Ski, Kotelnica Białczańska, Kozieniec Ski, and two large thermal baths—Termy Bania. In 2020, there were 97 tourist accommodation facilities here, including hotels, guesthouses, private accommodation, agrotourism lodgings, holiday centers, and hostels [67]. These facilities had a total of 5384 bed places. According to research carried out in the Tatra National Park, around 30% of the total touristic traffic in the Polish Tatras is concentrated on the route from Palenica Białczańska to Morskie Oko. In August 2009, on average, around 20,000 tourists entered the territory of the Tatra National Park every day [68]. Due to strong tourist pressure, there have been numerous attempts at limiting tourist traffic—for example, the online sale of tickets for the parking lot in Palenica Białczańska [74]. The analysis carried out by the service intermediating in booking accommodation—nocowanie.pl [75]—showed that tourists most often chose Małopolska as their place of rest in 2017. Zakopane dominated (44% of inquiries), followed by Białka Tatrzańska (8% of inquiries), with Bukowina Tatrzańska (4%) in the 6th position. On the other hand, during the winter holidays in 2019, Zakopane was chosen most often (13.5% of inquiries) in the entire country [76], while Białka Tatrzańska was ranked 4th (5.5% of inquiries) among the analyzed towns. Białka Tatrzańska is considered to be the fastest-growing town of Podhale, especially regarding the tourist aspect [70]. The development of mass tourism has caused the intense transformation of the natural environment and landscape of Białka Tatrzańska [71]. A large concentration of ski resorts in a relatively small space leads to extensive degradation of the slopes. The creation of new service facilities negatively impacts the environment and causes the devastation of the traditional aesthetics of the Białka Tatrzańska area [72]. Building on the valley floor also results in the disappearance of ecological corridors, limiting the free migration of animals in the catchment area. The annual influx of tourists during the summer and winter holidays is crucial for water management. Enlarging urbanization could be the cause of the increasing flood risk—impermeable surfaces such as roofs, roads, and parking lots are the cause of increasing runoff [73]. The well-developed tourist base is associated with the abstraction of water for economic purposes, including the snowing of slopes in winter [68]. The increase in the number of people in the tourist season (which now includes both winter and summer) combined with poorly developed infrastructure results in the strong pollution of the river waters. Selected issues related to the tourism are described in Sections 4.2 and 4.3.

4.2. Water Pollution from Point Sources

The increasing pressure resulting from tourism development in the catchment area necessitates water and sewage infrastructure development. In the territory of the Bukowina Tatrzańska community, the number of industrial and sewage treatment plants has doubled in recent years. In 2020, there were six such treatment plants [67]. Their capacity in 2014–2020 increased more than three times (Table 4). The number of residents who benefit from the treatment plant is also gradually increasing. The number of septic tanks (more than five times) and household sewage treatment plants (more than four times) has increased significantly in recent years. The municipality has also systematically recorded an increase in water consumption since 2014 (except for 2018 and 2020, when a decrease was noted compared to the previous year). A similar trend also occurs in the share of industry in overall water consumption [67]. Water quality monitoring is carried out by the Regional Inspectorate for Environmental Protection (WIOŚ), which has two measurement points on the Białka River—in Łysa Polana and Dębno (Białka estuary to the Czorsztyn reservoir) [81]. Information contained in the Classification and Assessment of the Condition of Surface Waters Bodies in 2019 (analysis based on data from 2014 to 2019) shows that the Białka River at the tested measurement points is characterized by a moderate ecological and good chemical state. The overall assessment of the surface water body indicates poor water conditions.

Table 4. Water and sewage management in the Bukowina Tatrzańska commune based on Statistics Poland [67].

Statistics Poland Data Subgroup	Unit	2014	2015	2016	2017	2018	2019	2020
Number of industrial and municipal sewage treatment plants	Number of units	3	3	4	4	4	6	6
Number of household sewage treatment plants		7	7	30	30	30	30	no data
Number of septic tanks		540	540	540	540	2804	2765	no data
People using the sewage treatment plants	people	6624	6690	7203	7453	7691	7713	7855
Capacity of industrial and municipal sewage treatment plants	m^3/d	835	1790	2150	2150	2150	2410	2410
Water consumption for the needs of the national economy and population during the year	dam^3	840.3	872.5	1001.3	2041.2	1152.9	1309.4	1021.2
Industry share in water consumption	%	64.1	65.8	66.9	83.2	69.0	72.2	70.3

Despite the measures taken, the intensive development of tourism adversely affects water quality, especially in terms of microbiology [77,78]. Along the course of the river, the impact of point and diffuse sources of water pollution increases. Moreover, sewage treatment in plants has low efficiency [79]. The water is treated mainly in terms of physicochemical indicators, while bacteriological contamination and high concentrations of antibiotics are observed below the discharges of treated sewage [80]. This problem is also the subject of public discussion in the media, which appears in the context of extreme events or the lack of an adequate sewage system [82–84]. There are reports in the local media about the contamination of the Białka River in the area of Białka Tatrzańska, where there is a lack of sewage, especially during peak tourism periods. The local government and the inhabitants of Białka Tatrzańska have drawn attention to the growing problem of illegal sewage discharge by, among others, owners of small- and medium-sized guesthouses or small farms with rooms for rent. On the other hand, the construction of the sewage system causes resistance from the owners of the plots through which the pipeline would run, which, in 2014, led to the blocking of local government activities leading to the sewerage of Białka Tatrzańska [85,86]. Another cause of water quality problems in the Białka River, which has been noticed by residents, tourists, and pro-environmental organizations, is the inefficient operation of the sewage treatment plant in Czarna Góra [87–89].

4.3. Use of Water in the Białka River Basin

One of the tools for managing water resources in Poland is water-legal permits—they are necessary to obtain a water law approval, which is a type of administrative decision authorizing the use of water or affecting the water environment [52]. Data on water-legal permits have been collected in the PGW WP databases since 2018. The current resources are not yet complete and require further arrangement with regard to the new provisions of the Water Law Act. In June 2021, information was obtained from PGW WP on all applicable water permits in the Białka River catchment area. The shared raw database was then prepared and developed for further work. As a result of the selection, those water-legal permits were rejected that did not apply to any water activity, water services, or use in the catchment area, but were only corrections, instructions, remissions, etc. A total of 203 water-legal permits were finally used for further analyses. They all have the status of writing as up to date. For 83 water permits (41%), water users are private persons, while the remaining are private and public enterprises, companies, and local governments. Water permits in the Białka catchment area were obtained for various types of activities, water services, or water use. They concern such categories as (some permits fall into more than one category): regulations (6), power plants (2), surface intakes (6), protection zones of surface intakes (1), groundwater intakes (62), protection zones of intakes underground (26), sewage treatment plants (10), wastewater discharges (62), pre-treatment facilities (2), crossing by watercourses (44), periodic surface intakes (1), fish ponds (1). A total of 38 water permits were issued between 1998 and 2010 and are still valid today. In the last decade (2011–2020), 165 such permits were issued, the most (20 and more permits) in 2016, 2017, 2019, and 2020. There was also a sharp increase in wastewater discharge permits granted between 2011 and 2020. During this period, 59 of the 62 permits currently in force were issued. Table 5 shows the general distribution of water-legal permits granted and in selected categories in 1998–2020.

According to the analysis of the database, it should be assumed that at least 32 permits in the detailed description have direct reference to the tourist function (guesthouse, hotel, restaurant, recreational function, ski, sports and recreation resort, etc.). Seven water-legal permits concerning the abstraction of surface waters (including one periodical) apply currently in the area of the Białka River Basin. Four permits were issued to entrepreneurs for snowmaking on ski slopes, while the others were issued for the needs of small hydroelectric power stations and for supplying waterworks. Water intake for skiing purposes exceeds 15,000 m^3 per day (Table 6). The permits provide for the abstraction of water from the Białka River in the following locations: Czarna Góra Grapa, Bukowina Tatrzańska, Białka

Tatrzańska, and Jurgów. Out of all water-legal permits related to wastewater discharge, only 14 were classified as discharges of treated domestic sewage or domestic sewage, with information on the permissible discharge. In total, on the basis of the information contained in the obtained database of the permits in force, their discharge is allowed in the amount of $Q\acute{s}r = 231$ m^3/d (Table 7).

Table 5. Number of water-legal permits issued in 1998–2020 in the Białka River Basin.

Year	Number of Water-Legal Permits Issued (Currently Valid)	Type of Water-Legal Permit	
		Surface Water Abstractions (Including Periodic Intakes)	Sewage Discharges
2020	23	-	4
2019	20	-	7
2018	14	1	6
2017	21	2 (1)	10
2016	21	1	6
2015	19	-	7
2014	17	-	5
2013	10	-	5
2012	11	-	7
2011	9	-	2
2010	7	1	1
2009	1	-	-
2008	7	-	-
2007	3	-	-
2006	7	1	1
2005	5	1	1
2004	2	-	-
2003	1	-	-
2002	3	-	-
2001	1	-	-
2000	0	-	-
1999	0	-	-
1998	1	-	-
TOTAL	203	7 (1)	62

Table 6. Purposes of water abstractions.

Purpose	Acceptable Quantity	
	Q_{mean} m^3/d	Q_{max} m^3/s
Artificial snowmaking of slopes (the Białka River)	16,950	-
Small hydroelectric power stations	-	2.3
Waterworks	no data	

Table 7. Sewage discharges.

Sewage Discharge Site	Acceptable Quantity
	Q_{mean} m^3/d
Czerwonka	67.8
Bryjów Potok	60
Rybi Potok	9.5
Unnamed stream	51.3
To the ground	42.3
Total	230.9

Over 40% of all water-legal permits in the catchment area were issued for Białka Tatrzańska (84 water-legal permits). According to the information obtained, 52% of users are private individuals (44 permits). The issued decisions concern, among others, wastewater discharge (35 permits), underground intakes (27), protection zones of underground intakes (11), exceedances (11), treatment plants (6), and surface intakes (1). According to the data, wastewater (including treated domestic sewage, rainwater, and snowmelt) is most often discharged to the ground, mainly using absorbent wells and sewage outlets to the Czerwonka stream. In the other major towns located in the catchment area (Czarna Góra, Bukowina Tatrzańska, Brzegi, Trybsz, Jurgów), 103 water-legal permits apply in total, out of which 36 permits (35%) are for private individuals. The permits issued in these towns include wastewater discharges (21 permits), exceedances (30), underground intakes (31), surface intakes (4, including one periodical), and sewage treatment plants (2). Wastewater is most often discharged into ditches and streams without names, representing the tributaries of other rivers, and to the ground using absorbent wells. Apart from water permits, which specify the volume of abstraction, discharge, or other environmental effects, information about the volume of water resources and how it changes over time in different parts of the catchment area is an integral part of water resource management. Currently, apart from two water gauges on the Białka River, the Institute of Meteorology and Water Management—National Research has a water gauge on the Morskie Oko Lake. Between 1967 and 1979, there was also a water gauge on the Wielki Staw Lake (Roztoka) in the Valley of the Five Polish Ponds (Table 8).

Table 8. IMGW-PIB measurement network in the Białka River Basin.

Type	Gauge Name	Date of Starting the Measurements	Date of Termination of the Station
Lake	Wielki Staw (Roztoka)	1967	1979
Lake	Morskie Oko	1951	Active
River	Białka	1917	Active
River	Białka	1994	Active
River	Młynówka (Białka)	1942	2000

4.4. Flood Risk

Floods in the Białka River Basin are a permanent manifestation of its hydrological regime [73]. This is related to the mountainous nature of the catchment area—the diverse morphology of the terrain and the significant height differences, which, combined with high rainfall, often of a torrential nature, result in the rapid development of floods, especially in the southern part of the catchment area. In addition, the spatial distribution of the population is uneven and concentrated in the central and northern parts of the catchment area, in the valley axis—along rivers and streams. This is another characteristic of this area that contributes to the development of floods, threatening the health and lives of residents. The first mentions of floods in Podhale date back to the early 19th century. The floods in 1934, 1970, 1997, and 2010 are considered extremely catastrophic events [90]. According to the inhabitants of the Białka catchment area [91], starting from 1997 and in the years 2001, 2002, 2004, 2005, 2008, 2010, 2014, and 2018, as a result of the floods, numerous roads, bridges, power transmission lines, and water supply systems were destroyed; moreover, drinking water intakes, sports fields, and houses were flooded. In 1965–2008, the bridge over the Białka River between Nowa Biała and Krempachy was broken three times. Flood losses between 2007 and 2012 in the community of Bukowina Tatrzańska amounted to over PLN 10 million [92]. The issue of flood risk, particularly the regulation of the Białka river bed, is the subject of a conflict between stakeholders representing the local and regional authorities, the inhabitants of the area, and the Regional Directorate for Environmental Protection in Cracow. This issue is reflected in the local media publications, similarly as in the case of water pollution [63,93–105]. At the lowest level, the inhabitants of the area demand effective regulation of the river bed through the construction of embankments and weirs and consent to the deepening and thus narrowing of the river bed, also in the area of a Natura 2000 nature reserve. Representatives of the local administration, such as the major of Nowy Targ, present a similar position. At the regional level, there is a conflict between the above stakeholders, the conservator of nature protection, and environmental organizations, which indicate that the proposed solutions are short-term and do not consider the nature of the Białka River [106]. In such circumstances, conflicts escalate, including protests. Flood risk is currently under discussion between the public administration and the local community [107]. Representatives of the local administration (commune heads, district governors, village chiefs), as part of consultations on regulatory works carried out on the Białka River in 2018, indicated that the protection of residents against the effects of floods is one of the management priorities in districts, communes, and towns. According to the information posted on the National Water Holding Polish Waters (PGW WP) website, in 2009, the first attempt was made to implement a flood protection project, taking into account the protection of the unique nature of the Białka River environment [108]. However, the proposed solutions met with the reluctance of the local community. Work is currently underway on a multidirectional flood protection project from the border of the Tatra National Park to the mouth of the Białka River to the Czorsztyn Reservoir. The Program for Białka is intended to protect valuable habitats and ensure corridors of free migration for aquatic and water-dependent organisms.

4.5. The Białka River Basin in the Context of Climate Change

When analyzing climate change, we will use two projects' results based on scenario analyses that form its basis. The scenarios relate to the specific emission of gases that retain solar energy in the Earth's atmosphere, known as greenhouse gases, and contribute to positive radiative forcing. In such a situation, there is an increase in energy absorbed by the climate system, which leads to climate warming. The scenarios included in the Special Report of Emission Scenarios (SRES) [109] are defined by families (A1, A2, B1, and B2) designed on the basis of a set of consistent assumptions. In turn, the Representative Concentration Pathways (RCP) scenarios [6] are classified according to the change in radiative forcing (+2.6 to +8.5 W/m^2) that will occur by 2100. The four proposed scenarios are called RCP2.6, RCP4.5, RCP6.0, and RCP8.5. They estimate the approximate radiative forcing in 2100 against 1750. The year 1750 is the reference year as the conventional end of the pre-industrial era. In the second half of the 21st century, the SRES A2 has a similar trajectory to RCP8.5. Both trajectories will reach around 8 W/m^2 by 2100. SRES A2 is also similar to RCP8.5 in terms of changes in mean global temperature [110].

Taking into account the described similarity of the A2 scenario to the RCP8.5 scenario, to illustrate the changes in water resources, the figures below show the percentage change in the mean annual flow and mean seasonal flows in rivers between 2071 and 2100 compared to the period 1961–1990 for the A2 scenario [110,111]. For the Tatra Mountains and Podhale region, the forecasted flow change is within the range of -5 to 5% in terms of mean annual flows (Figure 3).

In turn, a detailed analysis of the results of the Chase-PL project shows that, in the south of Poland, the number of hot days has increased, and winters are becoming milder—the number of very cold and extremely cold days is decreasing. In high mountain areas (such as the southern part of the Białka River Basin), a decrease in annual precipitation is observed, while, in the foreland of the Tatra Mountains, the opposite trend is observed. In the Podhale region, the ratio of winter precipitation to summer precipitation changes [112]. Both in the station located in the high mountain region and in the foreground of the Tatra Mountains, the mean seasonal precipitation decreases in winter and summer, while, in spring and autumn, it increases. A particularly unfavorable phenomenon is the change in the type of precipitation from snow to rain in winter. As a result, the water retained by plants during the summer quickly escapes beyond the catchment area. Additionally, the lack of snow cover may reduce the retention capacity of the soil as a result of soil freezing. The climate scenarios based on the radiative forcing in the RCP 4.5 and 8.5 variants for the time horizons 2021–2050 and 2071–2100 for the empirical–statistical downscaling (ESD) and dynamic downscaling (DD) models clearly forecast an increase in air temperature [113]. The DD projections indicate that the mountainous areas of Southern Poland will be exposed over a longer time horizon to temperatures above 2 °C per year in the RCP 8.5 scenario. The most significant change will take place for the winter months (both RCP 8.5 and 4.5). According to the DD method, precipitation in both time horizons and emission scenarios will increase from a few to several percent per year (Table 9). The highest increases are forecast for the winter months. In the case of ESD projections for the Podhale region, a decrease or a slight increase in annual precipitation is forecast. In winter and autumn, a weaker decrease for the RCP 4.5 scenario and stronger for RCP 8.5 (multi-year 2021–2050) is expected. The decrease in precipitation total is carried over to the spring and summer months for a longer time horizon. In addition, the authors of the study point out that the appearing divergences in the projection results for various types of models introduce high uncertainty of the obtained results. The results of the SWAT model for RCP 4.5 and 8.5 in the time horizons 2024–2050 and 2074–2100 for the value of river outflow do not show statistical significance of the projected changes [114]. Climate change will also affect aquatic organisms in the Białka River. Okruszko et al. [115] analyzed three groups of fish species: settled, partially migratory, and migratory. For the first group, the impact of the change on the rivers in the Podhale region will be medium for both RCP 4.5 and RCP 8.5 in the near and far future. Similarly, for the second and third groups, there will be slight deviations towards high and low impact. The IHA index [116] was used for the analyses, including identifying changes in parameters affecting fish habitat conditions. These included features such as water temperature, flow velocity, and vegetation changes.

4.6. Legal Forms—Polish Water Law in the Context of the Stakeholders' Participation

The current legal framework for water management in Poland is defined in the new Water Law Act, which entered into force in 2018 [52]. According to its content, water management in Poland should be carried out according to the principle of sustainable development, particularly regarding the development and protection of water resources, water use, and water resource management, including a wide and open process of consultation. For water management purposes, a system of management units with different spatial resolutions has been established. At the highest level of generality, the Białka River Basin is classified as the Vistula river basin, then to the Upper-Western Vistula water region. In the current text of the Act, the minister responsible for water management, the President of Polish Waters, the directors of the regional water management board in Cracow of Polish Waters, the management of the water catchment in Nowy Sącz, the head of the Water Supervision Zakopane, the Małopolska voivode, the governor of the Tatra and Nowotary district, and commune mayors (Bukowina Tatrzańska, Nowy Targ, Łapsze Niżne) are mentioned as the governing bodies in the context of the Białka catchment area [118]. It is understood through the principle of common interests based on the cooperation of various stakeholders groups to obtain the maximum benefit with minimum environmental costs. According to this Act, stakeholder participation at the national level of responsibility is assured by the State Council for Water Management. For a specific case, the President of Polish Waters also has the opportunity to appoint a consultative team consisting of experts and representatives of the public administration. However, the Act does not further define which rights such teams would have. The new Act has a direct impact on the management of water resources at lower levels of responsibility, through the limited possibilities of stakeholder participation. The previous water act established "water region councils", whose members were stakeholders representing economic, agricultural, fisheries, and community organizations and representatives of water users [117]. The current Act has abolished such councils.

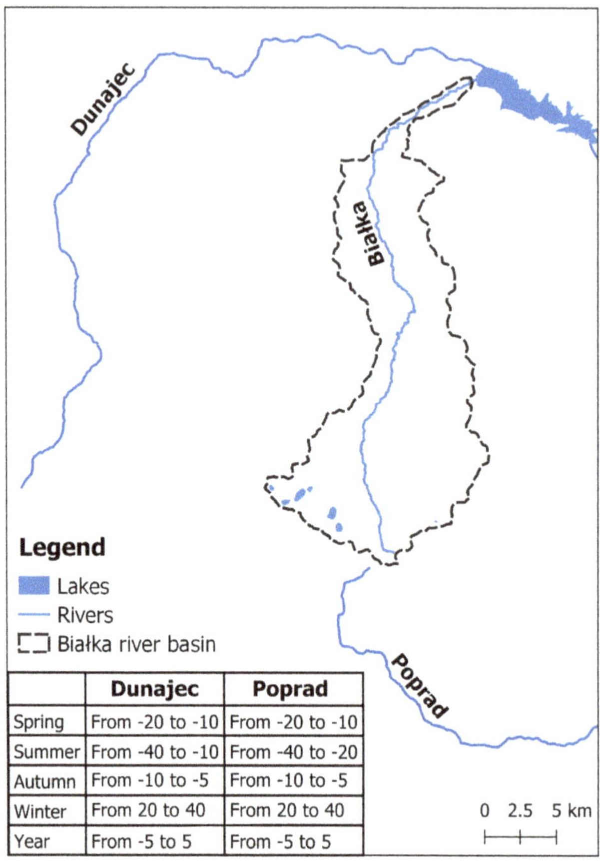

	Dunajec	Poprad
Spring	From -20 to -10	From -20 to -10
Summer	From -40 to -10	From -40 to -20
Autumn	From -10 to -5	From -10 to -5
Winter	From 20 to 40	From 20 to 40
Year	From -5 to 5	From -5 to 5

Figure 3. Projected changes (%) in the annual flow for the time horizon 2071–2100 in relation to the multi-year period 1961–1990 based on Majewski and Walczykiewicz [111].

Table 9. Projected changes in annual precipitation for the Podhale region—modeled by empirical–statistical downscaling (ESD) and dynamic downscaling (DD) based on Mezghani et al. [113].

Downscaling Method		RCP Scenario	Range of Changes (%)
DD	2021–2050 2071–2100	4.5	from 0 to 5 from 5 to 10
	2021–2050 2071–2100	8.5	from 0 to 5 from 5 to 10
ESD	2021–2050 2071–2100	4.5	from 0 to 5 from 0 to 5
	2021–2050 2071–2100	8.5	from −5 to 0 from 0 to 5

4.7. Survey Results

A total of 371 respondents took part in the survey. People with a Master's degree and secondary education prevailed. This trend was maintained for all age groups (except for "up to 18"). The older the age category, the fewer people participated in the study. In total, 63.6% of the respondents came from the Nowy Targ commune, 17.3% from Bukowina Tatrzańska, and 12.4% from Łapsze Niżne. The remaining 7.0% were respondents from other communes. A significant fraction of the respondents—85.2%—declared that they were associated with the Białka River Basin as a resident, and approximately 14% were entrepreneurs (Table 10). Other responses did not exceed 10%.

Table 10. Links between respondents and the Białka River Basin.

Type of Connection	Number of Answers	%
I live here	316	85.2
I run a business	53	14.3
I work in a non-governmental organization (NGO)/association	21	5.7
I work in public administration	28	7.5
Tourism	18	4.9
Other	14	3.8

4.7.1. Knowledge of Water Management and the Environment in the Białka River Basin

More than 80% of respondents indicated internet portals as the primary source of knowledge on water management. Radio and television were ranked lower—37%—while the remaining sources did not exceed 35% of all respondents. In their own responses, the respondents most often cited their experience and observation of the environment as the source of knowledge (20 responses—5.3% of all respondents). The survey respondents' knowledge of planning documents was relatively low—63.6% answered that they were not familiar with the planning documents related to water management (Figure 4). Awareness of documents did not exceed 30% of all respondents. Understanding the definition of water management was different—almost 60% of the respondents answered this question correctly, choosing the answer based on the definition of water management from the Water Law Act [52].

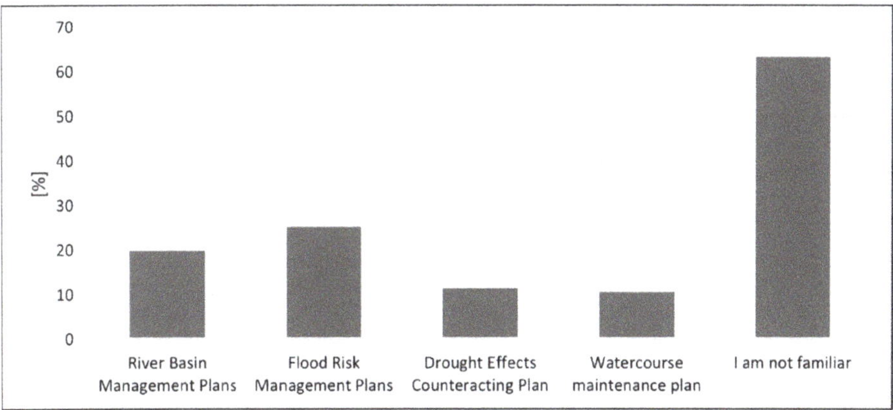

Figure 4. Knowledge of planning documents.

The high percentage (20.5% of respondents) of responses identifying the regulation of rivers and streams with the process of water management is also puzzling. The respondents noticed the variability of the water flow in the Białka River Basin during the year. Overall, 74.1% of people believed that the quantitative status of the waters in the basin had deteriorated. The answer to the question regarding what forms of nature protection exist in the Białka River Basin was quite surprising. Only 15.1% of people indicated the national park (a significant part of the basin is the Polish and Slovak Tatra National Park).

The vast majority (73.6%) pointed to Natura 2000 sites. Almost half of the respondents (49%) pointed to nature reserves and approximately 7.0% replied that they did not know the forms of nature protection in the Białka River Basin. Approximately 12.0% pointed to a landscape park but, in the Białka River Basin, such a form of nature protection does not exist.

4.7.2. Issues in the Białka River Basin

The survey included two questions on the issues surrounding the Białka River Basin and their causes. The number of responses to the questions varied from 330 to 361 for the question concerning the problems and from 288 to 342 for the question related to the causes of the issues. More than one answer was possible for the question, but few respondents made such a choice—in Tables 9 and 10, they are marked as "other". When asked about the issues in the Białka River Basin, the respondents indicated the flood risk and poor quality of water in rivers and streams as huge problems. In both cases, the responses amounted to more than 50% (Table 11). The most frequently cited cause of flood risk issues was the variability of the course of the river bed throughout the year. Subsequently, building on floodplains was mentioned as the cause (Table 12). Although water shortage was not the most crucial problem, many respondents believed that excessive water abstraction for tourist purposes, artificial snowmaking of slopes, and a lack of precipitation were significant causes of issues in the Białka River Basin. More than 40% of respondents identified this issue as very important. For the problem of poor water quality in rivers and streams, the respondents' highlighted the following as essential reasons: discharge of untreated sewage to surface water bodies, lack of sewage system, leaky septic tanks, excessive discharge of untreated sewage during the tourist season. For all four of these reasons, the answer "very important" was chosen by more than 60% of the respondents.

Table 11. Issues related to water in the Białka River Basin.

Scale of the Issue	Flood Risk (n = 361)	Water Shortage (n = 330)	Poor Water Quality in Rivers and Streams (n = 347)
		%	
Huge issue	53.7	14.8	56.2
Big issue	19.7	20.3	25.4
Moderate issue	18.0	37.3	12.7
Very small issue	3.6	15.2	3.2
There is no such issue	2.2	9.4	1.4
Other	2.8	3	1.2

Both questions enabled respondents to enter their own statements. Since the first question concerned the identification of issues in the Białka River Basin, and the second concerned the causes of these issues, respondents gave similar answers to both questions in the "other" category. Therefore, it was decided to group thematically the answers assigned to this category for both questions together. The result of grouping responses was to create eight categories of issues, the order of which was determined based on the most frequent responses among respondents. The results of this analysis are presented in Table 13. The comments included in the "other" category show that "poor water quality" (33.0%) is the most significant issue in the Białka River Basin. Among the other responses of respondents were "riverbed regulations" (18.0%), "formal and legal conditions, education" (14.0%), "digging up gravel and stones from the river bed " (10), and "flood risk" (9.0%). Problems related to "protected areas" (6.0%) and "water abstraction" (2.0%) were the least relevant, according to respondents.

Table 12. Causes of water issues in the Baiłka River Basin.

Cause	Total Number of Responses	Very Important	Important	Moderately Important	Unimportant	Negligible	I Don't Know	Other
	n				**%**			
Building on floodplains	288	26.4	28.8	18.1	14.2	8.3	2.8	1.3
Variability of the course of the riverbed during the year	342	52.3	21.6	13.5	7.6	2	0.9	2.1
Excessive water abstraction for tourist purposes	305	36.7	18.0	19.0	16.7	8.2	1.0	0.3
Artificial snowmaking of slopes	314	44.9	16.2	17.5	12.7	7.0	1.3	0.3
Discharge of untreated sewage to surface water	328	76.2	13.4	4.3	1.2	1.2	3.0	0.6
No precipitation, drought	297	23.6	30.0	24.6	13.1	5.7	1.7	1.2
No sewage system	333	76.6	12.0	3.6	2.1	1.2	3.6	0.9
Leaky septic tanks	320	64.7	15.9	7.2	3.8	3.4	4.4	0.6
No water supply network	295	41.4	18.0	16.9	9.2	9.5	4.1	1.0
Excessive discharge of sewage during the tourist season	336	76.8	8.6	4.8	2.4	2.1	4.8	0.6

4.7.3. Social Potential

When asked about the local community's interest in maintaining the quality and quantity of water in the Białka River Basin, 36% of respondents said that the local community was very interested in such activities, and 21% stated that this was moderately the case. The remaining 43% responded "Yes, slightly" (25%), "No" (17%), and "I don't know" (1%). In terms of the question regarding the selection of social groups that should be involved in the water management process, representatives of the residents were dominant (88%). Other social groups questioned, i.e., "entrepreneurs" and "representatives of organizations and associations related to nature protection", received 55% and 50% of responses, respectively. When asked about their willingness to cooperate in water management, 53% of respondents expressed their willingness to get involved, 17% stated that they would not like to get involved, while the rest did not have an opinion. In terms of the activities that the respondents wished to undertake, the most dominant was participation in the identification of water quality endangering sites (53%). Meanwhile, 20% wished to be involved in co-organizing training and workshops related to water management, while a similar number wished to help in carrying out measurements (23%), while assistance in the preparation of training and promotional materials was the least popular option.

The final part of the survey concerned entrepreneurs. The number of people who defined themselves as entrepreneurs (110 responses) differed from those who introduced themselves as a person running a business in the river basin at the beginning of the survey (53 people). The first question aimed to identify the possibility of incurring the costs of maintaining the measurement and observation infrastructure to support business activity. The number of affirmative responses was 50%, 21% were not interested in maintaining the measuring devices, and 29% had no opinion. Moreover, among entrepreneurs, 14% answered that they maintained the measurement infrastructure, while 50% of the respondents indicated that they would be willing to incur such costs. The last issue concerned establishing a hierarchy of water users—41% of the responses agreed to the creation of such a structure, while 36% disagreed, and others did not have an opinion.

Table 13. Respondents' own statements.

Issue Category	Description of the Category
Poor water quality	Direct and indirect responses related to poor water quality: • information on discharge of pollutants to surface water/water bodies from tourist centers in Białka Tatrzańska, Bukowina Tatrzańska, and Czarna Góra. • high-temperature water discharge from thermal baths. • illegal discharge of pollutants by residents. • information relating to the broadly understood contamination of the river basin, e.g., garbage and debris removal to the river, landfills along the riverbed. • no sewage system in Białka Tatrzańska.
Riverbed regulation	Responses concerning the issues related to the lack of regulation of the Białka River: • information on the lack of reinforcements of the banks and the river bed in critical places, i.e., in the sections of the river's regular flooding during floods.
Formal and legal conditions, education	Responses on formal, legal, and educational aspects related to the proper understanding of water management: • information on the lack of solutions regarding charges for land taken up by the flood. • improper river basin management, lack of knowledge of water retention, no compromise between the needs of the environment and local communities, improperly functioning administration. • lack of cooperation between the river basin managers and the inhabitants of towns threatened by the river, improper management of financial resources.
Digging up gravel and stones from the river bed	Responses to the illegal digging up of gravel and stones from the Białka river bed and the resulting problems: rising of the river bed due to the accumulation of transported rock material in the downstream river, thus contributing to flooding.
Flood risk	Responses concerning: • information on the frequency of floods and related damage. • activities related to reducing of flood risk, the impact of an anastomosing river on the flood risk.
Other	Responses of respondents that could not be assigned to any of the other categories. The statements were in the form of deliberations, assessments, or statements: • "Why and with whom the Slovaks agreed that they cut off one tributary of this stream" • "Rushing flow" "The problem is people who have nothing to do with our region and are fighting against the regulation of the river, allegedly ... ".

5. Discussion

The results of the analyses show that the Białka River Basin requires a multifaceted and comprehensive approach to water management. It is an area of interest for many groups of stakeholders due to the region's high natural value. It is a place of residence for many people and, additionally, a zone of development of intensive tourism, both summer and winter. This phenomenon is confirmed by the significant number of water-legal permits granted to tourist facilities. Due to the valuable mountain and river ecosystems in the catchment area, there are many forms of nature protection, including water-oriented ones, which remain an integral part of the water management process in this area. The Białka River, because of its mountainous nature—high flow dynamics and the flood-like nature of the river—affects various aspects of the functioning of the communities associated with the catchment area. Due to the fact that the Białka River Basin is a transboundary area, it may require cooperation between the Polish and the Slovak governments in the future. As Hussein [119] points out, transboundary issues in the context of water resources are a common problem in regions where there are frequent water shortages, and the group of stakeholders at the country level is large. However, the current situation in the management of the Białka River's resources is not of strategic importance for these countries and is not the subject of conflict.

5.1. Water Issues in the Białka River Basin

Analyses of the obtained questionnaire surveys, the information contained in the literature on the subject, and the media indicated that the main risks related to water management in the catchment area are the phenomena of flooding and water pollution. The causes of these problems were most often indicated as inadequately organized water and sewage management and, in the event of flood risk, the variability of the course of the river bed. Much less frequently, the respondents indicated the development of floodplains. The tourist significance of the region means that stakeholders related to the accommodation network, ski slopes, and restaurants have and will have strategic importance in water management in the Białka River Basin. Anthropopressure related to tourism has a negative impact on the quality of the water in the area. Due to the region's attractiveness, it is burdened with a year-round influx of tourists, which also results in an increased supply of sewage to the waters of both Białka and its tributaries. According to the permits, there are sites of discharge of domestic, treated domestic, rainfall, and snowmelt sewage in the catchment area. They are discharged into rivers, streams, or with absorbent wells to the ground. The occurrence of negative phenomena related to inefficient water and sewage management is confirmed by research conducted by A. Lenart Boroń and her team [77–80] in the field of microbiology. They indicate that the water from the treatment plant itself is not sufficiently purified. These studies also confirm the observations of the inhabitants of the Białka River Basin expressed in the survey, as well as the information available in the media. However, it should be remembered that this is a subjective assessment, often of people not involved in water management. The activities of entrepreneurs in the tourism industry now also generate increased water abstraction throughout the catchment area. Abstraction for snowmaking of the ski slopes in Białka Tatrzańska, Bukowina Tatrzańska, Jurgów, and Czarna Góra is becoming particularly significant. Based on the data from the permits, it appears that they can collect in total approximately 0.2 m^3/s (while the minimum flow in the multi-year 2001–2020 in Trybsz is 0.9 L/s—taking into account that this measurement is burdened with the abstractions mentioned above). As in the case of water quality, the location of IMGW-PIB measurement points does not allow for the monitoring of water abstraction because the distance between the stations is too large. However, it should be remembered that the Water Law in pp. 316–320 requires [52] that the method of monitoring the quantitative and qualitative parameters of the abstracted water is specified in the water law consent. Nonetheless, there is no publicly available database that contains data on water abstraction and the quality and quantity of discharged sewage. As a result, there is no information flow in this regard between the stakeholders. As a consequence, those who abstract water for snowmaking also do not know its quality parameters.

Due to the increase in the number of permits issued for the abstraction of surface waters (four permits have been issued for snowmaking of ski slopes since 2010—a two-times increase compared to previous years), the uncertainty of the results obtained from the above-mentioned measurement networks is also increasing. Climate change analyses do not clearly indicate an increase or decrease in water resources. In the Białka River Basin, a change in the precipitation structure is expected during the year. It will increase the time uncertainty, which may necessitate appropriate adaptation measures. The climatic scenarios for the southern regions of Poland indicate an increase in rainfall at the expense of snowfall, which may disrupt the river's natural regime and water availability in different seasons of the year. At the same time, the divergent results of climate models in terms of the various components of the water balance indicate that they are fraught with uncertainty, which significantly hinders the development of an appropriate water management strategy. Many authors point out that the uncertainty of climate projections is an important challenge for entities responsible for managing various areas of life, including water resources [120–125]. This uncertainty hampers policy-making, reduces public and management confidence in research results, and may increase investment costs. Moreover, according to Refsgaard et al. [126], uncertainty is a major obstacle to the efficient management of water resources and a key obstacle to the development of adaptation measures.

5.2. Barriers to Water Management in the Białka River Basin

5.2.1. Involving Stakeholders

It should be emphasized that even the full availability of data and the full ability to forecast changes in the aquatic environment are not enough to ensure the correctness of the water management process. It requires a socialization aspect of the decision-making process by ensuring stakeholder participation. This is a challenging task that requires involved stakeholders to influence different elements of the decision-making process in line with the common objective of adequate water management, taking into account needs, uncertainties, and conflicts [127,128]. It is crucial to move from imposing optimal solutions developed by experts to supporting experts in developing appropriate solutions considering stakeholders [129]. Public consultation in Sweden showed that the public was interested in participating in the water management process [130]. However, non-expert stakeholders indicated that they were overwhelmed by an overload of information and expressed doubts as to whether they should be involved at all phases of the process due to a poor level of knowledge. Moreover, many of them indicated that they would be more willing to engage in the problems of "their own backyard" than the entire catchment area, the impact of which on their lives they do not see. In the Białka River Basin, social potential in terms of willingness to participate in the water management process was also observed as a result of the survey. At least half of the respondents were interested in participating. As with the Swedish consultations, respondents claimed that they would prefer to be involved in local issues such as the localization of sources of pollution, while participatory activities, such as cooperation in the information exchange process, aroused much less interest among them. It is worth noting that, in the studies by Jacobs et al. [131], it is confirmed that positive effects of these activities can be observed, especially in terms of consensus building and conflict resolution in the catchment areas of Mexico, the USA, Brazil, and Thailand, where stakeholders were involved in the water management process. At the same time, the authors note that participatory processes generate huge financial and time costs.

5.2.2. Dialogue and Knowledge

Nevertheless, given the respondents' interest in participation, it can be assumed that implementing this type of management would improve the quality of water management processes in the Białka River Basin. Social participation allows for the development of thoughtful, open solutions that consider the change in knowledge concerning the water system in the future. The need to implement a new management model in the Białka River Basin is also reflected in the low communication assessment between stakeholders. Both the survey results and the literature query indicate that, currently, in the Białka River Basin, there is no proper dialogue between the public administration, stakeholders using water, and the inhabitants of the catchment area. This is reflected in a social sense of poor resource management (both in media reports and the survey, there are accusations that water management is insufficiently conducted). The implementation of a participatory management model would facilitate the flow of information and make stakeholders aware of the difficulties in reconciling all sides of the issues. These communication problems also translate into the level of knowledge. The survey results partially confirm this. Every fifth respondent believed that water management is based on the regulation of rivers and streams. Although the respondents perceived the river's variability throughout the year, they did not know about the forms of nature protection in the catchment area or about planning documents related to water management. Respondents also identified the internet as their main source of knowledge. The PGW WP is responsible for disseminating information on water management, whose principal medium of communication is the website www.wody.gov.pl (accessed on 15 October 2021). By analyzing the respondents' responses, it was found that the existing methods of communication used by public institutions do not meet the needs of society.

A worrying phenomenon is the increasing negative assessment of the Natura 2000 program and legal regulations in environmental protection, particularly regarding flood protection. In many cases, the respondents perceived nature protection as a tool for the unfounded repression of the local inhabitants. The situation is aggravated by the fact that the responsible institutions are unable to implement the assumptions of the environmental protection program, taking into account the needs of the inhabitants of the Białka River Basin (in particular, the Nowa Biała and Krempachy regions). This points to an imperfection in the functioning of the water management process (especially in terms of communication and knowledge transfer), which is intended to ensure the common interest of the whole of society. According to the inhabitants, the strong and fast current of the Białka River, during heavy rainfall, brings huge amounts of rubble, which causes the bed's grade line to rise continuously. Presumably, the grade line has been increased in some places by at least 2 m for 80 years, and thus the Białka river bed, in some sites, reaches 200–250 m wide [91].

In the survey and media reports, people associated with the region indicated that the most appropriate solution is artificially deepening the river bed and building flood embankments. Significantly, such activities would lead to the destruction of the unique character of the Białka River Basin, and, from a broader perspective, it would not improve flood safety. This problem reveals the low level of knowledge of society—in this case with regard to the development of the river bed and natural floodplain terraces. At the same time, a certain contradiction appeared in the results of the survey. Some respondents expressed a desire to deepen the riverbed on their own, while a large group indicated that one of the area's problems is the digging up of gravel and stones from the Białka River's bed. This is defined as an unequivocally negative phenomenon. There are also no legal permits for this type of activity in the catchment area. Flood risk is a sphere where the poor quality of communication plays a very negative role. Residents do not know which public administration bodies they should address. They acquire knowledge about flood protection from uncertain sources, which often do not consider the broader perspective. One example is the perception of an anostomotic river bed as a flood risk factor. This leads to conflicts between the residents, the administration, and organizations dealing with nature protection. Purkey et al. [132] indicate that conflicts of stakeholder needs and different perceptions of reality are premises for implementing a participatory management model. This is one of the more complex elements of the process because reaching a consensus is long and arduous. It is impossible to satisfy all parties to the conflict entirely. Furthermore, developing optimal solutions through social participation requires a willingness to cooperate, thoughtful and reasonable actions, precise and efficient communication, and the building of positive relations based on respect between stakeholders [129]. An essential element of this model is the exchange of knowledge and information, which is most effective through social learning, which is crucial for initiating changes and building and maintaining water management systems' adaptive capacity [133]. Effective social learning leads to new knowledge, a common understanding of the processes taking place in the environment, the transparent exchange of information between stakeholders, and increased trust in the managing authorities [134]. As a result, there is a change in practices and behaviors, the system of values is restructured, institutional changes take place, and the policy is adapted to the needs of the water management process.

5.2.3. Policy

The wide range of stakeholders and barriers and conflicts related to the management of water resources in the Białka River Basin requires the implementation of appropriate operational rules. It is worth noting that despite the low level of knowledge in nature protection forms and planning documents, a significant percentage of respondents were interested in cooperating in the development of the water management process, especially in identifying the risks and issues related to water resources. This potential can be used to implement the IWRM's principles in the studied area. IWRM's objectives are complementary at the national level, river basin level, and sub-basin levels [135]. It is optimal in this respect to achieve such a balance in activities that support the IWRM process from all parties involved. A holistic, integrated objective means that all aspects of water management, soil maintenance, spatial planning, land use, agriculture, transport, urban development, and nature conservation should be considered at the appropriate scale and administrative level [136]. Within a river basin or sub-basin, the integration of water management with spatial planning is not an easy process because aspects of spatial planning are related to, among others, agriculture, urban policy, transport, and industry, supervised by various administrations guided by their own policies [137]. Depending on the level of activities in the IWRM, three levels can be distinguished. At a local level, problems in the catchment area, water supply, and water protection plans are analyzed. Second, the implementation level covers the river basin scale or a separate administrative unit. Third, the political level is where national and international problems are resolved, and legislation is created to regulate water management issues. This requires the creation of an appropriate management structure with a network of connections, which will include public structures, including ministry offices responsible for water management in a strategic dimension; organizations, agreements, and agencies operating at the river basin level or its parts; local authorities and local governments; associations of communes and catchment unions; associations of water users; and non-governmental organizations. The local level is mainly responsible for the practical implementation of all measures while being their direct beneficiaries. At this level, the real, local problems of water management are known. The local level should form the IWRM based on the correlation between two complementary activity groups [138]. The first group should focus on the development of natural resources to ensure, among others, economic development, while the second group includes activities in the field of resource management, protection, and restoration. Both groups of activities require the participation and interaction of the operational level (users and society) to ensure balance and correctness in the management process. Given the numerous identified areas of uncertainty in water management in the Białka River Basin and the low level of public knowledge about the water management process, it seems that a necessary complement is the implementation of the Adaptive Water Management (AWM) principles into the IWRM. The AWM rules aim to build a resilient management system based on always incomplete information about the system, considering the uncertainty of its results [139]. AWM considers the complexity of the managed systems and the limitations in anticipating and controlling them. It assumes a comprehensive approach to all issues and the relations between them. The key tool for developing AWM is social learning, which should include the cooperation and exchange of knowledge of laypeople with experts and scientists, developing an understanding of key issues related to water resource management. AWM strives to build capacity through training and information distribution at every stage of the management process and shape stakeholders' conscious attitudes in the water resource management process. Properly built public awareness is intended to help broaden knowledge of the system and reduce uncertainty. From this perspective, AWM ensures greater resistance to unexpected and uncontrolled conditions in the system, reducing the negative environmental impact of its activities, building and strengthening the dialogue between stakeholders and area managers, including building positive communication relationships. Considering the barrier to the development of adequate water resource management, which is the low level of knowledge and measurement uncertainty, the AWM concept is probably the best way to achieve balance, resolve conflicts, and deal with the low spatial resolution of measurement networks and the uncertainty of climate projections for the Białka River Basin.

Both the IRWM and AWM need to develop an appropriate legal framework that will support the transition of all management stages to a local scale. Do the changes to the Water Law introduced in Poland in 2018 create the conditions for such support? The Act, in its current form, has maintained an opinion-making and advisory body at the national level as the State Water Management Council. It is of strategic importance, and its role is mainly based on issuing opinions on strategic documents or formal and legal solutions on a national scale. The water region councils have been abolished, and advisory committees have been proposed instead. Pursuant to Article 250 para. 1 of the above Act, the President of Polish Waters appoints consultative committees as opinion-making and advisory teams composed of governors, voivodeship marshals, representatives of the local government of the Joint Government and Local Government Commission, and directors of inland navigation offices. These committees are established for one or more water regions. However, is such a solution sufficient from a local point of view? Certainly not, because the act should also allow for the formal appointment of river basin committees and their operation at the local level, which is significant because of the aforementioned interactions. Thus, it is a barrier to the implementation of both IWRM and AWM. Table 14 shows how the current legal framework in Poland fits into the IWRM concept. The ideas of the concept are reflected at the political and implementation level. The legislation has the appropriate solutions here. On the other hand, the problem is the local level, theoretical assumptions of which are presented in the table, and which, in the current legislation, has not been defined (except for the spatial development plan). Local stakeholders, such as inhabitants and entrepreneurs associated with the Białka River Basin, are not included in the water management process. Their only tool is applications and petitions to the public administration managing the catchment area. The local level is crucial for AWM. The implementation of social learning must involve the exchange of knowledge between all stakeholders. Moreover, there are conceptual inconsistencies in the current Water Law, which make it difficult to understand the issue of water resource management. According to the first article of the current Water Law (in p. 1) [52], water management consists of water resource management, water development and protection, and water use. On the other hand, Article 10 of the Water Law (in p. 3) defines the following elements of water resource management: meeting the needs of the population and economy, the protection of waters, and the protection of the environment associated with these resources [52]. The emerging inconsistencies and lack of appropriate tools in the Water Law are some of the main barriers to the implementation of AWM, as well as IWRM. This is a basic problem, but the fragmentation of the management stages into many stages will generate further issues that the current legislation cannot minimize. For example, Saravanan et al. [140] indicate that although the decentralization of management is necessary, there is a particular risk of uncontrolled behavior, such as unfair selection of stakeholders for political reasons. It also confirms that a significant obstacle to engaging society in the water management process in many countries is unfavorable legal solutions that do not provide tools for building social participation.

Table 14. A concpetual framework for catchment management in the context of IWRM for the Białka River Basin.

Assumption/Premises of IWRM	Political Level	Implementation Level	Local Level
The type of river basin organization	International River Basin Committee	PGW WP National Water Management Board, regional water management boards, committees, associations, etc.	Local group, an association of communes, catchment union, the association of catchment users
Strategies and plans for the river basin	Agreement for the international catchment, management plan	River basin management plan	Local water plan, land use plan, local water and sewage management plan, local flood and drought protection plan
Decision-making level	Highest political level	Voivodeship, district, commune	Local administration, user associations, producer associations
The existing system of natural resources	A delimited geographical area, river basin, or part thereof, lake	A regional ecological system, catchment, groundwater reservoir, aquifer	Areas with relatively uniform ecological and hydrological conditions

6. Conclusions

The water resource management in the Białka River Basin is facing many difficulties. The most significant are river basin issues such as flood risk and water pollution, and, to a lesser extent, the risk of water scarcity. The causes of these problems are complex. They result from natural conditions (the seasonal variability of the hydrological regime of the Białka River, the flood-like nature of the river) and anthropogenic conditions (intense tourist pressure resulting in excessive discharge of municipal sewage and increasing water abstraction). The low emphasis on education in terms of water management results, among others, in the emergence of conflicts between the stakeholders and institutions responsible for water resource management. The low level of knowledge also leads to a considerable diversification of attitudes towards current methods of nature protection, and, in some cases, a strongly anthropocentric approach, especially in terms of the area's flood safety. Poor communication between stakeholders is the cause of, and, at the same time, the solution to this issue. The hydrological and socio-economic complexity of the river basin requires appropriate management methods such as IWRM, which will strive to preserve the unique value of the area, taking into account the needs of its inhabitants. Due to the low level of knowledge, high uncertainty of forecasts of changes in water resources, and uncertainty of measurements, the Białka River Basin needs a solution that will focus on the social aspects of management in order to reduce the negative effects of the uncertainty of system elements. The AWM seems to be the answer to these needs, which emphasizes social learning and knowledge exchange in adapting to changes in the water and economic system. In order for IWRM and, in the broader context, for AWM to be able to exist at all levels within the river basin, appropriate legislation is necessary. The current legal framework in Polish law covers the political and implementation level at the public administration level. However, at the local level, which is crucial for AWM, there is no defined framework for action and functioning in the context of developing the water management process.

Author Contributions: Conceptualization, M.B., T.W., M.S., M.Ż.; Methodology, M.B., T.W., M.S., M.Ż.; Software, M.B., T.W., M.S., M.Ż.; Validation, M.B., T.W., M.S., M.Ż.; Formal Analysis, M.B., T.W., M.S., M.Ż.; Investigation, M.B., T.W., M.S., M.Ż.; Resources, M.B., T.W., M.S., M.Ż.; Data Curation, M.B., T.W., M.S., M.Ż.; Writing—Original Draft Preparation, M.B., T.W., M.S., M.Ż.; Writing—Review and Editing, M.B., T.W., M.S., M.Ż.; Visualization, M.B., T.W., M.S., M.Ż.; Supervision, M.B., T.W.; Project Administration, T.W.; Funding Acquisition, T.W. All authors have read and agreed to the published version of the manuscript.

Funding: Subsidy from the Minister of Education and Science, Government of Poland. Project DS-6, 2021 "Adaptive planning and management of water resources in the light of climate change".

Institutional Review Board Statement: Not applicable.

Informed Consent Statement: Not applicable.

Data Availability Statement: The data that support the findings of this study are available on request from the corresponding author.

Acknowledgments: In this section, you can acknowledge any support given which is not covered by the author contribution or funding sections. This may include administrative and technical support, or donations in kind (e.g., materials used for experiments).

Conflicts of Interest: The authors declare no conflict of interest.

References

1. Vörösmarty, C.J. Global water assessment and potential contributions from Earth Systems Science. *Aquat. Sci.* **2002**, *64*, 328–351. [CrossRef]
2. Vörösmarty, C.J.; McIntyre, P.B.; Gessner, M.O.; Dudgeon, D.; Prusevich, A.; Green, P.; Glidden, S.; Bunn, S.E.; Sullivan, C.A.; Reidy Liremann, C.; et al. Global Threats to Human Water Security and River Biodiversity. *Nature* **2010**, *467*, 555–561. [CrossRef]
3. Parish, E.S.; Kodra, E.; Steinhaeuser, K.; Ganguly, A.R. Estimating Future Global per Capita Water Availability Based on Changes in Climate and Population. *Comput. Geosci.* **2012**, *42*, 79–86. [CrossRef]

4.   Vörösmarty, C.J. Global Water Resources: Vulnerability from Climate Change and Population Growth. *Science* **2000**, *289*, 284–288. [CrossRef]

5.   Kundzewicz, Z.W.; Mata, L.J.; Arnell, N.W.; Döll, P.; Jimenez, B.; Miller, K.; Oki, T.; Şen, Z.; Shiklomanov, I. The implications of projected climate change for freshwater resources and their management. *Hydrol. Sci. J.* **2008**, *53*, 3–10. [CrossRef]

6.   Pachauri, R.K.; Allen, M.R.; Barros, V.R.; Broome, J.; Cramer, W.; Christ, R.; Church, J.A.; Clarke, L.; Dahe, Q.; Dasgupta, P.; et al. *Climate Change 2014: Synthesis Report. Contribution of Working Groups I, II and III to the Fifth Assessment Report of the Intergovernmental Panel on Climate Change*; Intergovernment Panel on Climate Change (IPCC): Geneva, Switzerland, 2014.

7.   Azhoni, A.; Holman, I.; Jude, S. Adapting water management to climate change: Institutional involvement, inter institutional networks and barriers in India. *Glob. Environ. Chang.* **2017**, *44*, 144–157. [CrossRef]

8.   Dai, A.; Trenberth, K.E.; Qian, T. A global dataset of Palmer Drought Severity Index for 1870–2002: Relationship with soil moisture and effects of surface warming. *J. Hydrometeorol.* **2004**, *5*, 1117–1130. [CrossRef]

9.   Sivakumar, M.V.K. Agricultural Drought—WMO Perspectives. In *Agricultural Drought Indices, Proceedings of the Expert Meeting, Murcia, Spain, 2–4 June 2010*; Sivakumar, M.V.K., Motha, R.P., Wilhite, D.A., Wood, D.A., Eds.; World Meteorological Organization: Geneva, Switzerland, 2011; pp. 22–34.

10.   Sheffield, J.; Wood, E.F.; Roderick, M.L. Little change in global drought over the past 60 years. *Nature* **2012**, *491*, 435–438. [CrossRef]

11.   Spinoni, J.; Naumann, G.; Carrao, H.; Barbosa, P.; Vogt, J. World drought frequency, duration, and severity for 1951–2010. *Int. J. Climatol.* **2014**, *34*, 2792–2804. [CrossRef]

12.   Gustafson, E.J.; Sturtevant, B.R. Modeling forest mortality caused by drought stress: Implications for climate change. *Ecosystems* **2013**, *16*, 60–74. [CrossRef]

13.   Ghimire, Y.N.; Shivakoti, G.P.; Perret, S.R. Household-level vulnerability to drought in hill agriculture of Nepal: Implications for adaptation planning. *Int. J. Sustain. Dev. World Ecol.* **2010**, *17*, 225–230. [CrossRef]

14.   Udmale, P.D.; Ichikawa, Y.; Kiem, A.S.; Panda, S.N. Drought impacts and adaptation strategies for agriculture and rural livelihood in the Maharashtra State of India. *Open Agric. J.* **2014**, *8*, 41–47. [CrossRef]

15.   Sugg, M.; Runkle, J.; Leeper, R.; Bagli, H.; Golden, A.; Handwerger, L.H.; Magee, T.; Moreno, C.; Reed-Kelly, R.; Taylor, M.; et al. A scoping review of drought impacts on health and society in North America. *Clim. Chang.* **2020**, *162*, 1177–1195. [CrossRef]

16.   Mehta, L. Whose scarity? Whose property? The case of water in western India. *Land Use Policy* **2007**, *24*, 654–663. [CrossRef]

17.   Petersen-Perlman, J.D.; Veilleux, J.C.; Wolf, A.T. International water conflict and cooperation: Challenges and opportunities. *Water Int.* **2017**, *42*, 105–120. [CrossRef]

18.   Ahmed, S.A.; Hoque, B.A.; Mahmud, A. Water management practices in rural and urban homes: A case study from Bangladesh on ingestion of polluted water. *Public Health* **1998**, *112*, 317–321. [CrossRef]

19.   Ashraf, M.A.; Maah, M.J.; Yusoff, I.; Mehmood, K. Effects of polluted water irrigation on environment and health of people in Jamber, District Kasur, Pakistan. *Int. J. Basic Appl. Sci.* **2010**, *10*, 37–57.

20.   Cazcarro, I.; López-Morales, C.A.; Duchin, F. The global economic costs of the need to treat polluted water. *Econ. Syst. Res.* **2016**, *28*, 295–314. [CrossRef]

21.   Hussein, H.; Menga, F.; Greco, F. Monitoring Transboundary Water Cooperation in SDG 6.5.2: How a Critical Hydropolitics Approach Can Spot Inequitable Outcomes. *Sustainability* **2018**, *10*, 3640. [CrossRef]

22.   Marsalek, J. Overview of flood issues in contemporary water management. In *Flood Issues in Contemporary Water Management*; NATO Science Series: Series 2, Environment Seciurity; Marsalek, J., Watt, W.E., Zeman, E., Sieker, F., Eds.; Springer: Dordrecht, The Netherlands, 2000; Volume 71, pp. 1–16. [CrossRef]

23.   Zhang, S.; Li, Y.; Ma, M.; Song, T.; Song, R. Storm water management and flood control in sponge city construction of Beijing. *Water* **2018**, *10*, 1040. [CrossRef]

24.   Huntjens, P.; Pahl-Wostl, C.; Rihoux, B.; Schlüter, M.; Flachner, Z.; Neto, S.; Koskova, R.; Dickens, C.; Nabide Kiti, I. Adaptive water management and policy learning in a changing climate: A formal comparative analysis of eight water management regimes in Europe, Africa and Asia. *Environ. Pol. Gov.* **2011**, *21*, 145–163. [CrossRef]

25.   Martínez-Santos, P.; Varela-Ortega, C.; Hernández-Mora, N. Making Inroads Towards Adaptive Water Management Through Stakeholder Involvement, the NeWater Experience in the Upper Guadiana basin, Spain. In Proceedings of the International Conference on Adaptive and Integrated Water Management (CAIWA), Basel, Switzerland, 12–15 November 2007.

26.   King, A.B.; Thornton, M. Staying the course: Collaborative modeling to support adaptive and resilient water resource governance in the Inland Northwest. *Water* **2016**, *8*, 232. [CrossRef]

27.   Mutahara, M.; Warner, J.F.; Wals, A.E.; Khan, M.S.A.; Wester, P. Social learning for adaptive delta management: Tidal River Management in the Bangladesh Delta. *Int. J. Water Resour. Dev.* **2018**, *34*, 923–943. [CrossRef]

28.   Tran, T.A.; Pittock, J.; Tuan, L.A. Adaptive co-management in the Vietnamese Mekong Delta: Examining the interface between flood management and adaptation. *Int. J. Water Resour. Dev.* **2019**, *35*, 326–342. [CrossRef]

29.   McLoughlin, C.A.; Thoms, M.C.; Parsons, M. Reflexive learning in adaptive management: A case study of environmental water management in the Murray Darling Basin, Australia. *Riv. Res. Appl.* **2020**, *36*, 681–694. [CrossRef]

30.   Mysiak, J.; Henrikson, H.J.; Sullivan, C.; Bromley, J.; Pahl-Wostl, C. *The Adaptive Water Resource Management Handbook*; Routledge: London, UK, 2010; pp. 17–32. [CrossRef]

31. Meran, G.; Siehlow, M.; von Hirschhausen, C. Integrated Water Resource Management: Principles and Applications. In *The Economics of Water: Rules and Institutions.*; Springer International Publishing: Cham, Switzerland, 2021; pp. 23–121. [CrossRef]
32. The Need for an Integrated Approach. 2020. Available online: https://www.gwp.org/en/About/why/the-need-for-an-integrated-approach/ (accessed on 26 August 2021).
33. Akhmouch, A.; Clavreul, D. Stakeholder Engagement for Inclusive Water Governance: "Practicing What We Preach" with the OECD Water Governance Initiative. *Water* **2016**, *8*, 204. [CrossRef]
34. Dillon, P.; Bellchambers, R.; Meyer, W.; Ellis, R. Community perspective on consultation on urban stormwater management: Lessons from Brownhill Creek, South Australia. *Water* **2016**, *8*, 170. [CrossRef]
35. Mott Lacroix, K.E.; Megdal, S.B. Explore, synthesize, and repeat: Unraveling complex water management issues through the stakeholder engagement wheel. *Water* **2016**, *8*, 118. [CrossRef]
36. Megdal, S.B.; Eden, S.; Shamir, E. Water governance, stakeholder engagement, and sustainable water resources management. *Water* **2017**, *9*, 190. [CrossRef]
37. Van der Keur, P.; Lloyd, J.G. Integrated Water Resources Management (IWRM). In *Adaptive Water Resource Management Handbook*; Mysiak, J., Henrikson, J.H., Sullivan, C., Bromley, J., Pahl-Wostl, C., Eds.; Earthscan: London, UK, 2010; pp. 4–6.
38. Mehta, L.; Movik, S.; Bolding, A.; Derman, B.; Manzungu, E. Introduction to the Special Issue: Flows and practices—The politics of Integrated Water Resources Management (IWRM) in southern Africa. *Water Altern.* **2016**, *9*, 389–411.
39. Allouche, J. The birth and spread of IWRM—A case study of global policy diffusion and translation. *Water Altern.* **2016**, *9*, 412–433.
40. Michalak, D. Adapting to climate change and effective water management in Polish agriculture—At the level of government institutions and farms. *Ecohydrol. Hydrobiol.* **2020**, *20*, 134–141. [CrossRef]
41. McMillan, H.K.; Westerberg, I.K.; Krueger, T. Hydrological Data Uncertainty and Its Implications. *WIREs Water* **2018**, *5*, e1319. [CrossRef]
42. Sordo-Ward, Á.; Granados, I.; Martín-Carrasco, F.; Garrote, L. Impact of Hydrological Uncertainty on Water Management Decisions. *Water Resour. Manag.* **2016**, *30*, 5535–5551. [CrossRef]
43. Loch, A.; Adamson, D.; Mallawaarachchi, T. Role of Hydrology and Economics in Water Management Policy under Increasing Uncertainty. *J. Hydrol.* **2014**, *518*, 5–16. [CrossRef]
44. Pandey, C.L.; Bajracharya, R.M. Climate adaptive water management practices in small and midsized cities of Nepal: Case studies of Dharan and Dhulikhel. *Sustain. J. Rec.* **2017**, *10*, 300–307. [CrossRef]
45. Cosens, B.; Brian, C.; Chaffin, B.C. Adaptive Governance of Water Resources Shared with Indigenous Peoples: The Role of Law. *Water* **2016**, *8*, 97. [CrossRef]
46. Casiano Flores, C.; Vikolainen, V.; Bressers, H. Water governance decentralisation and river basin management reforms in hierarchical systems: Do they work for water treatment policy in Mexico's Tlaxcala Atoyac Sub-basin? *Water* **2016**, *8*, 210. [CrossRef]
47. Rouillard, J.J.; Spray, C.J. Working across scales in integrated catchment management: Lessons learned for adaptive water governance from regional experiences. *Reg. Environ. Chang.* **2017**, *17*, 1869–1880. [CrossRef]
48. Ashofteh, P.S.; Bozorg-Haddad, O.; Loáiciga, H.A. Role of adaptive water resources management policies and strategies in relieving conflicts between water resources and agricultural sector water use caused by climate change. *J. Irrig. Drain. Eng.* **2017**, *143*, 02516004. [CrossRef]
49. Pahl-Wostl, C. Adaptive and sustainable water management: From improved conceptual foundations to transformative change. *Int. J. Water Resour. Dev.* **2020**, *36*, 397–415. [CrossRef]
50. Derepasko, D.; Peñas, F.J.; Barquín, J.; Volk, M. Applying Optimization to Support Adaptive Water Management of Rivers. *Water* **2021**, *13*, 1281. [CrossRef]
51. Pahl-Wostl, C.; Craps, M.; Dewulf, A.; Mostert, E.; Tabara, D.; Taillieu, T. Social learning and water resources management. *Ecol. Soc.* **2007**, *12*, 5. [CrossRef]
52. Act of 20 July 2017, Water Law (Journal of Laws 2017, Item 1566). Available online: http://isap.sejm.gov.pl/isap.nsf/download.xsp/WDU20170001566/U/D20171566Lj.pdf (accessed on 15 October 2021).
53. Directive 2000/60/EC of the European Parliament and of the Council of 23 October 2000, Establishing a Framework for Community Action in the Field of Water Policy. Available online: https://eur-lex.europa.eu/resource.html?uri=cellar:5c835afb-2ec6-4577-bdf8-756d3d694eeb.0004.02/DOC_1&format=PDF (accessed on 15 October 2021).
54. Kubiak-Wójcicka, K.; Machula, S. Influence of climate changes on the state of water resources in Poland and their usage. *Geosciences* **2020**, *10*, 312. [CrossRef]
55. Więckowski, M. Tourism development in the borderlands of Poland. *Geogr. Pol.* **2010**, *83*, 67–81. [CrossRef]
56. Mapa Podziału Hydrograficznego Polski w Skali 1:10 000 (MPHP10k), Państwowe Gospodarstwo Wodne Wody Polskie. Available online: https://dane.gov.pl/pl/dataset/2167,mapa-podzialu-hydrograficznego-polski-w-skali-110 (accessed on 15 October 2021).
57. Kulesza, K.; Woźniak, Ł.; Walczykiewicz, T. Czynniki Determinujące Funkcjonowanie Korytarza Ekologicznego Doliny Białki na tle Możliwych Zmian Klimatu. In *Nauka Tatrom, Tom III—Człowiek i Środowisko*; Chrobak, A., Zwijacz-Kozica, T., Eds.; Wydawnictwa Tatrzańskiego Parku Narodowego: Zakopane, Poland, 2015; pp. 79–88.

58. Wrzesiński, D. Identyfikacja cech reżimu odpływu rzek w Polsce na różnych poziomach grupowania. *Bad. Fizjogr.* **2017**, *68*, 265–278. [CrossRef]
59. SDF of the Natura 2000 Białka River Valley. Available online: http://n2k-ws.gdos.gov.pl/wyszukiwarkaN2k/webresources/pdf/PLH120024 (accessed on 29 October 2021).
60. Hełdak, M. Rzeka Białka w krajobrazie wsi podhalańskiej. *Archit. Kraj.* **2010**, *2*, 39–46.
61. Commission Decision of 12 December 2008, Adopting, Pursuant to Council Directive 92/43/EEC, a Second Updated List of Sites of Community Importance for the Alpine Biogeographical Region, (2009/91/EC). Available online: https://eur-lex.europa.eu/LexUriServ/LexUriServ.do?uri=OJ:L:2009:043:0021:0058:EN:PDF (accessed on 15 October 2021).
62. Council Directive 92/43/EEC of 21 May 1992, on the Conservation of Natural Habitats and of Wild Fauna and Flora. Available online: https://eur-lex.europa.eu/legal-content/EN/TXT/PDF/?uri=CELEX:31992L0043&from=PL (accessed on 15 October 2021).
63. Bożek, K. Jak Udrożnić Rezerwat. Dzikie Życie 3/177. 2009. Available online: https://dzikiezycie.pl/archiwum/2009/marzec-2009/jak-udroznic-rezerwat (accessed on 29 October 2021).
64. Kondracki, J. *Geografia Regionalna Polski*; Wydawnictwo Naukowe PWN: Warsaw, Poland, 2002.
65. Solon, J.; Borzyszkowski, J.; Bidłasik, M.; Richling, A.; Badora, K.; Balon, J.; Brzezińska-Wójcik, T.; Chabudziński, Ł.; Dobrowolski, R.; Grzegorczyk, I.; et al. Physico-geographical mesoregions of Poland: Verification and adjustment of boundaries on the basis of contemporary spatial data. *Geogr. Pol.* **2018**, *91*, 143–170. [CrossRef]
66. Copernicus Land Monitoring Service, CORINE Land Cover (CLC). 2018. Available online: https://land.copernicus.eu/pan-european/corine-land-cover/clc2018 (accessed on 22 October 2021).
67. Statistics Poland (GUS), Local Data Bank. 2021. Available online: https://bdl.stat.gov.pl/BDL/start (accessed on 15 October 2021).
68. Krąż, P. Problemy w zarządzaniu systemem krajobrazowym Doliny Białki. *Probl. Ekol. Kraj.* **2012**, *33*, 229–234.
69. Krzesiwo, K.; Mika, M. Ocena atrakcyjności turystycznej stacji narciarskich w świetle zagadnienia ich konkurencyjności-studium porównawcze Szczyrku i Białki Tatrzańskiej. *Prace Geogr.* **2011**, *125*, 95–110.
70. Kruczek, Z. Wykorzystanie Wód Geotermalnych w Polsce w Celach Rekreacyjnych i Uzdrowiskowych. Studium Przypadku-Białka Tatrzańska. In *Wybrane Aspekty Zarządzania Zakładem Uzdrowiskowym*; Szromek, A., Ed.; Proksenia: Kraków, Poland, 2016; pp. 157–173.
71. Duda, E.; Ziaja, W. Wpływ turystyki i rekreacji na środowisko przyrodnicze i krajobraz białki tatrzańskiej. *Problemy Ekol. Kraj.* **2010**, *27*, 131–140.
72. Heldak, M.; Szczepanski, J. Wpływ rozwoju turystyki na transformację krajobrazu wsi Białka Tatrzańska. *Infrastrukt. I Ekol. Teren. Wiej.* **2011**, *1*, 151–161.
73. Janecka, K.; Lempa, M.; Czajka, B. Przyczyny i Skutki Powodzi w Tatrach i Na Podhalu. In *Z Badań Nad Wpływem Antropopresji na Środowisko*; Machowski, R., Rzętała, M.A., Eds.; Studenckie Koło Naukowe Geografów UŚ, Wydział Nauk o Ziemi UŚ: Sosnowiec, Poland, 2014; Volume 15, pp. 111–119.
74. Tatra National Park—Online Ticket Website. Available online: https://tpn.pl/zwiedzaj/e-bilety (accessed on 22 October 2021).
75. Stępień, K. Małopolskie Interesuje Turystów Najbardziej. Available online: https://www.nocowanie.pl/malopolskie-interesuje-turystow-najbardziej.html (accessed on 22 October 2021).
76. Ferie Zimowe 2019 w Polsce—Raport Nocowanie.pl. Available online: https://www.nocowanie.pl/ferie-zimowe-2019-w-polsce--raport-nocowanie-pl.html (accessed on 22 October 2021).
77. Lenart-Boroń, A.; Wolanin, A.; Jelonkiewicz, Ł.; Chmielewska-Błotnicka, D.; Żelazny, M. Spatiotemporal variability in microbiological water quality of the Białka River and its relation to the selected physicochemical parameters of water. *Water Air Soil Pollut.* **2016**, *227*, 1–12. [CrossRef]
78. Bojarczuk, A.; Jelonkiewicz, Ł.; Lenart-Boroń, A. The effect of anthropogenic and natural factors on the prevalence of physico-chemical parameters of water and bacterial water quality indicators along the river Białka, southern Poland. *Environ. Sci. Pollut. Res.* **2018**, *25*, 10102–10114. [CrossRef]
79. Lenart-Boroń, A.; Prajsnar, J.; Krzesiwo, K.; Bojarczuk, A.; Jelonkiewicz, Ł.; Jelonkiewicz, E.; Żelazny, M. Diurnal variation in the selected indicators of water contamination in the Białka river affected by a sewage treatment plant discharge. *Fresenius Environ. Bull.* **2016**, *25*, 5271–5279.
80. Lenart-Boroń, A.; Prajsnar, J.; Guzik, M.; Boroń, P.; Chmiel, M. How much of antibiotics can enter surface water with treated wastewater and how it affects the resistance of waterborne bacteria: A case study of the Białka river sewage treatment plant. *Environ. Res.* **2020**, *191*, 110037. [CrossRef]
81. Program Państwowego Monitoringu Środowiska Województwa Małopolskiego na lata 2016–2020. WIOŚ, Kraków, Poland, 2015. Available online: http://www.krakow.pios.gov.pl/Press/monitoring/pms/wpms_2016_2020.pdf (accessed on 22 October 2021).
82. Kuraś, B. Białka Tatrzańska. Turystów Przybywa, a Kanalizacji Jak Nie było, Tak Nie Ma. Trza Skończyć z Tym Szambem". 2020. Available online: https://krakow.wyborcza.pl/krakow/7,44425,25782184,trza-skonczyc-z-tym-szambem.html (accessed on 15 October 2021).
83. Będzie Oczyszczalnia, Kanalizacja i Nowa Droga. 2021. Available online: https://podhale24.pl/aktualnosci/artykul/75699/Bedzie_oczyszczalnia_kanalizacja_i_nowa_droga.html (accessed on 15 October 2021).
84. Starmach, P. Podhalański Śmierdzący Problem. Turystów Przybywa, Sieci Kanalizacyjnej Nie Ma. 2021. Available online: https://zakopane.wyborcza.pl/zakopane/7,179294,27098979,podhalanski-smierdzacy-problem-turystow-przybywa-sieci-kanalizacyjnej.html (accessed on 15 October 2021).

85. Podhale. Kiedy Ruszy Budowa Kanalizacji w Białce Tatrzańskiej? 2020. Available online: https://gazetakrakowska.pl/podhale-kiedy-ruszy-budowa-kanalizacji-w-bialce-tatrzanskiej/ar/c15-14888421 (accessed on 15 October 2021).
86. Zubek, A. Białka Tatrzańska na Kanalizację Musi Poczekać Do 2027 Roku? Pytamy w Gminie. 2021. Available online: https://podhaleregion.pl/index.php/z-podhala/33168-bialka-tatrzanska-na-kanalizacje-musi-poczekac-do-2027-roku-pytamy-w-gminie (accessed on 15 October 2021).
87. Białka Tatrzańska: Płyną Ścieki Do Rzeki. 2014. Available online: https://24tp.pl/?mod=news&id=21987 (accessed on 15 October 2021).
88. Włodarczyk, K. Czarna maź w Białce Tatrzańskiej Wypływa z Oczyszczalni Ścieków w Czarnej Górze. Czy ABW też Będzie Badało tą Sprawę? 2019. Available online: https://gazetaplus.pl/czarna-maz-w-bialce-tatrzanskiej-wyplywa-z-oczyszczalni-sciekow-w-czarnej-gorze-czy-abw-tez-bedzie-badalo-ta-sprawe/8695/ (accessed on 15 October 2021).
89. Podhale. Ekolodzy w Szoku! Wodę w Rzece Białce Zanieczyściły bobry? 2019. Available online: https://zakopane.naszemiasto.pl/podhale-ekolodzy-w-szoku-wode-w-rzece-bialce-zanieczyscily/ar/c8-7436941 (accessed on 15 October 2021).
90. Stoffel, M.; Wyżga, B.; Niedźwiedź, T.; Ruiz-Villanueva, V.; Ballesteros-Cánovas, J.A.; Kundzewicz, Z.W. Floods in Mountain Basins. In *Flood Risk in the Upper Vistula Basin. GeoPlanet: Earth and Planetary Sciences*; Kundzewicz, Z.W., Stoffel, M., Niedźwiedź, T., Wyżga, B., Eds.; Springer: Berlin, Germany, 2016; pp. 23–37. [CrossRef]
91. Waniczek, K. Petycja w Sprawie Białki. 2019. Available online: https://krempachy.pl/2019/03/20/petycja-w-sprawie-bialki/ (accessed on 15 October 2021).
92. Odpowiedź Podsekretarza Stanu w Ministerstwie Administracji i Cyfryzacji na Zapytanie nr 3074 w Sprawie Określenia Wysokości Środków Finansowych Przekazanych Do Tej Pory na Usunięcie Skutków Powodzi, z Uwzględnieniem Szacunków i Wyceny Zgłoszonych i Udokumentowanych Strat, Jakie Powstały w Wyniku Powodzi w Latach 2007–2012 w gm. Bukowina Tatrzańska. 2013. Available online: https://www.sejm.gov.pl/sejm7.nsf/InterpelacjaTresc.xsp?key=09288B00&view=null (accessed on 15 October 2021).
93. Słowik, J. Wody Białki Opadły. Pozostało Oburzenie. 2008. Available online: https://gazetakrakowska.pl/wody-bialki-opadly-pozostalo-oburzenie/ar/30317 (accessed on 15 October 2021).
94. Ludzie Przegrali z Wodą. 2008. Available online: https://dziennikpolski24.pl/ludzie-przegrali-z-woda/ar/2474868 (accessed on 15 October 2021).
95. Rzeka Białka Znowu Postraszyła. "Na Spiszu" 2(75). 2010. Available online: http://kacwin.com/nowe/naspiszu/2010_2/?rok=2010&nr=2&wydanie=75&strona=44,data (accessed on 15 October 2021).
96. Kaźmierczak, P. Powódź Przez Urzędników? (updated). 2016. Available online: https://interwencja.polsatnews.pl/reportaz/2011-02-08/powodz-przez-urzednikow_853131/ (accessed on 15 October 2021).
97. Para, M. Jurgów: Zdążyć Przed Powodzią. 2012. Available online: https://24tp.pl/n/17151 (accessed on 15 October 2021).
98. Giżycka, A.; Szkaradzińska, B. Ochrona Siedzib Ludzi, a Ochrona Przyrody. 2013. Available online: https://podhaleregion.pl/index.php/z-podhala/3936-ochrona-siedzib-ludzi,-a-ochrona-przyrody (accessed on 15 October 2021).
99. Ciężki Sprzęt Dewastuje Rezerwat Przełom Białki! 2018. Available online: https://www.wwf.pl/aktualnosci/ciezki-sprzet-dewastuje-rezerwat-przelom-bialki (accessed on 15 October 2021).
100. Breczko, B. Mieszkańcy Kontra Ekolodzy Kontra Państwo. A Rzeka Płynie Jak Chce. 2018. Available online: https://tech.wp.pl/mieszkancy-kontra-ekolodzy-kontra-panstwo-a-rzeka-plynie-jak-chce-6327138770130561a?nil=&src01=f1e45&src02=isgf (accessed on 15 October 2021).
101. Pałahicki, M.; Potocka, J.; Wiosło, M. Protest Ekologów Przeciwko Pracom Na Chronionej Rzece Białka. 2018. Available online: https://www.rmf24.pl/fakty/polska/news-protest-ekologow-przeciwko-pracom-na-chronionej-rzece-bialka,nId,2731131#crp_state=1 (accessed on 15 October 2021).
102. Rzeka Białka Niszczona Ciężkim Sprzętem. 2018. Available online: http://eko.org.pl/index_news.php?dzial=2&kat=20&art=2206 (accessed on 15 October 2021).
103. Topiński, P. Piotr Topiński: Protest nad Białką. 2018. Available online: https://studioopinii.pl/archiwa/190156 (accessed on 15 October 2021).
104. Wody Polskie o Regulacji Białki: "Chronimy Mieszkańców przed Powodzią". 2018. Available online: https://podhale24.pl/aktualnosci/artykul/60844/Wody_Polskie_o_regulacji_Bialki_quotChronimy_mieszkancow_przed_powodziaquot.html (accessed on 15 October 2021).
105. Słowik, J. Białka Znowu Groźna. Mieszkańcy Chcą Udrożnienia Koryta i Zabezpieczenia Brzegów. 2020. Available online: https://podhale24.pl/aktualnosci/artykul/71175/Bialka_znowu_grozna_Mieszkancy_chca_udroznienia_koryta_i_zabezpieczenia_brzegow.html (accessed on 15 October 2021).
106. Skowron, G. Kraków. Protest Przeciwko Regulacji Rzeki BIAŁKI. 2018. Available online: https://dziennikpolski24.pl/krakow-protest-przeciwko-regulacji-rzeki-bialki/ar/c3-13741932 (accessed on 15 October 2021).
107. Wody Polskie Zabezpieczają Białkę Przed Powodzią. 2018. Available online: https://www.kzgw.gov.pl/index.php/pl/aktualnosci/699-wody-polskie-zabezpieczaja-bialke-przed-powodzia (accessed on 15 October 2021).
108. Ochrona Przeciwpowodziowa Doliny Rzeki Białki w Zgodzie z Przyrodą. 2021. Available online: https://krakow.wody.gov.pl/aktualnosci/1837-ochrona-przeciwpowodziowa-doliny-rzeki-bialki-w-zgodzie-z-przyroda-2 (accessed on 15 October 2021).
109. Nakicenovic, N.; Alcamo, J.; Davis, G.; Vries, B.; Fenhann, J.; Gaffin, S.; Gregory, K.; Grubler, A.; Jung, T.Y.; Kram, T.; et al. *Special Report on Emissions Scenarios*; Cambridge University Press: Cambridge, UK, 2000.

110. Melillo, J.M.; Richmond, T.C.; Yohe, G.W. (Eds.) *Climate Change Impacts in the United States: The Third National Climate Assessment*; U.S. Global Change Research Program: Washington, DC, USA, 2014. [CrossRef]

111. Majewski, W.; Walczykiewicz, T. (Eds.) *Zrównoważone Gospodarowanie Zasobami Wodnymi Oraz Infrastrukturą Hydrotechniczną W Świetle Prognozowanych Zmian Klimatycznych*; IMGW-PIB: Warsaw, Poland, 2012.

112. Pińskwar, I.; Kundzewicz, Z.W.; Choryński, A. Projections of changes in heavy precipitation in the northern foothills of the Tatra Mountains. *Meteorol. Hydrol. Water Manag.* **2017**, *5*, 21–30. [CrossRef]

113. Mezghani, A.; Parding, K.M.; Dobler, A.; Benestad, R.E.; Haugen, J.E.; Piniewski, M. Projekcje Zmian Temperatury, Opadów i Pokrywy Śnieżnej w Polsce. In *Zmiany Klimatu i ich Wpływ na Wybrane Sektory w Polsce*; Kundzewicz, Z.W., Hov, Ø., Okruszko, T., Eds.; Narodowe Centrum Badań i Rozwoju: Poznań, Poland, 2017; pp. 94–118.

114. Piniewski, M.; Szcześniak, M.; Kardel, I.; Marcinkowski, P.; Okruszko, T.; Kundzewicz, Z.W. Zasoby wodne. In *Zmiany Klimatu i ich Wpływ na Wybrane Sektory w Polsce*; Kundzewicz, Z.W., Hov, Ø., Okruszko, T., Eds.; Narodowe Centrum Badań i Rozwoju: Poznań, Poland, 2017; pp. 131–147.

115. Okruszko, T.; O'Keeffe, J.; Utratna, M.; Marcinkowski, P.; Szcześniak, M.; Kardel, I.; Kundzewicz, Z.W.; Piniewski, M. Prognoza wpływu Zmian Klimatu na Środowisko Wodne i Mokradła w Polsce. In *Zmiany Klimatu i ich Wpływ na Wybrane Sektory w Polsce*; Kundzewicz, Z.W., Hov, Ø., Okruszko, T., Eds.; Narodowe Centrum Badań i Rozwoju: Poznań, Poland, 2017; pp. 148–167.

116. O'Keeffe, J.; Piniewski, M.; Szcześniak, M.; Oglęcki, P.; Parasiewicz, P.; Okruszko, T. Index-based analysis of climate change impact on streamflow conditions important for Northern Pike, Chub and Atlantic salmon. *Fish. Manag. Ecol.* **2019**, *26*, 474–485. [CrossRef]

117. Act of 18 July 2001, Water Law (Journal of Laws 2001, No. 115, Item 1229). Available online: https://isap.sejm.gov.pl/isap.nsf/download.xsp/WDU20011151229/U/D20011229Lj.pdf (accessed on 23 November 2021).

118. ISOK. Hydroportal. Available online: https://wody.isok.gov.pl/imap_kzgw/ (accessed on 15 October 2021).

119. Hussein, H. Yarmouk, Jordan, and Disi basins: Examining the impact of the discourse of water scarcity in Jordan on transboundary water governance. *Mediterr. Polit.* **2018**, *24*, 269–289. [CrossRef]

120. Jones, R.N. Managing uncertainty in climate change projections–issues for impact assessment. *Clim. Chang.* **2000**, *45*, 403–419. [CrossRef]

121. Webster, M.; Forest, C.; Reilly, J.; Babiker, M.; Kicklighter, D.; Mayer, M.; Prinn, R.; Sarofim, M.; Sokolov, A.; Stone, P.; et al. Uncertainty Analysis of Climate Change and Policy Response. *Clim. Chang.* **2003**, *61*, 295–320. [CrossRef]

122. Lawler, J.J.; Tear, T.H.; Pyke, C.; Shaw, M.R.; Gonzalez, P.; Kareiva, P.; Hansen, L.; Hannah, L.; Klausmeyer, K.; Aldous, A.; et al. Resource management in a changing and uncertain climate. *Front. Ecol. Environ.* **2010**, *8*, 35–43. [CrossRef]

123. Todd, M.C.; Taylor, R.G.; Osborn, T.J.; Kingston, D.G.; Arnell, N.W.; Gosling, S.N. Uncertainty in climate change impacts on basin-scale freshwater resources–preface to the special issue: The QUEST-GSI methodology and synthesis of results. *Hydrol. Earth Syst. Sci.* **2011**, *15*, 1035–1046. [CrossRef]

124. Kundzewicz, Z.W.; Krysanova, V.; Benestad, R.E.; Hov, Ø.; Piniewski, M.; Otto, I.M. Uncertainty in Climate Change Impacts on Water Resources. *Environ. Sci. Policy* **2018**, *79*, 1–8. [CrossRef]

125. Underwood, B.S.; Mascaro, G.; Chester, M.V.; Fraser, A.; Lopez-Cantu, T.; Samaras, C. Past and present design practices and uncertainty in climate projections are challenges for designing infrastructure to future conditions. *J. Infrastruct. Syst.* **2020**, *26*, 04020026. [CrossRef]

126. Refsgaard, J.C.; Arnbjerg-Nielsen, K.; Drews, M.; Halsnæs, K.; Jeppesen, E.; Madsen, H.; Markandya, A.; Olesen, J.E.; Porter, J.R.; Christensen, J.H. The Role of Uncertainty in Climate Change Adaptation Strategies—A Danish Water Management Example. *Mitig. Adapt. Strateg. Glob. Chang.* **2012**, *18*, 337–359. [CrossRef]

127. Gallego-Ayala, J.; Juízo, D. Integrating stakeholders' preferences into water resources management planning in the Incomati river basin. *Water Resour. Manag.* **2014**, *28*, 527–540. [CrossRef]

128. Basco-Carrera, L.; Warren, A.; van Beek, E.; Jonoski, A.; Giardino, A. Collaborative modelling or participatory modelling? A framework for water resources management. *Environ. Model. Softw.* **2017**, *91*, 95–110. [CrossRef]

129. Langsdale, S.; Beall, A.; Bourget, E.; Hagen, E.; Kudlas, S.; Palmer, R.; Tate, D.; Werick, W. Collaborative Modeling for Decision Support in Water Resources: Principles and Best Practices. *J. Am. Water Resour. Assoc.* **2013**, *49*, 629–638. [CrossRef]

130. Jonsson, A. Public participation in water resources management: Stakeholder voices on degree, scale, potential, and methods in future water management. *AMBIO* **2005**, *34*, 495–500. [CrossRef] [PubMed]

131. Jacobs, K.; Lebel, L.; Buizer, J.; Addams, L.; Matson, P.; McCullough, E.; Garden, P.; Saliba, G.; Finan, T. Linking Knowledge with Action in the Pursuit of Sustainable Water-Resources Management. *Proc. Natl. Acad. Sci. USA* **2016**, *113*, 4591–4596. [CrossRef] [PubMed]

132. Purkey, D.R.; Escobar Arias, M.I.; Mehta, V.K.; Forni, L.; Depsky, N.J.; Yates, D.N.; Stevenson, W.N. A Philosophical Justification for a Novel Analysis-Supported, Stakeholder-Driven Participatory Process for Water Resources Planning and Decision Making. *Water* **2018**, *10*, 1009. [CrossRef]

133. Pahl-Wostl, C.; Sendzimir, J.; Jeffrey, P.; Aerts, J.; Berkamp, G.; Cross, K. Managing change toward adaptive water management through social learning. *Ecol. Soc.* **2007**, *12*, 30. [CrossRef]

134. Lebel, L.; Grothmann, T.; Siebenhüner, B. The role of social learning in adaptiveness: Insights from water management. *Int. Environ. Agreem.* **2010**, *10*, 333–353. [CrossRef]

135. GWP; INBO. *A Handbook for Integrated Water Resources Management in Basins*; Global Water Partnership and International Network of Basin Organisations; Elanders: Mölnlycke, Sweden, 2009.
136. Konieczny, R.; Walczykiewicz, T. *Wstępna Analiza Proponowanych Rozwiązań Zawartych w Projektowanej Dyrektywie Powodziowej i ich Wpływu na Rozwiązania Prawne i Organizacyjne w Polsce*; (Unpublished Work, Typescript); Institute of Meteorology and Water Management, National Research Institute: Kraków, Poland, 2005.
137. Drużyńska, E.; Nachlik, E. (Eds.) *Podstawy Metodyczne i Standardy Zintegrowanego Planowania w Gospodarce Wodnej*; Praca zbiorowa, Politechnika Krakowska im, Tadeusza Kościuszki, Monografia 341; Seria Inżynieria Środowiska: Kraków, Poland, 2006.
138. Dourojeanni, A. Public Policies for Integrated Watershed Management. In *Integrated River Basin Management: The Latin American Experience*; Tortajda, C., Biwas, A., Eds.; Oxford University Press: Oxford, UK, 2001; pp. 58–73.
139. Van der Keur, P.; Jeffrey, P.; Boyce, D.; Pahl-Wostl, C.; Hall, A.; Lloyd, J.G. Adaptive Water Management in Terms of Development and Application within IWRM. In *Adaptive Water Resource Management Handbook*; Mysiak, J., Henrikson, J.H., Sullivan, C., Bromley, J., Pahl-Wostl, C., Eds.; Earthscan: London, UK, 2010; pp. 7–9.
140. Saravanan, V.S.; McDonald, G.T.; Mollinga, P.P. Critical review of integrated water resources management: Moving beyond polarised discourse. *Nat. Resour. Forum* **2009**, *33*, 76–86. [CrossRef]

Article

Water Management Balance as a Tool for Analysis of a River Basin with Conflicting Environmental and Navigational Water Demands: An Example of the Warta Mouth National Park, Poland

Dorota Pusłowska-Tyszewska

Chair of Environment Protection and Management, Faculty of Building Services, Hydro and Environmental Engineering, Warsaw University of Technology, Nowowiejska 20, 00-663 Warszawa, Poland; dorota.puslowska@pw.edu.pl

Abstract: Allocating finite water resources between different water uses is always a challenging task. Searching for a solution which satisfies the water needs (requirements) of all water users without compromising the water requirements of river ecosystems calls for analyzing different water management options and their expected consequences. Water management balances are usually used for comparison of water resources with the needs of water users. When aquatic and water dependent ecosystems are considered in a similar manner as other users, searching for the optimum water resources allocation, without neglecting requirements of the natural environment, is possible. This paper describes basic modeling assumptions and methodological solutions, which allow for taking into account some tasks related to the protection of aquatic and water dependent ecosystems. The water balance model, developed for a catchment comprising the Warta Mouth National Park, was applied to find out whether supplying adequate amounts of water for conservation (or restoration) of wet meadows and wetland habitats in the area is possible, while still satisfying the demands of other water users.

Keywords: water requirements of aquatic and water dependent ecosystems; water resources allocation; water balance model

Citation: Pusłowska-Tyszewska, D. Water Management Balance as a Tool for Analysis of a River Basin with Conflicting Environmental and Navigational Water Demands: An Example of the Warta Mouth National Park, Poland. *Water* **2021**, *13*, 3628. https://doi.org/10.3390/w13243628

Academic Editors: Andrzej Walega and Tamara Tokarczyk

Received: 26 October 2021
Accepted: 14 December 2021
Published: 16 December 2021

Publisher's Note: MDPI stays neutral with regard to jurisdictional claims in published maps and institutional affiliations.

1. Introduction

The issue of allocating sufficient volumes of water for aquatic and water-dependent terrestrial ecosystems has been analyzed for many years, but, in 2000, the Water Framework Directive (WFD [1]) introduced new and ambitious objectives to protect and restore these ecosystems as a basis for ensuring the long-term sustainable use of water for people, businesses, and nature. The key objective of the WFD is to achieve a good status for all water bodies. This comprises the objectives of good ecological and chemical status for surface waters and good quantitative and chemical status for groundwater. This becomes a priority task of water management. Maintenance of the appropriate environmental flows is mentioned often as one of the basic conditions to achieve good status of surface water bodies [2]. There are many methods of defining the environmental flows required for an aquatic environment—at least 200 of them have been identified [3–5]. These methods differ considerably from one another as regards the method of determination, scope of application, the hydrological regime elements taken into account [6–8], interactions with groundwater [9,10], and the socio-economic objectives of water use [11–13]. The flow magnitude and characteristics of the hydrological regime, such as variability of flows, their distribution during high- and low-water periods, duration, and frequency of occurrence, are treated as the key parameters [7]. Other parameters, such as water velocity and depth of the stream, river bed morphology, and connection with floodplains, are also mentioned in the context of quantitative requirements of river ecosystems [14–16]. For water dependent ecosystems, including a variety of wetlands, water requirements pertain to hydrological feeding types, time distribution and dynamics of water level changes, soil moisture content, and frequency of droughts [17,18].

Identification of water requirements of ecosystems or protected organisms is the first basic condition of their protection. Other conditions are related to the ability of meeting these requirements—in view of the existing socio-economic tasks of water management [19,20]. The search for a compromise fits into the concept of sustainable development. What has fundamental importance is the possibility of analyzing the potential alternatives of water resources allocation in a specific location and time [21,22].

At the current level of economic development, in Poland and Europe alike, and with the effected anthropogenic changes of the environment as a whole, one can hardly approve the idea of preservation or reconstruction of the natural hydrological conditions that originally formed the existing aquatic and water-dependent ecosystems. One should focus instead on allocation of sufficient (or appropriate) volume of water, which—in a specific situation as regards to water use, anthropogenic transformation of the basin, and social expectations, e.g., those related to flood risk—at least partly meet the ecosystems' requirements and secure a sufficient level of protection [23]. One of the tools for analyzing the water resources allocation alternatives is the model of water management balance (e.g., [24]). Since the water management balance means comparison of water resources with the needs of their users, both the resources and the needs should be described with sufficient precision. A dynamic balance takes into account data that change over time and the calculations are based on a simulation of the functioning of the water management system, usually a river basin [25,26].

In the Warta Mouth National Park (WMNP) there are conflicting objectives of water resources management: agricultural areas located in the park require at least periodical drainage, while protection of the Park's natural values requires maintenance of high humidity of habitats. The use of water resources of the Warta River to improve habitats' moisture conditions is limited due to the necessity to provide adequate flows for inland navigation. The aim of the study was to answer the question if it was possible for effective protection of wetland habitats, navigation, and agriculture to coexist in the area. In order to answer this question, a water management balance of part of the Warta catchment was performed, in which tasks related to maintaining appropriate moisture conditions in the WMNP area were taken into account. In this paper special emphasis is put on methodological solutions for these elements of water balances, which are of crucial value for adequate representation of the quantitative requirements of water dependent ecosystems. The balance model, which takes into account the specific features of a catchment comprising The Warta Mouth National Park, was applied to find out whether supplying an adequate volume of water for conservation or restoration of the marshy meadow ecosystems is possible, while still satisfying the demands of other water users. The present balance model assumptions, methodological solutions, and calculations are the effects of a study: "Water management optimization model for the Warta Mouth National Park" [27] undertaken within the preparations of a draft protection plan for the Park and Natura 2000 site PLC 080001, implemented by MGGP S.A. in 2013.

In the following part of the article a short description of the Warta Mouth National Park is presented, the applied water management balance methodology is discussed, and then the way in which the specific uses of the studied area were included in the balance model is described. The results of simulation calculations are presented on the example of a selected habitat that is protected in the WMNp. In the discussion, attention is paid to possible sources of uncertainty of the obtained results. Conclusions formulated in the final part concern both the usefulness of the applied approach in assessing the possibility of obtaining a compromise in case of conflicts between water management tasks and the scope of information necessary for an adequate description of tasks related to ecosystem protection.

2. Materials and Methods

2.1. Study Area

The Warta Mouth National Park (WMNP) lies in the lower part of the Warta River basin and comprises the right-hand part of the Warta valley and a fragment of the area between the Odra and Warta rivers. The WMNP covers an area of 8037.6 hectares. It is one of the most important refuges of water birds and marsh birds, as well as birds of prey, both in Poland and in Europe [28]. Due to its natural values, it has been entered on the list of the RAMSAR Sites and included in the Natura 2000 network (PLC 080001). The prevalent land cover consists of meadows with various moisture contents, some of them being used for agricultural purposes. The southern part of the park is regularly flooded during the spring freshets of Warta and the swelling of its waters at the mouth of the Odra River. The northern, right-hand part of the valley, located behind flood dikes, is not hydrologically connected with the Warta River today. The water conditions are shaped by small watercourses flowing down from the edge of the valley, and—first of all—by a system of drainage canals and ditches, as well as pumping stations, that drain water from the area [28]. The nature of the WMNP water conditions is one of the key elements for the protection of open meadow and marsh habitats, as well as nesting and resting areas of valuable bird species.

Besides meeting water needs related to the WMNP protection, the water management tasks in the Warta mouth catchment include: maintenance of the environmental (hydrobiological) flows in Warta and its tributaries; ensuring navigation flows in the Warta River (II class navigable route stretch); and water supply to the existing agricultural users.

2.2. Water Management Balance

The water management balance of surface water is a comparison of water resources with the needs of water users, which takes into account the requirements of the natural environment, the hierarchy of users, the effects of hydrotechnical facilities, and the impact of water abstractions and wastewater discharges on the volume of surface water resources, as well as the interactions with groundwater [29,30]. The balance calculations are performed as a simulation of water resources allocation among the users, for all time steps of the selected multi-annual period, taking into account the time variability of the input data (water resources, water needs and wastewater discharges, operation rules of hydrotechnical facilities, etc.). Simulation analyses shall cover the longest possible period for which reliable data on resources and needs are available. The allocation of water resources is carried out according to the adopted hierarchy of water use, which represents the priorities prevailing in the analyzed area and denotes the order in which users receive access to water. Water abstraction for a user placed lower in the hierarchy must not cause the occurrence or worsening of the deficit of the more important user. The comparison of water resources and water users' needs is carried out at control cross-sections, which are important for determining the quantity of water at main rivers above and below the mouth of significant tributary; at tributaries above the mouth to a higher-order river; at locations of significant water abstraction and sewage discharge, or hydrotechnical facilities (storage reservoirs, transfer channels); and at places important for the assessment of the amount of water resources due to protected ecosystems/habitats.

Time series of mean periodic flow (weekly, 10-day, monthly) at water gauge cross-sections are the basis for determining surface water resources. The flow series should be continuous, synchronous, and homogeneous, and should be free from the water use impact. Ensuring the last condition can be achieved, subject to data availability, by naturalizing water gauge flows (e.g., [31,32]). Flows at control cross-sections are computed by interpolation and extrapolation methods on the basis of water gauge observations, or, results of a hydrological model can be imported.

The needs of water users are represented by time series of average water demands (e.g., municipal or industrial users), or flow requirements at specific river cross sections (environmental flows, navigation flows, etc.). However, for water users capable of retaining water, such as fishponds, irrigated facilities, or certain nature conservation tasks, whose needs depend on the current water retention (including the amount of water supplied in previous time steps) and current hydrometeorological conditions, they are calculated during balance analyses. This approach allows for considering the build-up of demand volumes that have not been met in previous time steps. The simulation of users retaining water is carried out in two steps: first, user needs are calculated based on retention volumes and hydrometeorological conditions. Then, after the allocation of water resources in a given time step, the final state of retention is calculated based on the allocated water. This retention becomes the initial state in the next time step of simulation.

The wastewater discharges of groundwater users represent an additional source of water in the river. Discharges (return flows) of surface water users are calculated during water resources allocation, based on the amount of water allocated to the user.

The impact of groundwater use on river flows is described by pseudo-users of surface water, whose needs represent the reduction in groundwater discharge to rivers due to groundwater use. The volumes of pseudo-user needs are determined at the balance cross-sections either on the basis of the results of a groundwater model, or in a simplified way, according to the assumption that the reduction of groundwater discharge to a river is proportional to the area of groundwater filtration to the wells located in the catchment. However, the possibility to take into account the impact of groundwater use depends on the availability of results of hydrogeological analyses and groundwater use data.

The water system under study is modeled as a flow network of arcs and nodes. It reflects the spatial structure of the system: the layout of the river network, the routes of water transfer, the location of hydrotechnical structures, and the points of water intake and sewage discharge. The nodes of the network correspond to control cross-sections, water users, and hydrotechnical structures and the arcs represent the routes of water movement between the nodes: along river or water transfer stretches and between rivers or hydrotechnical structures and water users.

The basic task of the model is the multi-period simulation of the allocation of water resources between users. The flows calculated in the network arcs for each simulation time step must satisfy two basic conditions: flow compliance with the arc constraints (e.g., the water intake for a user must not exceed the amount of needs and must be a non-negative value) and preservation of the mass balance at the nodes (the sum of water inflows to a node must be equal to the sum of outflows). Allowing variability of flows within the constraints indicates that many different combinations that satisfy the constraints are possible. If in a time step there is a surplus of resources over demand, the solution is to assign to the intake arcs a flow equal to demand; the flows in the other arcs result from a simple summation (balance). In case of water scarcity, a combination of flows corresponding to the adopted hierarchy of water resource use is determined. The criterion for optimizing flows in the network is to minimize the sum of losses caused by failure to satisfy the needs of water users or to provide the required flows in river sections. The values of unit loss coefficients for water users and river reaches represent the water use hierarchy. The Out-of-Kilter network programming algorithm [33] is used to solve the water allocation task thus defined. The results of the calculations consist of the time series of: water intakes and wastewater discharges by users, volumes of water in storage reservoirs and at users that retain water, and flows in transfer channels and in all river reaches. From these, assessment criteria, such as time reliability, volume guarantee, maximum depth, maximum volume, and maximum duration of continuous deficit, are calculated. Criteria for users retaining water are usually based on the frequency of occurrence of a given retention condition [22,32]. Moreover, reserves of available water resources with assumed guarantees of occurrence are determined in all control cross-sections.

Balancing calculations often consider several water management variants, i.e., system operation is simulated for different sets of input data and model parameters. These variants may include: the occurrence, parameters and water management principles of hydrotechnical facilities (storage reservoirs, transfer channels), environmental flows, and water needs of users, as well as the hierarchy of water resource use.

2.3. Model Concept

The developed balance model allows for the consideration of the specificity of the area and identifies water users and their water needs related to WMNP protection—the basic scheme used to construct the model is shown in Figure 1. The following has been taken into account to assure the appropriate water conditions in the WMNP area: (i) satisfying the Northern Polder's water needs from the Old Warta River (N Polder PU4); (ii) supplementary Warta water supplies to the Northern Polder (N Polder PU4*); (iii) appraisal of the volume and time distribution of Warta water reserves for potential supplementary supplies to the southern part of the WMNP (Słoński Basin). The most important assumptions pertain to the method of modeling water requirements of the Northern Polder, the estimation of water volumes available for supplementary supplies to the Słoński Basin, and the method of representing water needs for navigation purposes (PU5).

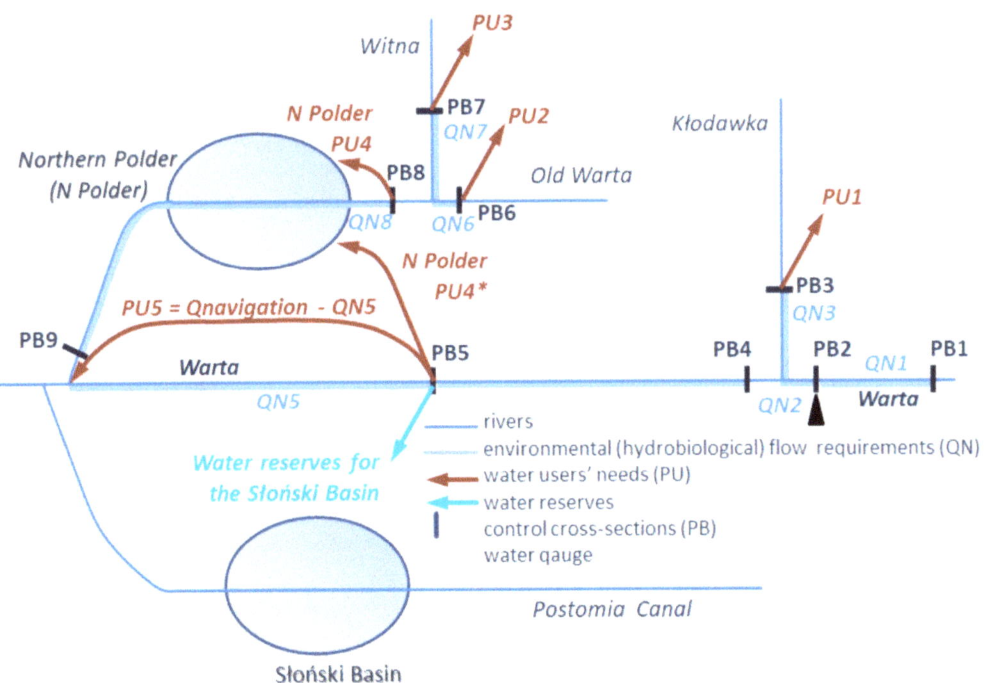

Figure 1. Water management system scheme.

Environmental (hydrobiological) flows (QN) were determined at control cross-sections (PB1-PB8) by the hydrological method defined by the Regional Water Management Authority, based on the method of Kostrzewa [34,35]. According to this method, the environmental flow is equal to the higher of two values: the product of the multi-year average of the annual minimum flows and the parameter of the method (k coefficient), or the lowest flow in the multi-year period. The k coefficient depends on the hydrological type of the river (lowland, transitional, mountain), which is selected on the basis of the average specific runoff and on the catchment area to the cross-section under consideration. The existing agricultural uses included fishpond complexes and areas of irrigated grassland, the needs of which were determined on the basis of water permits. The needs of these users were represented in the model as aggregated water demands PU1–PU3.

2.3.1. Northern Polder

The protection of the natural values of the Northern Polder WMNP aims to prevent the degradation of organic soils and vegetation and should, therefore, consist in maintaining the highest possible moisture content of the local hydrogenic habitats [28]. The habitats have been classified into three types based on their moisture conditions, and their respective uses have been defined:

- Marshy habitats, including reed fields not used for agricultural purposes, located in the southern part of the Polder—between Old Warta and the flood bank (area of marshy habitats Fmarsh = 500 hectares),
- Moist habitats, including extensively utilized once-mowed meadows (Fmeadow = 1050 hectares),
- Moderately moist habitats, including pastures (Fpasture = 1000 hectares).

The task of ensuring the habitats' high moisture content has been formulated as follows [28]:

- Admission of spring floods to the marshy and moist areas (until the end of June);
- Avoidance of excessive drainage when used for agricultural purposes (the groundwater level may be reduced to 50–60 cm below the ground from early June to mid-October in the case of pastures, and in July-August in the case of extensive meadows);
- Stopping of drainage and reconstructing of water retention in the soil profile after agricultural utilization ceases.

To maintain the habitats' high moisture content, the own waters and Old Warta's resources should be used first of all, with Warta waters used only in case of shortage of such resources. To represent water requirements of the Polder mentioned above, simplified water balance in the soil profile has been used to develop a model of the habitat's water needs, and the required parameters have been determined for each habitat moisture type (the desired water retention in the soil profile by seasons, the possibility of drainage or irrigation, and the occurrence of floods). According to the modeling method of water retaining user, the water needs of each habitats were calculated based on the soil water balance and the desired retention. Then, after solving the water allocation task, the final retention state was calculated, which became the initial state in the next simulation step.

2.3.2. Słoński Basin

No model of the Słoński Basin's water needs has been developed, due to insufficient exploration of the site and inventory/survey works carried out during the balance analyses. Instead, the volume of water resources (reserves) available for use as supplementary supplies for the area was estimated. The available reserves were determined based on the assumption that they are equal to the volume of water that remained after the needs of all users located downstream of the examined cross-section had been satisfied. Of crucial importance for the allocation of water for the potential supplementary supplies to the Słoński Basin, has been a discussion concerning satisfaction of the inland navigation water requirements.

2.3.3. Navigation

As follows from the information obtained from the Regional Water Management Authority in Poznań, the navigation season along the analyzed stretch of the Warta River comprises the whole year. The proper (standard) navigation conditions require water levels exceeding a specific threshold value (Hnav_stand). At the same time, navigation may take place, with some limitations, already at a specific lower water level (Hnav_min). Below that level, navigation is impossible. After preliminary balance analyses, it has been arranged with the Regional Water Management Authority in Poznań, that no resources are reserved for navigation purposes in the periods with water levels below Hnav_min. Therefore, current water requirements for navigation purposes (Qnavigation) have been modeled as:

- Equal to the navigable flow at water levels exceeding Hnav_stand,

- Equal to the actual flow within water level range ⟨Hnav_min–Hnav_stand⟩, and
- Equal to 0 at water levels below Hnav_min.

The water demand for navigation (PU5) is represented in the balance model as the excess flow over the environmental flow (PU5 = QNavigation − QN5).

2.4. Simulation

The balance analyses have been carried out in accordance with the methodology described in Section 2.2 by simulation of the catchment's functioning. The balance model was developed in an MS Excel workbook with Visual Basic Application macro support enabled. The model developed for the Warta mouth catchment consisted of 16 nodes and 81 arcs. The following hierarchy of water use was assumed in the balance calculations: maintaining environmental flows, maintaining navigation conditions, supplying existing agricultural users, and providing adequate moisture conditions for wetland habitats in the Northern Polder. The balancing covered the years 1984–2012, and the simulation based on 10 days' time steps. The interpolation and extrapolation method was used to determine the magnitude of flows at control cross-sections.

3. Results

On the basis of the balance simulation results, the criteria for assessing the degree to which users' water needs were met were calculated. In the system under analysis, water supply problems occurred in the basins of small watercourses—tributaries of Warta or Old Warta. With respect to those rivers, relatively low time reliability of maintenance of the environmental (hydrobiological) flows and satisfying water users demands were determined: QN8—58%, QN6 and QN7—60%, and PU1–PU3, respectively, 61%, 77%, and 23%. The volumetric guarantee, defining the ratio between the volume of water supplied and that required, was approximately 85% for maintaining environmental flows. The volumetric guarantee of water supply to agricultural users was in the range of 30–75%. Environmental flows in the Warta River, on the other hand, were 100% guaranteed, and navigable conditions occurred in 82% of the analyzed time steps (standard conditions—47%, minimum acceptable conditions—35%). The estimated water reserves are quite large and occur during periods when flows in the Warta River are greater than the environmental flow, but smaller than the minimum navigable flow, or they are above the standard navigable level. A considerable part of these reserves occurs in the spring period from March to May. The flow volumes determined in the control cross-section PB5, in which both environmental flow and the task of maintaining adequate navigable conditions were determined, are shown in Figure 2 (in hydrological years, that start in the 1 November). The volumes of water reserves that can be used for additional supply of the Słoński Basin are also included there. As can be seen from this figure, water reserves are not available all the time. They occur in about 50% of the time steps.

The water retention time series in the Northern Polder habitats demonstrate the habitats' satisfactory moisture content for most of the time. Considerable drying was found most often (18% of the time) in the moderately moist pasture areas. These are the driest of the habitats considered, with the smallest desired retention, which is related to their natural conditions and actual land use. Due to problems with maintaining appropriate humidity conditions, Figure 3 presents the water retention time series in this habitat in the studied multi-year period. Apart from the retention, the available water resources in the Old Warta River and water intakes to improve water conditions in this habitat are presented. Figure 4 shows the water retention and meteorological parameters for a selected year (2000), where relatively unfavorable moisture conditions were observed. Winter season retention was high, but drainage at the start of the grazing season, the subsequent period of high temperatures, lack of rainfall, and deficit of water resources for irrigation resulted in significant drying of the habitat, which lasted from early August to late October. In November the restoration of retention began.

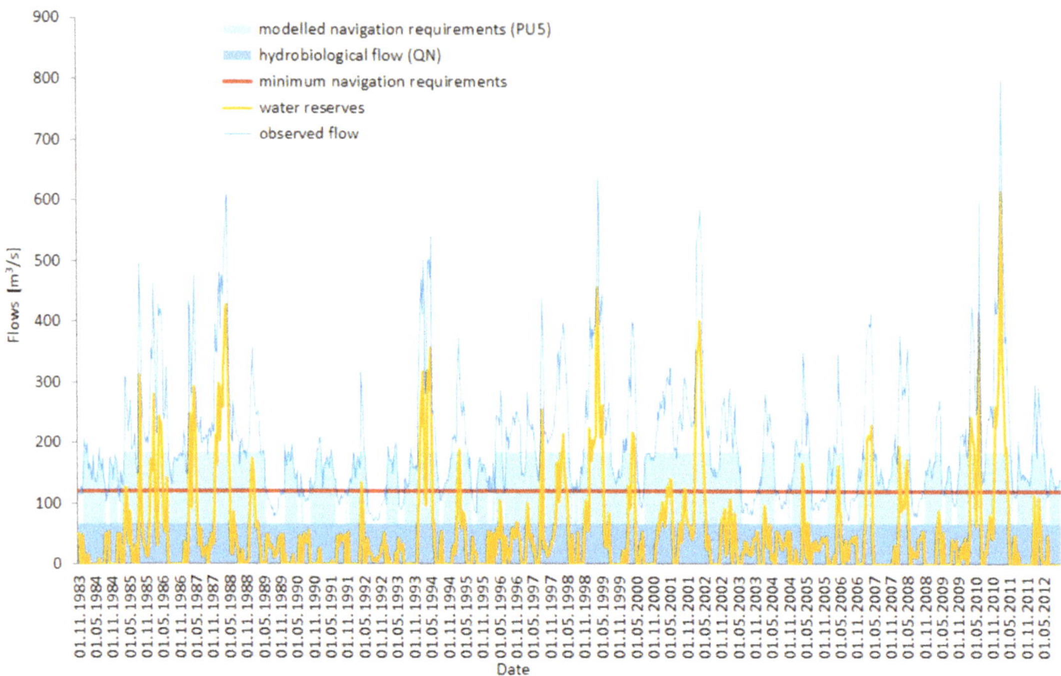

Figure 2. Balance flows of the Warta River at the cross-section PB5 with appraisal of the navigation conditions and water reserves.

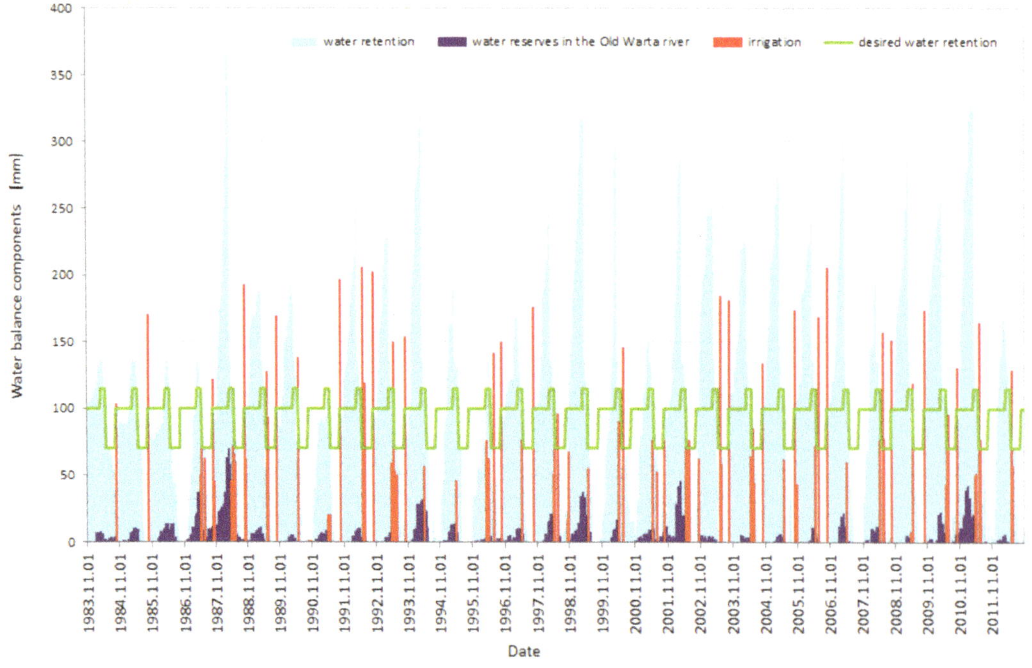

Figure 3. Water retention in moderately moist habitats in the analyzed multiple years' period.

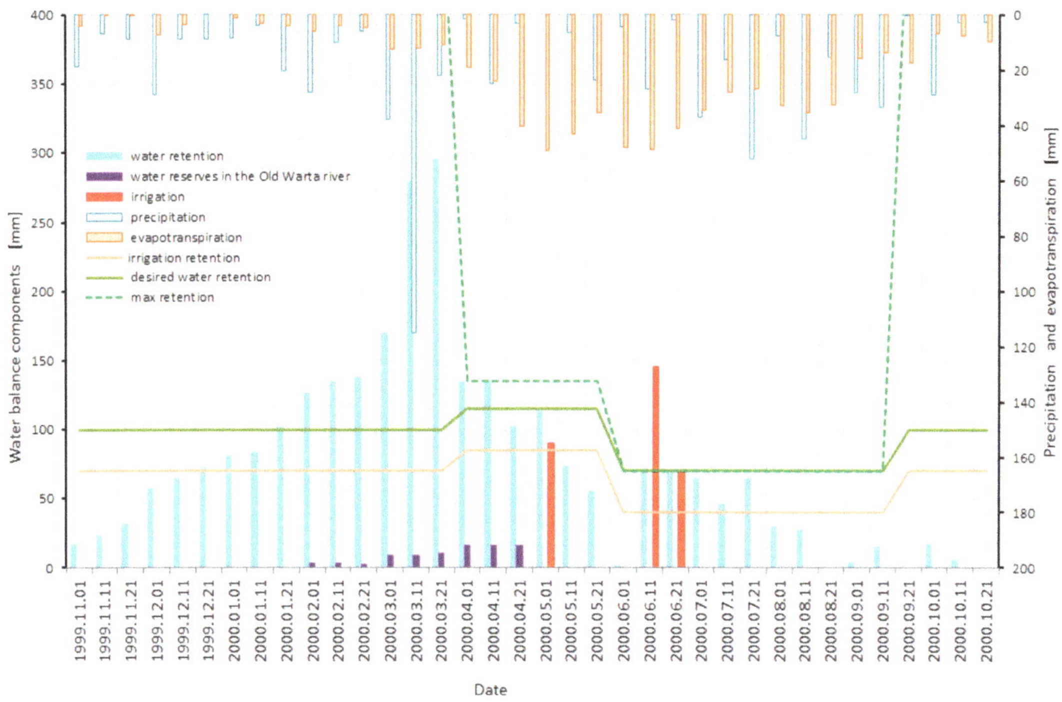

Figure 4. Water retention and other water balance elements in moderately moist habitats in 2000.

In moist habitats (meadows) water retention below the assumed irrigation threshold, i.e., overdrying threshold, was observed only 5% of the time. In marshy habitats, such a situation occurred more frequently, i.e., in about 12% of time steps, which is related to higher humidity of these habitats.

The resources of the Old Warta River were mainly used for irrigation. The task of providing suitable conditions for navigation limited the supply to the Polder from the Warta River.

4. Discussion

As identified within the water balance analyses, the problems with maintenance of the environmental hydrobiological flows—and, thus, with meeting the users' water requirements—in small water courses of the analyzed basin, are related to the high values of the required flows, determined in some of the still valid documents [34], much exceeding those determined in earlier studies [28]. Since maintenance of environmental flows was defined as the most important task in the modeled system, low values of its implementation criteria indicate the need for verification of the determined requirements. Verification, ideally preceded by research of the existing water ecosystems and definition of their specific quantitative requirements, would lead to a more reliable appraisal of any potential problems with maintaining appropriate flows, possibly threatening the good ecological status.

Furthermore, a more precise determination of the water resources of small watercourses would certainly contribute to a better recognition of the relevant catchment problems. The water flow data used in the balance calculations are subject to high uncertainty—the flows in all rivers were estimated on the basis of observations from the water gauge on the Warta River. For higher reliability of resource determination, it is advisable to establish at least periodic water gauges, which would provide data to improve the relationships used to transfer hydrological data, or to calibrate hydrological models [36]. Similar problems concerning availability, reliability of hydrological data, and the necessity to strive for their improvement, were also raised by other authors dealing with modeling for decision support in water management (e.g., [37–39]).

The problem of reliability of input data to the balance appears again when analyzing results for the habitats of the Northern Polder. In 7 years out of the 29 analyzed, water retention was not restored to the assumed optimum level during the winter season. Reasons for such results could be:

- too low values of groundwater recharge for habitats in the valley edge zone and of infiltration from the Warta River to habitats located near its bed, based on estimates and other studies' data [28] and
- the applied method of determining reference evapotranspiration (Penman's method), for which overestimation of calculation results was reported in other studies [40].

Field measurements and modeling aimed at identifying the best method of estimation of actual evapotranspiration from the area, and monitoring of groundwater levels permitting estimation of the inflow of waters from the upland and from the Warta River, would improve the accuracy of the habitats' water balance modeling.

In spite of the discussed inaccuracies in the description of some elements of habitats' water balance, it can be concluded from the results for the Northern Polder that a possibility to irrigate and retain water in the polder (prohibition of land drainage) in spring is of key importance for the occurrence of high moisture content in hydrogenic habitats. Water reserves of the Old Warta River might be used to ensure appropriate moisture conditions in the Northern Polder (Figures 3 and 4), however, due to their time distribution, the use of these reserves depends on the possibility of water retention in the area. The application of hydrotechnical solutions, e.g., trough damming devices, is one of the options, whose expediency and effectiveness should be further considered.

Another thing worth considering is the task of ensuring adequate navigation conditions. The proposed concept of giving up resources' preservation, in periods when the river flow is below the minimum navigable requirements, yields considerable volumes of water for other tasks. For the practice of water resources management, this way of meeting the navigation requirements, negotiated on the basis of the preliminary balance results, is an advantageous option for the environment. In the context of modeling the navigation requirements in balance analyses, the proposed approach is recommended where the needs depend on the defined threshold values and the current river flow.

5. Conclusions

This paper presents the application of water management balances to the search for a compromise between socio-economic water use and the tasks of protecting water and water-dependent ecosystems. The water management balance model proved to be a useful tool for such analyses. A necessary condition for including the tasks of protecting water and water-dependent ecosystems in the balance analyses is treating the water needs of these ecosystems as one of the water users. Only then can the impact of water management priorities on the amount of water available to both ecosystems and socio-economic users be analyzed. However, the possibility to model the water needs of ecosystems depends on the recognition of their water needs, which is necessary to define the model parameters.

For a complete description of the water needs of ecosystems it is necessary to provide not only the desired values that ensure optimum conditions for the development of ecosystems, but also the threshold values, beyond which significant changes in the ecosystems' functioning occur. The determination of desirable and threshold values has, for years, been an important research problem in the field of water management and protection of water-dependent ecosystems. The accuracy with which the requirements of aquatic and water-dependent ecosystems are represented in a water balance model depends on the recognition of their functioning and the role of flow for ecosystem sustainability and conservation.

The balance model was developed for the part of the Warta River catchment comprising the Warta Mouth National Park. The low availability of hydrological data and the resulting inaccuracy of water resources assessment, together with the uncertainty of input data for modeling the requirements of protected habitats in the WMNP, contributed to the limited reliability of the balance results. Nevertheless, it can be concluded that it is possible to satisfy both the needs of water users—agriculture and navigation—and, to a considerable extent, the requirements of protected wetlands. The abandonment of drainage in spring and the possibility of irrigation in late summer are both key to ensuring high moisture content of the Northern Polder habitats. As navigation requirements limit the use of the Warta River flows, and due to the unfavorable time distribution, the Old Warta River resources do not allow for fully meeting the water needs of the protected habitats, and the increase of water quantity for the Northern Polder would depend on the implementation of retention measures in the area. The developed balance model can be used to help determine the location and technical parameters of potential facilities.

The possibilities to improve the reliability of the balance results depend primarily on improving the quality of the input data. The coupling of water user models with the balance model allows for the correct determination of water needs and the proper assessment of the degree to which water needs are being met.

Funding: This research received no external funding.

Data Availability Statement: The source hydrological and meteorological data were collected by the Institute of Meteorology and Water Management National Research Institute and are available here: https://danepubliczne.imgw.pl/data/dane_pomiarowo_obserwacyjne/ (accessed on 26 October 2021). On their basis, the time series of mean 10-day values used in the study were calculated.

Acknowledgments: I would like to acknowledge T. Okruszko and S. Tyszewski for the work that fostered me to write this article. I would also like to thank the Regional Water Management Authority in Poznań and the Warta Mouth National Park staff for provision of necessary information and insight into respective water management issues as well as MGGP S.A. for the experience which I gained owing to cooperation. I would also like to thank the Anonymous Reviewers who have contributed to improving the quality of the article.

Conflicts of Interest: The author declares no conflict of interest.

References

1. *Directive 2000/60/EC of the European Parliament and of the Council of 23 October 2000 Establishing a Framework for Community Action in the Field of Water Policy, OJ L 327*; 2000; pp. 1–73. Available online: http://eur-lex.europa.eu/LexUriServ/LexUriServ.do?uri=CELEX:32000L0060:en:NOT (accessed on 26 October 2021).
2. Acreman, M.C.; Ferguson, A. Environmental flows and the European Water Framework Directive. *Freshw. Biol.* **2010**, *55*, 32–48. [CrossRef]
3. Tharme, R.E. A global perspective on environmental flow assessment: Emerging trends in the development and application of environmental flow methodologies for rivers. *River Res. Appl.* **2003**, *19*, 397–441. [CrossRef]
4. Acreman, M.; Dunbar, M.J. Defining environmental river flow requirements—A review. *Hydrol. Earth Syst. Sci.* **2004**, *8*, 861–876. [CrossRef]
5. Piniewski, M.; Acreman, M.C.; Stratford, C.S.; Okruszko, T.; Giełczewski, M.; Teodorowicz, M.; Rycharski, M.; Oświecimska-Piasko, Z. Estimation of environmental flows in semi-natural lowland rivers—The Narew basin case study. *Pol. J. Environ. Stud.* **2011**, *20*, 1281–1293.
6. Richter, B.D.; Baumgartner, J.V.; Wigington, R.; Braun, D.p. How much water does a river need? *Freshw. Biol.* **1997**, *37*, 231–249. [CrossRef]
7. Poff, N.L.; Allan, J.D.; Bain, M.B.; Karr, J.R.; Prestegaard, K.L.; Richter, B.D.; Sparks, R.E.; Stromberg, J.C. The natural flow regime. *BioScience* **1997**, *47*, 769–784. [CrossRef]
8. Poff, N.L.; Richter, B.D.; Arthington, A.H.; Bunn, S.E.; Naiman, R.J.; Kendy, E.; Acreman, M.; Apse, C.; Bledsoe, B.P.; Freeman, M.C.; et al. The ecological limits of hydrologic alteration (ELOHA): A new framework for developing regional environmental flow standards. *Freshw. Biol.* **2010**, *55*, 147–170. [CrossRef]
9. Hendriks, D.M.D.; Kuijper, M.J.M.; van Ek, R. Groundwater impact on environmental flow needs of streams in sandy catchments in the Netherlands. *Hydrol. Sci. J.* **2014**, *59*, 562–577. [CrossRef]
10. Streetly, M.J.; Bradley, D.C.; Streetly, H.R.; Young, C.; Cadman, D.; Banham, A. Bringing groundwater models to LIFE: A new way to assess water resources management option. *Hydrol. Sci. J.* **2014**, *59*, 578–593. [CrossRef]

11. Acreman, M.; Aldrick, J.; Binnie, C.; Black, A.; Cowx, I.; Dawson, H.; Dunbar, M.; Extence, C.; Hannaford, J.; Harby, A.; et al. Environmental flows from dams: The Water Framework Directive. *Eng. Sustain.* **2009**, *162*, 13–22. [CrossRef]

12. Shafroth, P.B.; Wilcox, A.C.; Lytle, D.A.; Hickey, J.T.; Andersen, D.C.; Beauchamp, V.B.; Hautzinger, A.; McMullen, L.E.; Warner, A. Ecosystem effects of environmental flows: Modelling and experimental floods in a dryland river. *Freshw. Biol.* **2010**, *55*, 68–85. [CrossRef]

13. Yin, X.A.; Yang, Z.F.; Petts, G.E. Optimizing environmental flows below dams. *River Res. Appl.* **2012**, *28*, 703–716. [CrossRef]

14. Verdonschot, P.F.M.; Nijboer, R.C. Towards a decision support system for stream restoration in The Netherlands: An overview of restoration projects and future needs. *Hydrobiologia* **2002**, *478*, 131–148. [CrossRef]

15. Gostner, W.; Parasiewicz, P.; Schleiss, A.J. A case study on spatial and temporal hydraulic variability in an alpine gravel-bed stream based on the hydromorphological index of diversity. *Ecohydrology* **2013**, *6*, 652–667. [CrossRef]

16. Parasiewicz, P.; Rogers, J.N.; Gortazar, J.; Vezza, P.; Wiśniewolski, W.; Comglio, C. The MesoHABSIM Simulation Model–development and applications. In *Ecohydraulics: An Integrated Approach*; Maddock, I., Harby, A., Kemp, P., Wood, P., Eds.; John Wiley & Sons Ltd.: Hoboken, NJ, USA, 2013; pp. 109–124.

17. Godyń, I.; Indyk, W.; Jarząbek, A.; Owsiany, M.; Pusłowska-Tyszewska, D.; Sarna, S.; Stańko, R.; Tyszewski, S. *Good Practices for Water Management Planning in the Sites of High Natural Values. Guidance*; Regional Water Management Authority: Kraków, Poland, 2011; p. 139. (In Polish)

18. Okruszko, T. *Hydrologic Criteria for Wetlands' Conservation*; Warsaw University of Life Sciences SGGW: Warsaw, Poland, 2005; p. 151. (In Polish)

19. Loucks, D.P.; Van Beek, E. *Water Resources Systems Planning and Management: An Introduction to Methods, Models and Applications*; UNESCO: Paris, France, 2005; p. 681.

20. Acreman, M. Linking science and decision-making: Features and experience from environmental river flow setting. *Environ. Model. Softw.* **2005**, *20*, 99–109. [CrossRef]

21. Holmes, M.G.R.; Young, A.R.; Grew, R. A catchment-based water resource decision support tool for the United Kingdom. *Environ. Model. Softw.* **2005**, *20*, 197–202. [CrossRef]

22. Bonenberg, J.; Drużyńska, E.; Kindler, J.; Nachlik, E.; Pusłowska-Tyszewska, D.; Tyszewski, S. Basis for planning and water resources allocation. In *Methodological Basis and Standards for Integrated Planning in Water Management*; Environmental Engineering Series, 341, Nachlik, E., Drużyńska, E., Eds.; Krakow University of Technology: Kraków, Poland, 2006; pp. 56–78. (In Polish)

23. Acreman, M.; Arthington, A.H.; Colloff, M.J.; Couch, C.; Crossman, N.D.; Dyer, F.; Overton, I.; Pollino, C.A.; Stewardson, M.J.; Young, W. Environmental flows for natural, hybrid, and novel riverine ecosystems in a changing world. *Front. Ecol. Environ.* **2014**, *12*, 466–473. [CrossRef]

24. Jha, M.K.; Das Gupta, A. Application of Mike Basin for Water Management Strategies in a Watershed. *Water Int.* **2003**, *28*, 27–35. [CrossRef]

25. Draper, A.; Jenkins, M.; Kirby, K.; Lund, J.; Howitt, R. Economic-engineering optimization for California Water Management. *J. Water Resour. Plan. Manag.* **2003**, *129*, 155–164. [CrossRef]

26. Dai, T.; Labadie, J. River basin network model for integrated water quantity/quality management. *J. Water Resour. Plan. Manag.* **2001**, *127*, 295–305. [CrossRef]

27. Pusłowska-Tyszewska, D.; Tyszewski, S.; Okruszko, T. *Water Management Optimization Model for the Warta Mouth National Park*; Project Report; MGGP S.A.: Kraków, Poland, 2013; p. 101. (In Polish)

28. Chormański, J.; Giełczewski, M.; Kardel, I.; Malinowski, R.; Szporak, S. *Concept of Revitalization of Meadow and Marsh Habitats in the Warta Mouth Park-Northern Polder*; Project Report; Nature Conservation Society Ptaki Polskie: Warsaw, Poland, 2009; p. 189. (In Polish)

29. Kloss, A.; Łaski, A.; Rutkowski, M.; Sokołow, W.; Tyszewski, S. *Methodology for Water Management Balances*; Hydroprojekt-Warszawa: Warsaw, Poland, 1992; p. 123. (In Polish)

30. Tyszewski, S.; Herbich, P.; Indyk, W.; Jarząbek, A.; Pusłowska-Tyszewska, D.; Rutkowski, M. *Methodology for Developing Conditions for Water Use in Water Regions and in Catchments*; Water Management Lab Pro-Woda–Guidance Document Prepared for the National Water Management Authority; Water Management Lab.: Warsaw, Poland, 2008; p. 66. (In Polish)

31. Stewardson, M.J.; Acreman, M.; Costelloe, J.F.; Fletcher, T.D.; Fowler, K.J.A.; Horne, A.C.; Liu, G.; McClain, M.E.; Peel, M.C. Chapter 3. Understanding Hydrological Alteration. In *Water for the Environment. From Policy and Science to Implementation and Management*; Horne, A.C., Webb, J.A., Stewardson, M.J., Richter, B., Acreman, M., Eds.; Academic Press: Cambridge, MA, USA; Elsevier: London, UK, 2017; pp. 37–64.

32. Pusłowska-Tyszewska, D.; Dybkowska-Stefek, D.; Relisko-Rybak, J. Analysis of Surface Water Resources Availability in the Area of the Planned Silesian Canal. Water-Management Balances of Surface Waters of the Ruda, Bierawka, Gostynia and Pszczynka Catchment areas. *Gospod. Wodna* **2021**, *13*, 10–24.

33. Ford, L.R.; Fulkerson, D.R. *Flows in Networks*, 1st ed.; PWN: Warsaw, Poland, 1969; p. 256. (In Polish)

34. Conditions for Water Use of the Warta Water Region, Official Journal of the Lubuskie Voivodeship No. 810 of 02.04.2014. Available online: http://dzienniki.luw.pl/WDU_F/2014/810/akt.pdf (accessed on 3 September 2014). (In Polish).

35. Kostrzewa, H. *Verification of Criteria and Magnitude of Environmental Flow for Polish Rivers*; IMGW Research Materials, Series: Water Management and Protection; IMGW: Warsaw, Poland, 1977. (In Polish)

36. Pusłowska-Tyszewska, D.; Tyszewski, S. Water management balance as a basis for conditions for water use in river catchments–the Jeziorka river case study. In *Monographs of the Committee on Water Resources Management of the Polish Academy of Sciences XX*; Banasik, K., Hejduk, L., Kaznowska, E., Eds.; KGWPAN: Warsaw, Poland, 2014; pp. 259–270. (In Polish)
37. Wurbs, R.A. Methods for Developing Naturalized Monthly Flows at Gaged and Ungaged Sites. *J. Hydrol. Eng.* **2006**, *11*, 55–64. [CrossRef]
38. Bangash, R.F.; Passuello, A.; Hammond, M.; Schuhmacher, M. Water allocation assessment in low flow river under data scarce conditions: A study of hydrological simulation in Mediterranean basin. *Sci. Total Environ.* **2012**, *440*, 60–71. [CrossRef] [PubMed]
39. Cai, X. Implementation of holistic water resources-economic optimization models for river basin management–reflective experiences. *Environ. Model. Softw.* **2008**, *23*, 2–18. [CrossRef]
40. Kasperska-Wołowicz, E.; Łabędzki, L. A comparison of reference evapotranspiration according to Penman and Penman-Montheith in various regions in Poland. *Water-Environ.-Rural Areas* **2004**, *11*, 123–136. (In Polish)

MDPI
St. Alban-Anlage 66
4052 Basel
Switzerland
Tel. +41 61 683 77 34
Fax +41 61 302 89 18
www.mdpi.com

Water Editorial Office
E-mail: water@mdpi.com
www.mdpi.com/journal/water